"Nico Stehr is one of the foremost expert thinkers on the intersection of science, politics, and society. This will be one of the foundational resources for navigating the political and economic orders in responding to climate change."

Charles Lemert, *University Professor and John C. Andrus Professor of Social Theory Emeritus at Wesleyan University, and author of* Americans Thinking America *(2025)*

"Nico Stehr's insights on science and policy are important and highly respected. In this book he builds on his well-established analyses but moves them in a new direction by focusing on risk, climate and democracy."

Tom Dietz, *Distinguished Professor Emeritus, Michigan State University, and author of* Decisions for Sustainability *(2023)*

"What matters more—finding ways of settling the convulsions of the global climate or preserving the political freedoms of the western democratic tradition? In *The World We Have Created*, veteran sociologist Nico Stehr answers this question: it is not that democracy is failing, but that we—the defenders of democracy—are failing to be democratic enough."

Mike Hulme, *Professor of Human Geography, University of Cambridge, and author of* Why We Disagree About Climate Change *(2009)*

"Nico Stehr, one of the co-authors of the influential Hartwell Paper, provides a well-documented account of climate change, including the challenge to democratic governance."

Stephen Turner, *Distinguished University Professor, University of South Florida*

THE WORLD WE HAVE CREATED

Addressing the complex interplay between climate change, democratic governance, and scientific knowledge in the Anthropocene era, this book focuses on how political systems respond to the global ecological crisis and to what extent they are structurally capable of doing so.

The World We Have Created: Climate, Democracy and Knowledge details how the de-democratization of ecological decision-making favors models of authoritarian governance, as democratic values such as transparency, participation, and accountability are continually undermined. Offering a robust account of the responses to the climate crisis and the necessary social and political changes posed by the scientific consensus, this book analyses the growing discrepancy between scientific knowledge about the causes and consequences of climate change and the lack of political and social willingness to implement necessary measures. Comprehensive social change is needed that focuses on social justice, intergenerational responsibility, and political participation, and it is only by strengthening democratic structures and integrating knowledge into responsible governance can we open paths to a resilient, just, and liveable future.

This sharp, interdisciplinary intervention bridges sociological, political, and philosophical thinking to underline the necessity of hope and utopian social imagination together with concrete changes to the constitution that guarantees a safe environment for future generations.

Nico Stehr is Karl Mannheim Professor of Cultural Studies Emeritus at the Zeppelin University, Friedrichshafen, Germany. He is a fellow of the Royal Society (Canada). He is one of the authors of the Hartwell Paper on climate policy. His recent books include *Knowledge Capitalism* (Routledge, 2022), *Money: A Theory of Modern Society* (Routledge, 2020), and *Knowledge: Is Knowledge Power?* (with Marion Adolf, Routledge, 2017).

A New Order of Social Things

Edited by Waverly Duck, (University of California, Santa Barbara), Kristin Plys, (University of Toronto), Charles Lemert (Wesleyan University) and Katharine Wallerstein, (University of California, Davis)

The social and human sciences may well be going through a sea change similar to what they experienced after the revolutionary moment of 1968. Today, it is fair to say, there are several obvious spheres of thought and research where this new order is beginning to show itself:

- The Global South
- Cultural and Critical Theories
- Multicultural Urban Ethnographies
- The Instability of Social Democracies
- Theorizing the Virtual World

To address these important questions we look to theorists of multiple geographies, genders, races, castes, and perspectives—established theorists who will levy this forum to take their existing work in new directions, along with early-career theorists whose ideas will chart new avenues for a renewal of critical theory. This breadth embraces both the plurality and ambiguity of theory today while also exploring bold solutions to the current impasse.

Titles in this series:

Americans Thinking America
Elements of American Social Thought
Charles Lemert

Social Thought From the Ruins
Quixote's Dinner Party
David A. Westbrook

The World We Have Created
Climate, Democracy and Knowledge
Nico Stehr

THE WORLD WE HAVE CREATED

Climate, Democracy and Knowledge

Nico Stehr

Routledge
Taylor & Francis Group
NEW YORK AND LONDON

Designed cover image: Maxim Tolchinskiy, Unsplash

First published 2026
by Routledge
605 Third Avenue, New York, NY 10158

and by Routledge
4 Park Square, Milton Park, Abingdon, Oxon, OX14 4RN

Routledge is an imprint of the Taylor & Francis Group, an informa business

ISBN: 978-1-041-15961-2 (hbk)
ISBN: 978-1-041-15958-2 (pbk)
ISBN: 978-1-003-68206-6 (ebk)

DOI: 10.4324/9781003682066

Typeset in Sabon
by Newgen Publishing UK

CONTENTS

PREFACE

> The ultimate hidden truth of the world is that it is something we make and could just as easily make differently.
>
> *Attributed to David Graeber*

> Based on a binary perspective, the choices open to human beings for social organization [...] is the choice between coordination by central authority, on one hand, and by mutual adjustment among those to be coordinated, on the other.
>
> *Charles Lindblom (1995: 688)*

The understanding of the fundamental drivers of climate change has evolved significantly. The climate crisis is a *global* as well as local and *ongoing* problem.[1] The internationally agreed-upon Paris 2015 (arbitrary) goal of keeping the global temperature increase below 1.5 degrees Celsius (2.7°F) compared to pre-industrial levels[2] is certainly not achievable.[3] A new report by the *World Meteorological Organization* (WMO)[4] has revealed that, for the first time, the data suggest a small likelihood of the world experiencing a year that is 2°C hotter than the pre-industrial era before 2030. Scientists have described this possibility as "shocking."[5] Many social, economic, and political pressures, such as *adaptation*, both reactive and proactive, are *local* and short-term. The often-overlooked issue of greenhouse gas *retention* in the atmosphere makes climate change a *persistent* problem. Climate change is the greatest *governance* challenge in history. Climate change is a disruption that should prompt action, but it is not fully recognized as a disorder.

By contrast, only recently have scholars begun systematically exploring the socio-political barriers to effective climate action. It is based primarily on contributions from the social sciences. Both scientific and political discussions on climate change are heavily focused on the future. This discourse is inherently difficult to navigate, yet it is crucial to engage with it. However, the future orientation of the climate change discourse is contingent upon the realities of the present. Niklas Luhmann (1976: 140) formulates this principle as follows: "all iteration of temporal forms has to have its base in a present." It is a future perspective that assumes the persuasiveness of societal influence on our environment and the notion that non-human forces act on society. However, what is already clear is that the political choice between centrality and democratic governance is a large, fateful one.

In developed democracies, the increasingly pessimistic discourse among climate scientists, activists, and institutions has begun to frame the relationship between knowledge and governance as one of "inconvenient democracy."[6] The discrepancy between the extent of our knowledge about climate change and the obligations of citizens to change their behavior (not in my backyard) is identified as an "*inconvenient reason.*" Politicians' reluctance to act on climate science has been labeled a failure of "inconvenient institutions." The inconvenient institutional arrangements are not confined to the national level, they extend to the global field. As some observers note, for example, Yasha Mounk (2018: 96),

> allowing each country to go its own way makes it impossible to confront a whole range of global challenges like climate change. In the end, it seems we must choose between achieving international cooperation on key issues by a troublingly undemocratic path—and not achieving it at all.

The broad secular societal trend away from democracy in many countries around the world is not only due to green issues. As surveys demonstrate, disenchantment with democracy is widely shared and based on multiple disappointments.[7] Prominent observers of the dangers of global warming also despair of democracy: For example, more than twenty years ago, former US Vice President Al Gore (2009: 303) referred to global warming as "the biggest failure of democratic governance in history" and philosopher Philip Kitcher (2011: 243), in his monograph *Science in a Democratic Society* calls the failure of climate policy "an enormous failure of worldwide democracy." Annika Stechemesser and her colleagues (2024: 891) who have examined existing climate emission reduction policies in 41 countries from six continents offer a sobering perspective on closing the emission gap between those policies and what would be required to reach the level necessary stipulated in the Paris Agreement of 2015; efforts would have to be quadrupled. Although there is some evidence that democracies emit less (e.g. Povitkina, 2018: 425), research on the impact of regime type on efforts to *reduce* emissions is inconclusive; some of the evidence obtained thus far indicates that regime type hardly matters (e.g. Chesler et al., 2023: 210).

Notwithstanding successful political assaults on the climate ambitions in some countries,[8] the existence of climate change is perceived as an existential emergency and a threat to human civilization if no action is taken, with the potential for future political struggles that resemble a state of war. Under anticipated conditions of profound uncertainty and extreme danger, the desire for prompt action, and ultimately a war narrative, becomes a crucial framework for coping. A state of war that, by definition, would allow for the implementation of effective measures against the threat regardless of resistance (McKibbon, 2016). It is argued that political action based on the principles of democratic governance[9] should be abandoned in favor of a more technocratic-authoritarian approach.[10] An approach favored by many Silicon Valley "tech actors" because "of their reliance on techno-fixes, representatives of Silicon Valley express an ideology of ecomodernism, which believes that human progress can be 'decoupled' from environmental decline" (Riemens, 2024: 294). The technocratic approach is favored because expert scientists are better able to make decisions on how to respond to the threat of climate change.[11] The populist counterargument is that democracy is incapable of deflecting a vigorous climate policy approach.[12] Right-wing populists hope that authoritarian political structures can successfully lead the fight against rapid decarbonization. Both sides support a strong state with large-scale planning as the best way to implement—or block—climate measures. The poles of the left and right political perspectives tend to converge in another respect, that is, in their critique of the capitalist economic order as a new form of "neo-feudalism," "digital feudalism," or "information feudalism" (cf. Morozov, 2022).

However, what is the optimal course of action? Will we be able to adapt, for example I argue that relying on a strong state to address future climate impacts is misguided.[13] A critical question is whether I argue that relying on a vision of the future dominated by a powerful state is flawed—and may even erode constitutional democracy. Whether societies can adapt to climate impacts within a time horizon that fits typical human social planning. This underscores the importance of time horizons in climate analysis.

The topic of the human time horizon intersects with sociology, economics, anthropology, psychology, and political science, and it is a theme deeply embedded in the history of the intellectual labor of social scientists; for example, Norbert Elias's ([1939] 1978]) work in *The Civilizing Process* and *Time* argues that modern societies have developed increasingly longer time horizons through social and institutional development; Pierre Bourdieu's ([1972] 1990) notion of *habitus* includes dispositions toward time—some social classes are socialized to think in short-term or long-term horizons;

Robert K. Merton's (e.g. 1957) work is foundational in framing how social structures shape temporal orientations. Barbara Adam in her *Timescapes of Modernity* focuses on how societies construct and govern futures, especially in relation to environmental and technological risks. Adam argues that modern societies need "futures literacy" to responsibly manage long-term outcomes. The classic observations of social sciences regarding the societal time horizon are extended by observers of modern society that our time horizon or attention span has shortened considerably in recent decades (e.g. Rosa, 2017).

The useful concept I will employ to capture the complexity of the varied human time horizons of a "future present," in contrast to the present past or the present,[14] is Niklas Luhmann's terminology (1976: 140–141):

> If we characterize processes or activities as beginning or ending, we use terminology that belongs to the present. When we use these terms to refer to distant dates – for example, the Roman Empire began to fall – we are referring to a past present or a future present. [...] we experience our future as a generalized horizon of surplus possibilities that have to be reduced as we approach them.[15]

Indeed, this is, as Niklas Luhmann described it, one of the central task and frame of my analysis.

A related, core narrative of my analysis is the idea that climate change is a *wicked problem*. That climate change is a wicked problem is a sober and a realistic diagnosis. Climate change is a multidimensional issue that can't be solved simply by reducing CO_2 emissions. Climate change is a societal problem that cannot be solved. It is hard to really relax. Suggesting that capitalism merely needs repair or reform is also an oversimplification. In addition, unlike in the case of past capitalism crises, "there are no lifeboats here for the rich and the privileged [...] Climate change is an unintended consequence of human actions and shows, only through scientific analysis, the effects of our actions as a species" (Chakrabarty, 2009: 221). In a self-critical comment, Dipesh Chakrabarty (2017: 341) explains that it

> is possible that the lifeboat metaphor was too cryptic (and it clearly misfired for some readers) but my point was that climate change, potentially, has to do with changes in the boundary conditions needed for the sustenance of human and many other forms of life.

In fact, the wealthier segments of the population can afford more strategies than the poor to cope with weather extremes such as heat. These households also have no incentive to internalize the cost of greenhouse gas emissions caused by the electricity needed to operate their cooling devices (Kahn, 2016: 171).

To the disappointment of climate experts, climate policies have not really progressed "over the last three decades. The dangers of global warming are much better understood, but nations have not adopted effective policies to slow the coming peril" (Nordhaus, 2020: 13). Considering Nordhaus's diagnosis of unsatisfactory conditions, is authoritarian climate governance perhaps legitimate (see Mittiga, 2022)?[16] Many notable observers of the ecological crisis agree with Mittiga's diagnosis:

> Why are we not doing what needs to be done, actions that many decision-makers deem necessary? At the heart of the problem lies democracy itself. The choices that we have to make will be painful and costly, in part because they directly affect our economic models and prosperity.
>
> *(Zuleeg, 2024)*

The common-sense attitude toward the future present is characterized by a mix of curiosity and denial. Today's social science discourse about the future is either absent or far too optimistic. Both attitudinal complexes are not helpful in this age. An age of anthropogenic climate change poses a fundamental problem of our attitude toward time: "It pits those who have long lived in a world of now against those who fear there may soon be no world for them to inhabit" (Barba-Kay, 2023: 85). However, we should not make the mistake of believing that the future will be better than the present just because it is the future, or that the future will be worse than the present just because the prospect of the future present is bleak. Neither overconfidence nor excessive pessimism is the best starting point. The intrinsic dilemma of time—the power of pressing circumstances—will inevitably confront us, giving rise to the dilemma of governance and the dilemma of democracy.

Notes

1 Elinor Ostrom (2009: 32) frames the challenge of climate change and the advancing of approaching climate policies very much in this spirit:

> Rather than only a global effort, it would be better to self-consciously adopt a polycentric approach to the problem of climate change in order to gain the benefits to multiple scales as well as to encourage experimentation and learning from diverse policies adopted at multiple scales.
>
> (also Ostrom, 2010)

2 The average temperature between 1850 and 1900 is used as the pre-industrial reference value. This leaves open the question of when exactly anthropogenic warming began. The climate researcher Benjamin Sander and his colleagues (2025) tried to answer this question. They link the beginning of global warming to the year 1885. In a thought experiment, using nine state-of-the-art climate models, the researcher concludes that, given the measuring devices now available, those available then would have yielded evidence of global warming in the year

1885. The study "provides strong evidence that a discernable human influence on atmospheric temperature has likely existed for over 130y."

3 See https://unfccc.int/process-and-meetings/the-paris-agreement. The Paris Agreement of the United Nations Framework Convention on Climate Change (UNFCCC) was adopted by 194 countries in December 2015 and entered into force on 4 November 2016 (ratified by 185 parties to date). Its main objective is to limit the global average temperature increase this century to well below 2°C (35.6°F). While this is intended as a long-term threshold, the latest five-year figure is the highest on record and shows that warming has already reached around 1.1°C (33.98°F).

4 WMO Global Annual to Decadal Climate Update (2025-2029); https://wmo.int/publication-series/wmo-global-annual-decadal-climate-update-2025-2029?access-token=pNLbdBu8q2rFHbkLrdh9YE5cold58Ic7lc47kQiUg4U.

5 Cf. Jonathan Watts, "Global temperatures could break heat record in next five years," *Guardian*, May 28, 2025; www.theguardian.com/environment/2025/may/28/global-temperatures-break-annual-heat-record-next-five-years-world-meteorological-organization.

6 Among the multitude of almost identical and skeptical observations in the last few decades on the inconvenient nature of democracy in the face of the historical challenge posed by the climate crisis, one could reference Claus Leggewie's (2010: 150) sober observation, for example,

> Authoritarian regimes may not only (as expected) be able to escalate global climate change with a top-down process of catch-up industrialization, but as green-enlightened oligarchies, they could also impose climate protection and energy transition on a scale and at a speed that Western democracies would not be willing or able to achieve.

In other words, "dictatorial politics is not *per se* an evil" (Leggewie, 2010: 151).

7 See Thomas Peterson, "Widerstandsreflex Ost [Resistance reflex East]," *Frankfurter Allgemeine Zeitung*, August 22, 2024, reporting on a representative nationwide survey in Germany: One of the indicators collected shows that more than half of the respondents agree with the statement that "we need a strong politician to lead instead of never-ending debates and compromises."

8 See David Wallace-Wells, "The world seems to be surrendering to climate change," *New York Times*, April 23, 2025; Cf. the declining importance of the topic of the climate crisis in the recent (2025) federal election in Canada: "Climate change, once a big issue, fades from Canada's election," *New York Times*, April 26, 2025; www.nytimes.com/2025/04/26/climate/canada-election-climate-change.html, www.nytimes.com/2025/04/23/opinion/climate-trump-world.html.

9 Governance, as opposed to government, is the broader term, intended to encompass and refer to a broader set of decision-making contexts than would be the case for formal (state) government.

10 Examples include the considerations of Yascha Mounk (2028: 243):

> Some elements of undemocratic liberalism are difficult to avoid. If we want to deal with climate change, we need international cooperation. And if we want to ensure that there are no dangerous chemicals in our food, we need to give considerable power to scientists and bureaucrats. An indiscriminate dash to return power to the people by scrapping independent agencies and abolishing international organizations would accomplish little.

11 Niklas Luhmann ([1992] 1998: 70, 95) strong objection is evident:

> The modern type of *expert*, that is, someone who, when asked questions he cannot answer, can be led back toward a mode of uncertainty [...] Today we can speak of the future practically only in terms of the probable or improbable that is, in terms of a fictively secured (duplicated by fictions) reality.

12 Even more ominous is the alleged "yearning for an authoritarian state" among significant segments of the East German public as discussed in a commentary in *Der Spiegel*: "Two-thirds of citizens in eastern Germany can imagine voting for extremists, and many would like to see a 'dictatorship of the majority'" (August 21, 2024; www.spiegel.de/kultur/wahlen-in-ostdeutschland-die-repraesentative-demokratie-ist-gefaehrdet-wie-nie-gastbeitrag-a-92aab600-bdb3-429d-91a0-33d872ccddfe).

13 The disconnect between science and politics has, of course, been framed in terms different from my emphasis on governance. It has been argued that the disconnect is due to moral deficiencies, lack of scientific literacy, capitalism, modernity as such, or the success of organized climate change denial. In the final analysis, the matter is complicated but boils down to governance.

14 I have examined a distinctive feature of the time horizon of the present as the simultaneity of the non-simultaneous (Stehr, 2022). The idea of the simultaneity (or the contemporaneity) of the non-simultaneous (non-contemporaneous). refers to layers of social phenomena. At a given time in the history of society (human time), the sum of the layers that make up the phenomenon of society consists of distinct attributes that emerged in different epochs, for example, different generations or indigenous and natural science knowledge. This is the normal situation in modern societies. Society lacks distinctiveness. What sounds like a deficit in some respects is however much more of a virtue.

15 Nikas Luhmann ([1976] 1990: 128–129) draws further attention to this and concedes that we can speak of an open future, which is "undoubtedly only a vague metaphor," as well as different degrees of openness: "If we accept the distinction between a present future and future presents, we can define an open future as the present future that has room for several mutually exclusive future presents."

16 Climate scientists have at times expressed frustration with the publics in democratic countries that have rejected climate policies, such as carbon taxes or fuel efficiency targets, through referendums or protests (see Andrea Kucera, "Die Wirtschaft ist der Klimapolitik voraus," *Neue Zürcher Zeitung*, May 22, 2022).

ACKNOWLEDGMENTS

Sections of the book arose, organically, from previous work. I have chosen not to acknowledge each gift. In fact, each loan is seldom a mere loan from an earlier reflection but a gain, as the gift often comes with substantial modifications. My basic plan, therefore, is to reciprocate by adding to the original gifts. I have translated the foreign quotations. "His" and "he" are used in a genetic sense. Due to the rapid evolution of issues related to climate, many citations of my analysis are taken from newspapers and other quick sources. References to items from newspapers and magazines are not listed separately in the reference section. A summary of the analysis of "A World of Our Making" forms the basis of a chapter in *Society and Knowledge in a Turbulent Era* (Stehr, 2025). My special gratitude, in the case of my reflections on climate, policy, and society, goes to Hans von Storch, my decades-long collaborator from the culture of the natural science community (e.g. Stehr and von Storch, 2023). Over many years, I owe thanks to William Catton, Aaron Cicourel, Ralph Matthews, and Volker Meja for their critical comments on my reflections on society and climate. The present work has benefited from the technical support of Marian Adolf, Dustin Voss, and Bruno Hopp. I would like to thank David Altheide, Jim Cohn, James Fleming, Steve Fuller, Ronald Glasman, John Hannigan, Mike Hulme, Andoni Ibarra, John Johnson, Myanna Lahsen, John Myles, Ortwin Renn, Göran Therborn, Wilhelm Vossenkuhl, and Rafal Wierzchosławski for their constructive comments.

INTRODUCTION

> A heat wave gripped Ontario and Quebec for a second consecutive day on Tuesday, while dangerously hot and humid conditions settled in over parts of Atlantic Canada. Environment Canada has warned of dangerously high temperatures this week – in Toronto, Ottawa and Montreal, daytime highs could hit 35 C, with the humidity making it feel like it's above 40.
>
> *The Globe and Mail, June 17, 2024*

> Five lead authors of IPCC reports told the *Guardian* that scientists should be given the right to make policy prescriptions and, potentially, to oversee their implementation by the 195 states signed up to the UN Framework Convention of Climate Change.
>
> *UNFCCC (as cited in the Guardian)*[1]

For centuries, the most powerful nation in the world tended to be a democratic state. Today, there has been a noticeable shift. The most powerful nation soon may no longer be a democratic nation. The secular trend represented by a shift toward the global ascendancy of autocratic political regimes is accompanied by a growing belief, which I would strongly question, that the "happy" unintended consequence of this change is a convergence between the trend away from democratic governance and the kind of political control needed to address the unique consequences of global warming. As a result of such expectations, the label of democracy will likely not disappear, just as it has not disappeared in autocratic forms of governance, but the way politics would then be conducted should bear little resemblance to democratic governance.

DOI: 10.4324/9781003682066-1

Daniel A. Bell (2015: 53), known for his relatively positive view of China's political system and state-controlled economy, which he calls "political meritocracy," is a prominent voice among social scientists promoting a perspective favoring a forceful disenchantment with democracy. Bell's endorsement of the complementary position of non-democratic governance, or what is often referred to as *authoritarian environmentalism*, first forcefully articulated by Richard Heilbroner (1974; Gilley, 2012),[2] is ultimately predicated on the enduring political influence of the Chinese Communist Party: "Perhaps the best reason to be optimistic is that the same party will likely still be in power several decades from now. The CCP is likely to stick to its long-term commitments." Daniel A. Bell (2015: 54) offers a defense of China's authoritarian model, arguing that political meritocracy can sometimes outperform democracy in governance while maintaining economic growth. In support of China's environmental policies, he stresses, only mildly reserved,

> I do not know if China will be able to develop an environmentally sustainable way of life, but so long as it does not adopt a democratic system that privileges the interests of the voting community at the cost of future generations, that's where I'd place my bet.

Indeed, Bell (2015: 52) predicts that non-democratic states may provide the most efficient policy response to global warming:

> the real player (among nondemocratic countries) in the climate-change debate is China, given its size and power. If we can't expect electoral democracies to take necessary measures to protect the interests of people several decades from now, is there any reason to be optimistic that China's unelected leaders can do better? From the planet's point of view, it is imperative that China set a good example. If China were to use as much energy per capita as the United States does, it would produce carbon emissions in such volume that our planet would be condemned to unmanageable climate change. So, clearly, China must embrace an outlook that puts resource management at the center of all policy making.

More recently, Peter Drahos (2021) has defended the same political perspective in his *Survival Governance*. Survival governance is a call for radical systemic change. To find ways to steer humanity away from catastrophe, especially to address existential threats such as climate change, pandemics, and nuclear war, we need a new paradigm of technological and social development aimed at restoring ecological systems—the bio-digital energy paradigm—and China is the world power best positioned to lead this change. Drahos argues that existing governance structures are ill-equipped to deal with these complex, interconnected crises that transcend national boundaries and require coordinated, long-term solutions. A key theme of Survival Governance is a critique of short-term political and economic thinking. Drahos shows how neoliberal capitalism, with its focus on profit maximization and deregulation, has exacerbated environmental degradation and inequality and undermined global stability. He calls for a shift to stewardship-based governance, where decision-making considers ecological limits and intergenerational equity. This involves rethinking property rights, corporate accountability, and international law to align them with planetary survival. It follows that no capitalist state "is in a strong position to deflect world capitalism from its current trajectory. China probably represents the best of a set of poor chances" (Drahos, 2012: 3).

In short, China's centralized system enables more effective long-term environmental planning than most Western democracies. Authoritarian environmentalism—often paired with "authoritarian capitalism"—underpins much of the analysis in this book. Authoritarian capitalism is an economic system in which a market economy and private property rights coexist with a powerful centralized government that maintains strict control over political and social affairs. It blends capitalism's free markets and private ownership with authoritarian rule marked by limited freedoms and restricted pluralism (see Witt and Redding, 2014). The definition of the concept highlights the degree to which such a policy approach rests on restrictions on the liberty of individual citizens and social collectives, let alone permissible participation *and* emotions:

> authoritarian environmentalism can be provisionally defined as a public policy model that concentrates authority in a few executive agencies manned by capable and uncorrupted elites seeking to improve environmental outcomes. Public participation is limited to a narrow cadre of scientific and technocratic elites, while others are expected to participate only in state-led mobilisation for the purposes of implementation. The policy outputs that result include a rapid and comprehensive response to the issue and usually some limits on individual freedoms.
>
> *(Gilley, 2012: 288)*

Populists—often backed by wealthy elites—support environmental authoritarianism because they reject climate science and its warnings. This is further reflected in opposition to multilateral agreements and foreign influence as unnecessary and suspicious, strong support for fossil fuel extraction and use, and the description of climate scientists and politicians advocating mitigation and adaptation as "corrupt elites" engaged in a conspiracy to defeat the will of most citizens (Fiorino, 2022). It is precisely this fundamental attitude, or even fear, toward international organizations and agreements that also applies to political action in Trump's second term.

A political strategy of the German populist party AfD in climate change discussions is to mimic

> scientific language to criticize the authority of scientific institutions and assumptions. [...] populists politicize such reflections: they decontextualize scientific expertise to point to scientific uncertainties and highlight the political interests of researchers in order to disqualify them and 'reveal' false evidence.
>
> *(Boecher et al., 2022: 835)*

Erik Swyngedouw (2022: 904) argues, contrary to my description, that "most mainstream as well as many radical climate discourses, practices, and policies are formally similar to populist arguments and should be considered as an integral part of a deepening process of post-politicization." The overlap between environmental and populist rhetoric stems from shared reactions to 'post-politicization'—a condition that obscures deeper social issues like class and capital. It is doubtful whether this discredited view adds to our understanding.

Populists are advocates of *political* authoritarianism (Applebaum, 2020; Frankenberg and Heitmeyer, 2024; Zürn, 2024). Hence also the appropriate designation of *authoritarianism populism* (Lockwood, 2018). Either support the *status quo* or argue for a *status quo ante* that defeats environmental authoritarianism. To create a theory of "autocratization," Michael Zürn (2024: 62; my emphasis) develops a typology of two forms of authoritarian political rule:

> The first is *populist authoritarianism*, which is ideologically linked to the upheavals in liberal democracies in the wake of globalization. The second is called *technocratic authoritarianism*, which defines the form of old autocracies that have managed to resist the spread of democracy.

Whereas Zürn sees populist authoritarianism as political, I classify environmental authoritarianism as technocratic.

The slogan "system change not climate change" (Beer, 2022) commonly seen on banners, placards, and signs at climate change protests, such as Fridays for Future or Extinction Rebellion, seems to unite environmental and political authoritarianism. Both movements share an essential disillusionment with current democratic practices and in making demands against the state and/or elites in the name of the members of the movements (cf. Caramani, 2017). Both radical views call for political orders that bypass democratic norms to enable rapid action, even against constitutional limits.

My emphasis in this book on the risks and dangers of climate change is less on the failure or promise of capitalism (Felli, [2016] 2021) as the culprit, but rather, as in the case of China, on populist critiques of climate policies on *governance*. My focus is on the potential of democratic governance to guide our response to climate-driven social change.[3] An ascending view in the form of, at times, still cautious predictions is, however, frequently, in both climate science and in the social sciences that "global climate change is likely to make it difficult for liberal democratic political systems to govern effectively and [...] is also likely to end the dreams of progress that liberal democracies have helped to produce." The future of democracy is seen as threatened. In practice, efforts to promote democracy internationally are weakening; as one observer notes, "democracy support is entering a much more uncertain, and potentially difficult, era" (Hobson, 2012: 977). This does not imply that the same critics emphasize that governance dominated by "capitalism may fail, or that totalitarian political regimes will flourish" (Edmundson and Levy, 2023: 33). Whether the Chinese government "seems to be taking steps in the right direction," with respect to responding to the climate change consequences, as Daniel A. Bell (2015: 52) implies, remains to be seen (also Gilley, 2012: 289–292).

Sunday, July 21, 2024, was the *world's* hottest ever recorded day. This record will likely be surpassed soon. According to preliminary data from the *Copernicus Climate Change Service*,[4] the average surface air temperature reached 17.09°C (62.76°F). The *Guardian* quotes the Copernicus director, Carlo Buontempo,

> what is truly staggering is how large the difference is between the temperature of the last 13 months and the previous temperature records. We are now in truly uncharted territory – and as the climate keeps warming, we are bound to see new records being broken in future months and years.[5]

As the *Washington Post* reported on August 1, 2024,

> ground temperatures in *East Antarctica* have soared more than 50 degrees (28 Celsius) above normal in the second major heat wave to afflict the region in the past two years. This historic warm spell could persist for another 10 days and is an ominous example of the major temperature spikes this polar climate could experience more frequently in a warming world.

The *World Meteorological Organization's* latest report calculates an 80% chance that at least one of the next five years will top 2024 as the warmest year on record.[6] It is happening faster. The trajectory of global warming shows it is getting hotter, faster. A report published in the journal *Earth Systems*

Science Data (Forster et al., 2025) in June 2025 found that human-caused global warming is now increasing by 0.27 degrees Celsius per decade. That rate was recorded at 0.2 degrees in the 1970s and has been growing since.

In addition, from such short-term views, a longer perspective on the impact of global warming also shows dramatic impacts; for example, "many glaciers in the tropics are probably now smaller than they have been in at least 11,700 years, making the tropics the first large region where this milestone has been documented" (Gorin et al., 2024) or, according to a report, named *Planetary Solvency*, by risk management experts of the *Institute and Faculty of Actuaries* (University of Exeter),[7] the prediction that the global economy could face a 50% loss in gross domestic product (GDP) between 2070 and 2090 unless immediate action is taken to decarbonize. Moreover, the report expects "at 3C or more of heating by 2050, there could be more than 4 billion deaths, significant sociopolitical fragmentation worldwide, failure of states (with resulting rapid, enduring, and significant loss of capital), and extinction events." The fragility of such predictions is illustrated by other observations, confidently claiming that the macroeconomic damage from future climate change is more likely to be mild to moderate (see Tol, 2024; Neal et al., 2025).

Launched by *Climate Central* in 2022, the *Climate Shift Index* (CSI) indicates how climate change has altered the frequency of *daily temperatures* at any location around the world. Changes in the index may therefore be more relevant to the perception of climate change in everyday life since

> people don't experience global average temperatures. Instead, we mainly experience climate change through shifts in the daily temperatures and weather patterns where we live. Bridging this gap, the *Climate Shift Index* is a system that quantifies the influence of climate change on local daily temperatures around the world.[8]

According to the index, Germany, for example, has lost an average of 18 winter days (now above-freezing days) over the past decade. As a result, winters that are not cold enough to keep snow on the ground are becoming increasingly rare. Compared with other regions, Europe is reported to have lost a particularly large number of winter days, especially in Denmark, Estonia, Latvia, and Lithuania. On average, the temperature is above zero degrees Celsius for at least three extra weeks each winter. In the United States, 28 states and approximately 63% (39 of 62) of the cities surveyed lost at least one week of winter days.

These long-term and short-term figures highlight the increasing rise in concerns about the future of democracy. The fates of climate change and democracy are intertwined. As Benjamin Barber (2012: 165–166) asks, how can democracy afford to be present in the face of the brutal reality of the global challenges it faces? His answer was pessimistic: "Democracy affords a politics of procrastination that is quite deadly in the environmental arena [and ...] democracy tethered to the sovereign national state, who borders are evermore irrelevant to the challenges of a borderless world." Climate change does not know or respect borders.

Today's critics of democracy both from the left and the right of the political spectrum agree that democracy is in danger. The general legitimacy of the democratic system is dwindling. As the weekly magazine *The Economist*[9] noted, "political scientists have been increasingly dissatisfied with the state of the world's democracies since the early 2010s." The term "*democratic backsliding*"[10] was used to describe the assault on the rules of constitutional democracy and the decline in political competition or civil rights in countries as Brazil, Hungary, the Philippines, Tunisia, and Turkey appeared 6.4 times more frequently in books in 2019 than in 2010, according to the Google NGram database. In general, **democratic backsliding**, born out of particular social conditions, refers to the gradual decline in the quality of democracy within a country. It is characterized by the weakening of democratic institutions, the erosion of political freedoms and civil liberties, and the undermining of checks and balances. Democratic backsliding can become manifest in various forms, for example such as: **undermining electoral integrity, attacking the independence of the judiciary, restricting civil liberties, consolidating executive power and eroding checks and balances**: "This process is propelled by the desire of a government to monopolize power and to remove obstacles to realizing its ideal policies" (Przeworski, 2019: 173). The narrative of democratic backsliding is sometimes embraced, and perhaps thereby fuelled, by democratic politicians who envisage the demise of democracy because of populist political movements gaining power.[11]

Scholars now speak of a "third wave of autocratization" that partially reversed the post-Cold War surge in democratization.[12] The erosion of democratic institutions has affected countries at relatively advanced levels of economic development, including those with long-standing democratic traditions. Part of the process of democratic backsliding in today's democratic countries is the support of significant parts of the voting public for authoritarian policies. The consensus among scholars about recessions in democracies is not unanimous, nor is their agreement about the timing or the reasons for the decline. In recent years, growing empirical evidence has strengthened the consensus that democracies are becoming less resilient. It's still unclear whether environmental problems are fueling reactionary trends—or if the two developments are merely coincidental.[13] A more radical view questions whether democracy is actually in decline at all.

In a decade-old article, Levitsky and Way (2015) maintain that the gloomy narrative about the decline of democracies around the globe is a myth. Specifically, the

> state of global democracy has remained stable over the last decade [2005–2014], and it has improved markedly relative to the 1990s. I argue, however, that perceptions of a democratic recession, are rooted in a flawed understanding of the events of the early 1990s. The excessive optimism and voluntarism that pervaded analyses of early post-Cold War transitions generated unrealistic expectations that, when not realized, gave rise to exaggerated pessimism and gloom. In fact, despite increasingly unfavorable global conditions in recent years, new democracies remain strikingly robust.
>
> *(Levitsky and Way, 2015: 45–46)*

Ekiert Grzegorz (2023: 1) noted that there is an often overlooked parallel political backsliding affecting hybrid and authoritarian regimes: this correlated process

> can be described as dictatorial drift and implies the transition from 'soft' forms of authoritarian rule to hard core authoritarian policies characterized by the concentration of executive power, the destruction of political institutions such as fair elections, independent judiciary, free media, and autonomous civil society organizations, and worsening political repressions.

Grzegorz explained that the origins and the societal processes that drive backsliding in democratic and autocratic political regimes are dissimilar. In democracies, backsliding is shaped by public demand and institutional weaknesses; in authoritarian regimes, it is imposed from the top down.

A survey conducted by the *International Institute of Democracy and Electoral Assistance* (IDEA), Stockholm, between July 2023 and January 2024, which included approximately 1,500 people in each of 19 countries including the United States, Brazil, Chile, Colombia, Gambia, Iraq, Italy, Lebanon, Lithuania, Pakistan, Romania, Senegal, Sierra Leone, South Korea, and Tanzania established that voters in many countries are sceptical of democracy. In many developed countries, distrust in democratic institutions has become the norm. The *Edelman Trust Barometer* from 2022 reported that there has been a collapse of trust in developed democracies; many failed to reach a score of 60 on the 2022 Trust Index, including Australia (53), France (50), Germany (46), UK (44), and the United States (43), which has dropped ten points since 2017. Contrast that to the trust jumps in non-democratic countries—China (83) and the UAE (76), which saw increases of 11 and nine points, respectively. The likely cause of this distrust? Most respondents in developed countries believe they'll be worse off financially in five years.[14]

As Eva Darian Smith (2022: 93) noted, "a feature of the global drift toward antidemocracy is the weaponizing of climate change." In many countries, active support at the ballot box for parties with anti-liberal, nationalist (anti-transnational political commitments), nativist, contempt for minorities, anti-elitist, and traditional values is growing, or the political platforms of traditionally democratic parties are drifting toward more populist policies. Populism, at least in the eyes of its detractors, is also characterized by a lack of respect for knowledge, scientific evidence, and expertise, and an overreliance on conspiracy theories (Conti, 2014). Anti-environmentalism is an important element in the mix of populist political agendas. In the United States, for example, fewer than half of the people believed that the 2020 presidential election was free and fair. Robust democracies are far from being established in many countries. In fact, evidence shows that democracies around the world have declined in number as well.[15]

In short, the optimism expressed by Fukuyama (1992) in the 1990s—that global democracy marked the end of major political upheavals—has now been replaced by growing fears that democracies are under siege. Fukuyama's view was not lonely. He was articulating the political conviction of his time. At the time, most political observers agreed that: "liberal democracy is the only political and economic system capable of generating sustained prosperity and political stability [...] These are remarkably rare features of human societies, both in history and today" (Wolf, 2004: 32–35). Presumably, as not only Futurama's claim implies, the remarkable political and economic period achieved by liberal democracies would be a permanent stage. But history has proven these hopes for lasting political stability and prosperity to be misguided.

A survey carried out in 2019 by the *Hansard Society* in the United Kingdom, known as the *Audit of Political Engagement*, showed that public opinion of the governing system was at its lowest point in the 15-year history of the study. The 2019 year's report contained several striking findings, including that 54% of respondents said that "Britain needs a strong leader willing to break the rules," 43% preferred political parties and leaders "with radical ideas for change who haven't been in power before," and 42% believed that "many of the country's problems could be dealt with more effectively if the government didn't have to worry so much about votes in Parliament." Furthermore, less than 35% of respondents were confident that MPs, parties, and the government act in the public interest, compared to over 60% for judges and the military.[16]

Worldwide crises such as the pandemic, neo-nationalism, financial miseries, and predicaments associated with globalization, and the evolution of capitalism may have contributed to and accelerated a robust disenchantment with democratic governance. The developed democracies of the world, for instance, the 34 member states of the OECD, all experience doubts expressed about democracy and its capacity to perform. In many countries, disappointing personal economic gains go hand in hand with a sense of disillusionment with democratic governance (cf. Heitmeyer, 2018). Nevertheless, I argue that democratic principles and processes are best suited to address the grand challenge of climate change—even in the face of unprecedented cognitive demands, diminishing national decision-making space, growing transnational challenges, and problems of political scale.

The role of knowledge is central to any debate about the nature of the future present. However, I will refrain from making or quoting (numerically) specific predictions based on scenarios developed, e.g. a decade or more into the future, about the future present affected by climate change,[17] e.g. the extent of heat-related deaths, the economic impact on global or national GDP, any intellectual impact,[18] the increase in wildfires,[19] or the spread of diseases. Exact predictions are impossible. Vaclav Smil's (2019: 425) cautionary note about quantitative predictions about the underlying causes of economic growth also applies to predictions about the future present of the consequences of climate change: "Broad qualitative conclusions explaining the roots of economic growth and the sources of productivity are thus preferable to allocations of shares attributable to specific causes." However, there is nonetheless a flood of such foresights. I will quote but one of these descriptions of the likely catastrophic future present by the British naturalist and broadcaster Sir David Attenborough who issued the warning to political leaders in 2021: "If we continue on our current path, we will face the collapse of everything that gives us our security: food production, access to fresh water, habitable ambient temperature, and ocean food chains," he said, adding "and if the natural world can no longer support the most basic of our needs, then much of the rest of civilization will quickly break down."[20] When predictions are made, for example, with respect to the size of the global economic surplus or deficit, they lack seriousness and are part of the political struggle over climate change.

However, discourses on climate in the social sciences in recent decades range from the complete absence or discussion of any such concerns to vigorous debates about the likely impacts of climatic conditions, of course, in response to the discovery of climate change in the natural sciences. The discourse on climate in the social sciences over the past few centuries has had radically different perspectives: on the one hand, many social, economic, and cultural conditions are attributed to climate, but climatic conditions are described as essentially fixed and unchangeable. In the tradition of this narrative, climate is about how climate makes us and our world. These perspectives, represented by several major contributors to whom I will refer in a later section, are summarized under the heading "climate determinism." More recent views, on the other hand, emphasize a changing climate and how a variable climate produced not least by human "effort" influences social, economic, and cultural conditions.

In contrast to a straightforward neoliberal policy agenda, especially its conviction that the state must get out of the way of economic affairs of society to allow for the unfettered pursuit of self-interest, the opposing call for a strong state remains a powerful voice in the discussion of how to navigate the difficult future present embedded in the challenges posed by climate change. Ultimately, the questions posed in this context revolve around how to govern in the face of rapidly changing environmental conditions. If urgent action and novel responses are essential for adaptation and mitigation, will democratic governance prove to be an asset or a danger?

In a spirit of self-criticism, I acknowledge that my analysis predominantly focuses on liberal democratic and Western institutional frameworks (e.g. the UNFCCC, US, and European environmental policy), overlooking insights from Indigenous, non-Western, or localized governance models that could already incorporate elements of reflexivity.

The following chapter outlines the urgent state of climate change, detailing its present and projected future impacts. It highlights rising global temperatures, shrinking ice sheets, more frequent and intense heatwaves, and the increasing prevalence of extreme weather events such as droughts and hurricanes. Greenhouse gas emissions, especially CO_2 and methane, are accelerating these changes, with significant impacts on human health,[21] ecosystems, and economic stability. Studies show that wealthier populations contribute disproportionately to emissions, while poorer communities suffer most from climate impacts. Despite growing renewable energy use, fossil fuel consumption remains high. Understanding what we currently know about climate change is crucial—not just for imagining the future, but for shaping it. This understanding, of course, extends to a rejection of the climate science community's consensus. Chapter 1 also explores controversial geoengineering proposals and emphasizes the role of adaptation in mitigating climate damage. It concludes by stressing the necessity of political will, global cooperation, and social justice to address this multifaceted "wicked problem." Technological solutions alone are insufficient; equitable and comprehensive adaptation strategies are essential for resilience in the "future present" shaped by accelerating climate change.

Notes

1 *The Guardian*, December 7, 2023: "We need power to prescribe climate policy, IPCC scientists say"; www.theguardian.com/environment/2023/dec/07/we-need-power-to-prescribe-climate-policy-ipcc-scientists-say. The five IPCC authors do

not represent the entirety of IPCC scientists, but the views of even a small group with such a strong political agenda are noteworthy.

2 Bruce Gilley's (2012: 300) affirmative summary of the effectiveness of authoritarian environmental protection—which he calls a useful and important counterweight to democratic environmental protection—is based primarily on the following characteristics: "Authoritarian environmentalism's merits are its ability to produce a rapid, centralised response to severe environmental threats, and to mobilise state and social actors."

3 I have had the opportunity to comment on and analyze the connection between cognitive, political, and climate change challenges. For example, I contributed to an edited volume by Glückler, Herrigel, and Handke (Stehr, 2020) and published two articles in 2015 (Stehr, 2015a; Stehr, 2015b). In this volume, I attempt to maintain and extend the theoretical position developed in these contributions.

4 https://climate.copernicus.eu/; Copernicus, previously known as GMES (Global Monitoring for Environment and Security), is the European Union's Earth observation programme.

5 Guardian, July 23, 2924; www.theguardian.com/environment/article/2024/jul/23/world-temperature-records-shattered-hottest-day-climate-crisis.

6 "WMO global annual to decadal climate update (2025–2029)"; https://wmo.int/publication-series/wmo-global-annual-decadal-climate-update-2025-2029?access-token=pNLbdBu8q2rFHbkLrdh9YE5cold58Ic7lc47kQiUg4U.

7 "Global economy could face 50% loss in GDP between 2070 and 2090 from climate shocks, say actuaries," *Guardian*, January 16, 2025; www.theguardian.com/environment/2025/jan/16/economic-growth-could-fall-50-over-20-years-from-climate-shocks-say-actuaries.

8 The *Climate Shift Index* ranges from –5 to +5. Positive levels indicate temperatures that are becoming more likely due to climate change (negative scores indicate conditions that are becoming less likely); www.climatecentral.org/climate-shift-index.

9 "Global democratic backsliding seems real, even if it is hard to measure," *The Economist*, September 12, 2023.

10 See Przeworski (2019: 172–1760); Levitsky, Ziblatt, and Daniel (2018); Bermeo (2016); Acemoglu and Robinson (2006).

11 In *How Democracies Die*, political scientists Steven Levitsky and Daniel Ziblatt (2018) explore the subtle, incremental ways that democracies can erode from within—not through violent coups, but via the rise of elected leaders who gradually subvert democratic norms and institutions. Drawing on global and historical examples, including Venezuela under Hugo Chávez and interwar Europe, the authors show that authoritarianism often emerges through legal mechanisms, cloaked in democratic legitimacy. The authors sometimes conflate norm-breaking political behavior with outright authoritarianism, which could blur distinctions between flawed democracy and democratic collapse.

12 See also the *Economist*'s status report: "Where democracy is most at risk," February 14, 2023. As "objective" as these attributions may be through the development of corresponding indicators, these status reports reflect an increasingly widespread assumption about the state of democracies among social science observers and in the media. In the long term, the impact of climate on people appears to be small, perhaps even negligible, and difficult to detect.

13 Günter Frankenberg and Wilhelm Heitmeyer (2014: 28) are however convinced that the "climate catastrophe is becoming a driver of authoritarianism in several respects: it promotes strategies of denial and, in the case of further policy failure, the ecological state of emergency."
14 The Trust 2020 survey is based on an online survey in 28 markets 34,000+ respondents' total. All fieldwork was conducted between October 19 and November 18, 2019 (www.edelman.com/trust/2022-trust-barometer).
15 The *Varieties of Democracy Institute* (V-Dem) at the University of Gothenburg in Sweden has been studying the declining democracy trend: Staffan I. Linberg, the institute's director reports, "A democratic decline has taken place globally, and an increasing number of people are living in closed autocracies [...] The level of democracy enjoyed by the average world citizen in 2022 is back to 1986 levels. This means that 72 percent of the world's population, 5.7 billion people, today live under authoritarian rule"; www.gu.se/en/news-and-events/find-news?hits=25.
16 www.hansardsociety.org.uk/'strong-leader'-helps-provoke-responses-to-2019.
17 A *Guardian* (May 8, 2024) survey of hundreds of the world's leading climate scientist finds that "80% of the respondents, all from the authoritative Intergovernmental Panel on Climate Change (IPCC), foresee at least 2.5C of global heating, while almost half anticipate at least 3C (5.4F). Only 6% thought the internationally agreed 1.5C (2.7F) limit would be met. Many of the scientists envisage a "semi-dystopian" future, with famines, conflicts and mass migration, driven by heatwaves, wildfires, floods and storms of an intensity and frequency far beyond those that have already struck" (www.theguardian.com/environment/article/2024/may/08/world-scientists-climate-failure-survey-global-temperature).
18 Park, Behrer, and Goodman (2021).
19 Despite its significant potential to cause damage, the impact of climate-induced wildfire smoke is rarely considered when estimating the societal costs of climate change. A study of pollution caused by wildfires in the United States based on "new causal evidence that exploits the natural experiment of drifting wildfire smoke [shows] that the relationship between air pollution exposure and health is concave: small increases in pollution cause significant health impacts, while larger increases yield diminishing additional effects" (Miller et al., 2024; Qiu et al., 2025). Significant also are the CO_2 emissions of wildfires. For example, Byrne et al. (2023) quantified the carbon emissions from the 2023 Canadian wildfires. They concluded that the magnitude of these emissions was 647 TgC (570–727 TgC), which is comparable to the annual fossil fuel emissions of large nations. Only India, China, and the USA release more carbon per year.
20 Source:https://news.un.org/en/story/2021/02/1085452.
21 For example, a study with data from the Middle East and North Africa by Wafa Abu El Kheir-Mataria and Sungsoo Chun (2025) found "a significant correlation [emphasis added] between prolonged exposure to high ambient temperatures and all four cancer types studied. Notably, the prevalence of breast, ovarian, and cervical cancers is markedly influenced by temperature increases."

1

THE CLIMATE CLOCK IS TICKING

It is certain, that of all Powers in Nature, Heat is the chief.

Francis Bacon *Sylva Sylvarum* (1627), Century X, §876[1]

And now we tend to view the atmosphere almost as something itself alive, because it is the product of the biota[2] and an essential channel by which elements of the great living animal communicate – it is indeed the milieu internal which is maintained by the biota as a whole for the wellbeing of its components.

Dian Hitchcock in a letter to James Lovelock, 1966[3]

I discuss the state of our knowledge about the nature and impact of climate change under two headings: (1) state of the art—the present—and (2) state of the art—the future present. In all instances, Francis Bacon's definition is appropriate; it underscores his view of heat as a fundamental force in nature. The knowledge acquired to date relates largely to the present state of nature and the impacts of climate change. More speculative and uncertain knowledge—such as future risks—falls under what I call the "future present" (Stehr, 2001). I list and comment on what can be considered robust knowledge about the future present. For example, data on mortality patterns associated with past and observed extreme weather events can be extended into the future present, assuming limited changes in the conditions that accompany extreme events.

State of the art—the present

While Earth's climate has always changed, it hasn't typically done so abruptly—or within human history. Satellite data show that the Earth's

DOI: 10.4324/9781003682066-2

albedo (the amount of solar energy reflected into space) has been decreasing since the beginning of the 21st century. The solar light that is not reflected is absorbed by the Earth system and contributes to the warming of the planet. The decreasing albedo is one of the reasons why the Earth's warming has accelerated to approximately 0.2°C per decade. One cause of "dimming" is air pollution, and the other is the melting of ice and snow, which exposes darker surfaces such as ocean water and land, which absorb more solar radiation.[4] However, CO_2 pollution, the release of other greenhouse gases into the atmosphere, and the strengthening of El Niño combined to cause longer and more intense heat waves and droughts[5] expose billions of people to hot days and dryness, such as like those in the spring and summer of 2024. During heat waves, chemicals such as formaldehyde and ozone are more likely to form in the air. Climate change and deteriorating air quality are happening together.[6]

Despite its severity, air pollution receives far less public and political attention than visible disasters like floods or storms. The complexity and interconnection of its causes are the main reasons. These causes include heat waves, wildfires, mobility, and industrial activities. Bad air is much more deadly than floods, fires, and heat waves. The danger is on a completely different scale. The World Health Organization is clear: air pollution kills seven million people every year.[7]

The summer of 2023 broke many records, only one year earlier. The geographers Jan Esper and colleagues (2024) have now shown this in the journal *Nature*. Researchers combined data from temperature measurements with tree rings and concluded that the summer of 2023 was the warmest in at least the last 2,000 years.[8] This does not refer to the entire globe but only to the Northern Hemisphere, excluding the tropics (Figure 1.1):

According to Esper's study, in the last 2,000 years, the year 246 stands out as the year with the warmest summer before the Industrial Age. At that time, the Roman Empire was in an imperial crisis. Today, summers are significantly hotter: of the past 28 years, 25 have exceeded the temperatures of the year 246. In 2023, preliminary results from an international team of researchers revealed that the amount of carbon absorbed by the land temporarily collapsed. The result was that forests, plants, and soil—as a net category—*absorbed almost no carbon.*[9] On the other hand, rivers and streams are a significant source of carbon, as they release carbon dioxide and methane into the atmosphere through their surface waters, until now, CO_2 and CH_4 emitted from rivers were thought to be predominantly derived from recent (sub-decadal) biomass production. A new study indicates, however, previously unrecognized release of old, pre-industrial-aged carbon to the atmosphere from long-term soil, sediment, and geologic carbon stores (Dean et al., 2025).

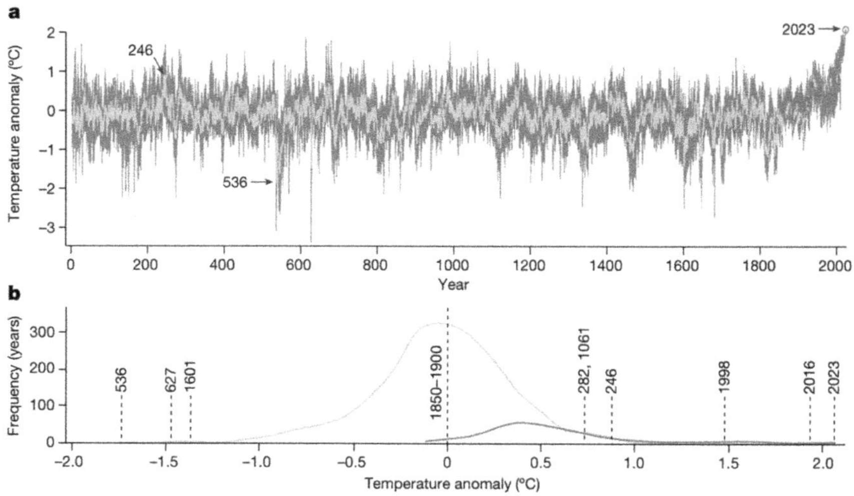

FIGURE 1.1 Reconstructed temperature extremes. Comparison of temperature anomalies in the four warmest summers (246, 282, 1061, 986 CE) and four coldest summers (535, 627, 1601, 1642 CE).

Source: Esper, Torbenson and Ulf Büntge (2024).

The consequences of global warming can be seen in many weather phenomena, according to the report "Global Climate Highlights 2024" from the European Union's *Copernicus Earth Observation System*. Most significantly, Europe is the fastest-warming continent. All four European seasons were warmer than average. Spring and summer were the warmest on record, with spring 0.36°C warmer than the previous warmest, in 2014, and summer at 0.20°C warmer than the previous warmest, in 2022. The European winter (December 2023 to February 2024) was the joint second warmest on record, at 1.44°C higher than the 1991–2020 average, and 1.4°C cooler than that of 2019/20, the warmest European winter. As fossil fuel companies reaped record profits, the average temperature for European autumn (September to November) 2024 was the third highest on record, at 1.25°C above the 1991–2020 average; 0.21°C cooler than the warmest autumn in 2020 (Figure 1.2).

Moreover, the more recent and more *regional* heat extremes are remarkable: In New Delhi toward the end of the month of May 2024, for example, official temperatures reached almost 50°C:

> The heatwave has been building up inexorably for weeks, but even so residents were shocked by conditions on Tuesday and Wednesday. People told of fingers being scorched from touching the steering wheel of a car, and tap water was coming out at boiling temperatures.

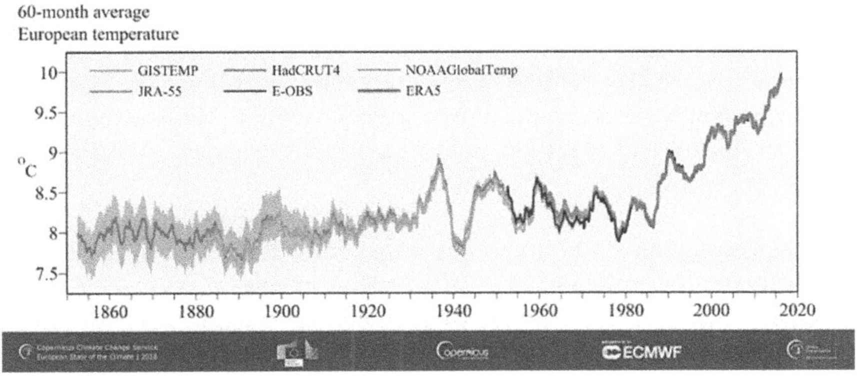

FIGURE 1.2 60-month average European temperature (up to 2019).
Source: Copernicus "*European State of the Climate 2018 report*"; *Copernicus Climate Change Service (C3S)/ECMWF.*

However, as Clarke and his colleagues (Clarke et al., 2022) stressed, unlike

> some other impacts of climate change, extreme weather events are manifesting on immediate timescales and changes in extremes are poorly described by the climatological means studied in many projections [...] As a result, there is no systematic basis to quantify the major contribution of human influence on extreme weather to the costs of climate change.

An analysis by Belloca, Poncela, and Ruiz (2024) about the impact

> of temperature exposure on the earnings per share of large European firms over the 21st century. Our findings reveal that earnings are sensitive to extreme temperatures across a large proportion of sectors. Depending on the sector and quarter, exposure to extreme temperatures can have either a positive or negative impact on profitability.

A new study (Wong et al., 2024) revealed that the world's oceans are currently facing a "triple threat" of extreme heating, a loss of oxygen, and acidification.[10] Acidification of the oceans is often called the "evil twin" of climate change. These conditions have intensified in recent decades, placing significant stress on the diverse marine life of the planet. These changes can pose a substantial threat to marine organisms and ecosystems, severely reducing the available habitable space.

To date, the major public and policy attention in mitigation efforts has focused on CO_2 emissions. However, another greenhouse gas, *methane*, long ignored despite being the most serious greenhouse gas,[11] is an equally or perhaps more serious problem. Methane gas is a by-product of the drilling and processing of fossil fuels. The methane emitted from livestock and rice production contributes. Rising global warming contributes to methane emissions through the decomposition of organic matter. Methane emissions are responsible for half of the global heating already experienced. In terms of methane emissions,

> the main component of natural gas is far more effective at trapping heat while it lasts. Over the first two decades after its release, methane is more than 80 times more potent than carbon dioxide in terms of warming the climate system.

Moreover, methane emissions have increased since 2006 and continue to do so. Carbon dioxide has a much longer retention time in the atmosphere than methane does. Hence, there would be a much more rapid response rate in stalling or reducing warming by removing methane or reducing methane emissions. Methane is referred to as a "super-pollutant."

Among the more specific risks of a warming planet that already exist, especially in recent years, are the number of fatalities caused by heat waves and lightning. An examination of European mortality figures for the year 2022 by Beck et al. (2024) concludes that the

> record-breaking temperatures in Europe during the 2022 summer were associated with over 60,000 heat-related deaths. By combining epidemiological models with detection and attribution techniques, we attribute half of this mortality burden (~56% [95% CI 39–77%]) to anthropogenic warming. Likewise, this applies to all sexes, ages, and heat-related mortality burdens during previous years (2015–2021).

However, caution is needed in discussions about heat death and lightning. A telling example of the one-sidedness of the discussion and climate protection efforts is the often clinically used term "heat-related deaths." People are only victims of nature and not victims of certain social conditions that expose those affected to extreme heat and do not protect them proactively. Heat-related deaths may ultimately protect only the municipalities, regions, or states that fail in their duty of care. The use of the term heat death almost guarantees that the underlying societal conditions and developments will repeat themselves owing to thoughtlessness and neglect.

Adaptation can be effective, but it is deeply shaped by social, political, and economic complexity.[12] For example, forward-looking measures are effective in preventing heat-related deaths. While a tragedy occurred in Chicago in mid-July 1995, with more than 700 "heat deaths," the so-called "Hot weather health warning and watch system" in the city of Philadelphia saved the lives of approximately 300 people that summer. Experiences with extremely high temperatures in Philadelphia in 1993 and 1994 triggered the development of an efficient warning system and social networks that benefited older and other vulnerable people. What does that mean? In fact, the isolation of older people in Chicago who did not know how to help themselves, or the much more concentrated poverty (and thus also helplessness) in these areas ten years ago, led to a high number of deaths (cf. Klinenberg, 2002).

On a global scale, as Ishan Nath (2025) has explored, shifts in productivity induced by climate change affect the global allocation of labor and production between agriculture and non-agriculture on a global scale. Climate change disproportionately harms agricultural productivity, particularly in hot, impoverished countries. Conversely, nonagricultural sectors such as manufacturing and services are much less sensitive to extreme temperatures. This disparity suggests that shifting labor and production toward nonagricultural sectors and increasing food imports could be powerful adaptation strategies. However, empirical evidence and model simulations show that this reallocation is limited by two major constraints: subsistence-driven consumer demand for domestic food production and high trade barriers in low-income countries. These constraints demonstrate the complexity of global adaptation measures, as they create what the author refers to as a "food problem": a situation in which vulnerable countries must continue agricultural production despite declining productivity, thereby exacerbating climate damage.

In recent years, due to global heat waves in many countries,[13] heat death and heat inequality have received renewed attention. The summer of 2025 in Europe signaled the change from near-perfect weather into a "paradise lost," as the *New York Times* reported: "Across Spain, Italy, Greece, France and beyond, sand-devouring storms, rising seas, asphyxiating temperatures, deadly floods and horrific wildfires have year after year turned some of the continent's most desired getaways into miserable locales to get away from."

Heat waves are the most lethal form of extreme weather. The number of heat deaths is difficult to track. There are no established and agreed-upon criteria for recording the contribution of extreme weather events to mortality statistics. The climate scientist Friederike Otto noted,

> heatwaves are the deadliest type of extreme weather, but they don't leave a trail of destruction or striking images of devastation. They kill poor, lonely people in rich countries and poor people working outdoors in developing countries. In the last 13 months, there will be thousands and thousands of stories of poor people dying in heat that will never be told.

According to a recent study by Howard et al. (2024), using data from the *Federal Center for Disease Control*, 2023 was the hottest year on record in the United States[14] that led to at least 2,325 heat-related deaths. According to the same study, more than 21,518 people have died from heat since 1999; most of the heat deaths have been recorded in the southwestern United States, and in the states of Arizona, California, Nevada, and Texas. This study "found that heat-related mortality rates in the US increased between 1999 and 2023, especially during the last 7 years" (Howard et al., 2024).[15]

The research on "excess" mortality patterns that I have cited is limited to a single point in time or a small range of points in time. Immediate deaths and damage can be observed immediately after a disaster. Delayed downstream outcomes caused indirectly by a disaster are difficult to trace back to the initial extreme event. Rachel Young and Solomon Hsiang (2024) studied complex chains of events triggered by natural disasters within human societies. More specifically, they have examined the long-term impact (for the 172 months = 14.3 years following a tropical cyclone) of 501 tropical cyclones (TCs), that is, hurricanes and tropical storms, on human mortality in the contiguous United States for all tropical cyclones between 1930 and 2015. Young and Hsiang observed "a robust increase in excess mortality that persists for 15 years after each geophysical event. We estimate that the average TC generates 7,000–11,000 excess deaths, exceeding the average of 24 immediate deaths reported in government statistics." The researchers conclude their assessment of the long-term impact of tropical cyclones, noting that the seriousness of the impact has not been realized. The findings

> identify the TC climate as a driver of broad public health outcomes. This suggests that critical healthcare needs of TC-affected populations are not being fully addressed, and many affected individuals probably do not realize the extent to which their own health was affected by a TC.

Globally, lightning flashes are responsible for a significant number of fatalities each year, estimated to be between 6,000 and 24,000, with injuries outnumbering fatalities by a factor of ten (Mishra et al., 2024). The range of fatalities reflects the variability in data reporting and the inherent challenges in accurately quantifying such incidents. Lightning fatality rates vary among developed, developing, and least-developed countries. Post-2001 data on lightning fatalities in India collected by Mishra and colleagues revealed a significant increase in the average annual fatality rate due to lightning. Lightning in India

> has emerged as one of the most lethal and challenging disasters among natural hazards. Over the past 54 years, lightning has caused a total of 101,309 fatalities. The data highlight an increasing trend, with the last two decades showing the greatest increase. [...] The rising trend of extreme

climate conditions is likely to exacerbate the situation in a country like India.

(Mishra et al., 2024)

According to researchers in the United States, the number of lightning strikes has increased by 12% with each additional degree of global warming (Romps et al., 2014). This increase is attributed to higher temperatures leading to more convective available potential energy (CAPE), which fuels thunderstorms and, consequently, lightning activity. As in the case of extreme heat adaptation, practicing lightning safety can significantly reduce fatalities.

In addition to the state of the art of climate research, which informs citizens and policymakers about the past development of key warming indices, emissions data, and temperature, as well as the probable future present of these indicators, another set of "things" needs to be part of a description and analysis of the state of the art of climate change, namely the range of major societal actors. This is not an exhaustive list, but I mention some of the major collective actors: the renewable energy movement and the fossil fuel industry, aware, as it were, as early as 1954 of the conditions for climate change and its dangers.[16] An emerging collective actor is the decarbonization industry.

Given the information on increasing warming and the likely increase in hot days and weeks around the globe, despite a record increase in the use of renewable energy in 2024, total fossil fuel consumption will continue to rise in 2024, because on the consumption side there is an unrelenting and persistent demand for cheap energy, reaching a new global record, according to an annual review of world energy by the Energy Institute. In advanced economies, there are signs of demand for fossil fuels peaking, in contrast to economies in the Global South, for which economic development and improvements in quality of life continue to drive fossil growth. At the same time, oil and gas corporations are expanding their activities, and institutional investors, banks, insurance companies, and state funds provide financing.[17] However, compare the observation about the expansion of the conventional activities of the large fossil fuel firms with the model about the "end of oil" by Ryan Kellog (2024). The strongest empirical support for Kellog's model anticipates the possibility "because oil extraction requires potentially long-lived investments in wells and other infrastructure, the opposite may occur: an anticipated demand decline reduces producers' investment rates, decreasing near-term oil production and CO_2 emissions."

According to the financial research paper "Investing in Climate Chaos" by the non-governmental organization *Urgewald*, investments in bonds and shares in coal, oil, and gas companies add up to 4.3 trillion dollars. The four U.S. asset managers, Vanguard (413 billion dollars), Blackrock (400 billion), State Street (171 billion), and Capital Group (165 billion) have invested the largest sums. A report issued by the *Americans for Financial Reform*

Education Fund Global Energy Monitor and Private Equity Stakeholder Project concludes that private equity funds

> have ploughed more than $1tn (£750bn) into the energy sector since 2010, often buying into old and new fossil fuel projects and, thanks to exemptions from many financial disclosures, operating them outside the public eye [... moreover] Public sector workers' money, through national, state, and retirement pensions, provides much of the capital for private equity firms' energy investments, but there is limited disclosure to the pension fund managers that the deferred earnings of their beneficiaries have potential climate impacts.[18]

Observations about the difference in the political and economic approaches to climate change typically invoke the dichotomy of a set of opposing twins of collective actors: corporations, legal frameworks, environmental norms, or everyday behaviors that are tied either to the fossil fuel industry or collective actors involved in the development of clean energy sources.

Recent climate projections (Greene et al., 2024) indicate with certainty that the Greenland ice shield will continue to lose ice mass throughout this century; however, "the timing, magnitude and character of future changes to the ice sheet will depend on the severity of climate forcing and the sensitivity that each glacier exhibits to changes in its environment." Nearly every glacier in Greenland has thinned or retreated over the past few decades. The *Nature* publication by Chad Greene and colleagues reiterates the warning of climate scientists that rapid sea level rise due to the warming of Polar Regions represents a significant threat. For example, over the last 15 years, the Arctic Circle region has been warming four times faster than the global average. The average temperature is now 3°C warmer than the temperature level in the 1980s: "Nearly every glacier in Greenland has thinned or retreated over the past few decades 1–4, leading to glacier acceleration, increased rates of sea-level rise and climate impacts around the globe" (Greene et al., 2024: 523).

There is a broad consensus in the scientific community that global emissions of greenhouse gases must be reduced dramatically in the coming decades to avoid some of the most serious impacts of climate change. Many countries have voluntarily and unilaterally reduced their emissions over the past two decades. Reducing emissions is a *public goods problem*. A public goods problem arises when a particular behavior provides benefits that are non-excludable and non-rivalrous. These characteristics mean that individuals cannot be effectively excluded from the benefits of reducing emissions, and one person's use does not reduce availability to others. As a result, voluntary reductions in greenhouse gas emissions are unlikely to be sufficient to address the consequences of climate change. Global cooperation is needed to effectively reduce emissions. I will return to this issue later when I discuss the scalability of climate policy.

On a global scale, approximately 50 billion tons of greenhouse gases are emitted annually. In recent years, annual CO_2 emissions have reached 35 billion tons (Figure 1.3):

Aside from the total annual CO_2 emissions between 1750 and 2022 worldwide and in different regions of the world, the changes in greenhouse gas emissions (CO_2eq)[19] in different economies and the rest of the world are of interest (Figure 1.4).

In 2022, the United States was responsible for 6 billion Gt CO2eq tons of greenhouse gases, China for 15.7 billion, and the European Union for 3.6 billion. Carbon emissions are not solely the responsibility of the nations within which they occur. The absolute statistics hide the significant differences among emitters within each region. Extremely wealthy individuals located

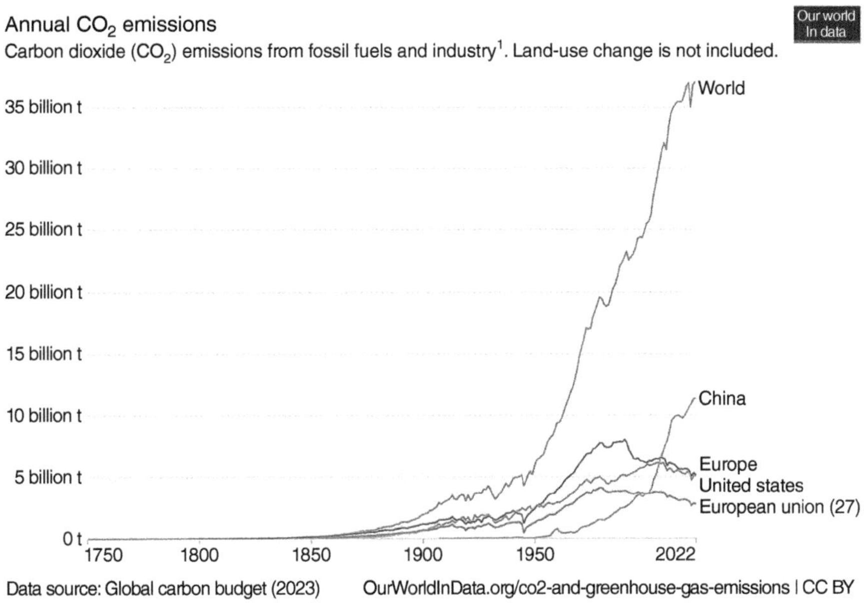

1. Fossil emissions: Fossil emissions measure the quantity of carbon dioide (CO_2) emitted from the burning of fossil fuels, and directly from industrial processes such as cement and steel production. Fossil CO_2 includes emissions from coal, oil, gas, flaring cement, steel, and other industrial processes. Fossil emissions do not include land use change, deforestation, soils, or vegetation.

FIGURE 1.3 Annual carbon dioxide (CO_2) emissions from fossil fuels and industry, 1750–2022.

Source: Our World in Data of the Global Change Data Lab: https://ourworldindata.org/explorers/co2?tab=chart&xScale=linear&yScale=linear&stackMode=absolute&endpointsOnly=0&time=earliest..latest&country=~World&region=World&Gas=All+GHGs+%28CO%E2%82%82eq%29&Accounting=Production-based&Fuel=Total&Count=Per+country&Relative+to+world+total=false.

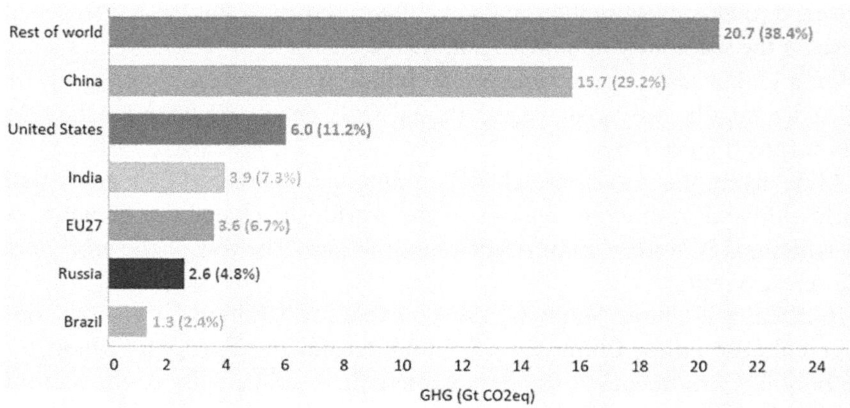

FIGURE 1.4 GHG emissions (in Gt CO_2eq) and contributions of major emitting economies and the rest of the world in 2022 (in Gt CO_2eq).

Source: Crippa et al. (2023).

Notes: In 2022, "Gt CO_2-eq" is composed of the following contributing gases: the majority of greenhouse gas emissions will be fossil CO_2, accounting for 71.6% of total emissions, with CH_4 [methane] contributing 21%, N_2O 4.8% [nitrous oxide], and F-gases 2.6% [fluorinated gases]. Global fossil CO_2 emissions have increased by more than 70% since 1990. CH_4 and N_2O emissions have increased at a slightly slower rate: CH_4 will increase by 32.4% and N_2O by 36.5% between 1990 and 2022, while F-gases will quadruple over the same period.

mainly in the United States are responsible for the highest carbon emissions. Between 2010 and 2018

> the 10 percent of the world's people responsible for the highest emissions emit on average 2.3 times the global average; together they account for 45 percent of global emissions. Of these emissions, North America represents 46 percent, Europe16 percent, and China12 percent.
>
> *(Piketty, [2019] 2020: 667)*

A 2023 Oxfam report, "Climate Equality: A planet for the 99%," refers to the massive personal inequality of emissions and probable consequences of unequal greenhouse gas emissions:

> The richest 1% of humanity is responsible for more carbon emissions than the poorest 66% [...] his elite group, made up of 77 million people including billionaires, millionaires and those paid more than US$140,000 (£112,500) a year, accounting for 16% of all CO_2 emissions in 2019 – enough to cause more than a million excess deaths due to heat, according to the report.[20]

Table 1. Shares in 2022 global emissions[12], yearly GHG emission relative changes[13] over the period 2019-2022 and the CAGR[14] in 1990-2022 (%)

	Share in global	Change 2019-2020	Change 2020-2021	Change 2021-2022	Change 2019-2022	Change 2020-2022	CAGR (1990-2022)
China	29.2%	1.9%	5.1%	0.3%	7.4%	5.4%	4.3%
United States	11.2%	-8.7%	5.5%	1.6%	-2.2%	7.2%	-0.1%
India	7.3%	-5.7%	6.7%	5.0%	5.7%	12.1%	3.2%
EU27	6.7%	-7.7%	5.6%	-0.8%	-3.4%	4.7%	-1.0%
Russia	4.8%	-3.9%	7.2%	-1.0%	2.0%	6.1%	-0.5%
Brazil	2.4%	-0.3%	5.1%	-2.4%	2.3%	2.6%	2.0%
Indonesia	2.3%	-4.9%	2.1%	10.0%	6.8%	12.3%	3.4%
Japan	2.2%	-5.3%	1.2%	0.6%	-3.6%	1.8%	-0.3%
Iran	1.8%	-1.6%	3.9%	1.6%	3.9%	5.6%	3.3%
Mexico	1.5%	-6.5%	3.5%	7.1%	3.7%	10.9%	1.8%
Saudi Arabia	1.5%	-0.8%	3.3%	3.9%	6.4%	7.3%	3.9%
Canada	1.4%	-8.2%	3.0%	3.2%	-2.4%	6.4%	0.8%
South Korea	1.3%	-4.3%	4.5%	-0.7%	-0.8%	3.7%	2.5%
Türkiye	1.3%	3.5%	8.5%	3.1%	15.8%	11.9%	3.5%
Australia	1.1%	-3.9%	-2.0%	1.7%	-4.1%	-0.3%	0.7%
South Africa	0.99%	-9.8%	-0.5%	-2.5%	-12.5%	-3.1%	0.8%
Global		-3.7%	4.8%	1.4%	2.3%	6.2%	1.5%
International Aviation	0.8%	-52.3%	15.4%	23.3%	-32.1%	42.3%	1.5%
International Shipping	1.4%	-8.5%	5.7%	5.7%	2.2%	11.7%	2.0%

Source: JRC, 2023

(12) In Table 1, countries are ranked by their GHG emission share in the global total (countries with share of more than 1% are shown, together with international shipping and aviation).

(13) It is important to acknowledge that year-to-year variations in emissions are estimated with an accuracy level of approximately ±0.5% (Olivier et al., 2016) when relying on robust statistical activity data (such as IEA energy balance data or CO_2 emissions from fossil fuel combustion for the period 1970-2020). For the data spanning 2021-2022, the accuracy can range up to ±2% (based on a Fast-Track approach), contingent upon regional, sectoral, and fuel-specific contributions. Emission magnitudes, on the other hand, have a range of accuracy that depends on the level of aggregation (for example global or country level, total emission, or specific sector, as detailed by Solazzo et al., 2021), as well as the substance, with N_2O in particular having higher levels of uncertainty, and CO_2 the least. Global total GHG emissions are estimated with around ±10% accuracy, while the range of accuracy for country level total CO_2 emissions is between ±4% and ±35% (95% confidence interval). Policy makers and the scientific community should consider these uncertainties when using these data for further analysis.

(14) Compound annual growth rate (CAGR) calculates annual changes over a specified number of years as if this change had happened steadily each year over that time period.

Moreover, the wealthy command resources to shelter themselves from the worst effects and harm of climate change in the present and in the future present.

No country, region, or continent can reduce emissions quickly enough to avoid climate change. In 2022, the six largest GHG emitters were China, the United States, India, the EU27, Russia, and Brazil. Together, they account for 50.1% of the world's population, 61.2% of the global GDP, 63.4% of the global fossil fuel consumption, and 61.6% of the global GHG emissions. In 2022, China, the United States, and India emitted more than they did in 2021. India had the largest increase (5%). The EU27 emitted 27% less than in 1990 at 3.59 Gt CO_2eq and 0.8% less than in 2021, representing 6.7% of global emissions. In 2022, international aviation and shipping emissions increased by 23% and 5.7%, respectively. During the pandemic, aviation emissions fell by almost half in 2020 but rebounded in 2021. Shipping emissions increased by more than twice the 2020 reduction, with 2022 emissions being 2.2% higher than those in 2019 (cf. Crippa et al., 2023) (Figure 1.5).

In addition to the widely researched and discussed emissions of Gt CO_2eq emissions, which are often neglected, factors other than carbon dioxide may have significant climate consequences. In particular, black carbon, which is the soot that results from the incomplete combustion of fossil fuels (petroleum, coal), biofuels, and biomass (wood, animal dung, etc.), in the atmosphere suggests that an emphasis on improving air quality and on reducing the emission of black carbon from burning biomass in households and from forest fires (hence forest issues) would produce much quicker physical reduction of the human influence on climate, especially in the high polar latitudes; and, in any case, it already has considerable support politically (Prins et al., 2010: 13).

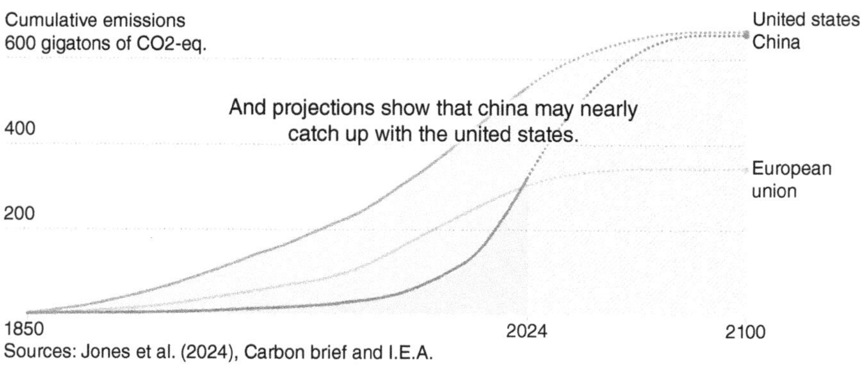

FIGURE 1.5 Cumulative CO_2 emissions by region and projected emissions until 2100.

Modern climate change has accelerated the melting of glaciers and polar ice sheets, leading to a rise in sea level. There is, however, considerable *uncertainty* about how and how quickly the ice breaks apart in Greenland and Antarctica under stress. Pole-to-equator mass transport has significantly increased the Earth's oblateness [Earth's oblateness, also known as its equatorial bulge, is a measure of how much the shape of Earth deviates from a perfect sphere to an oblate spheroid] and length of day (LOD) [Length of Day (LOD) refers to the duration of one complete rotation of the Earth on its axis, measured from one noon to the next] since 1900. Shahvandia, Adhikarib, Dumberryc, Mishrad, and Sojaa (2024) reported that the present rate of increase was higher than that at any point in the 20th century. In accordance with projections for elevated emissions, the climate-induced LOD rate is expected to persist in its upward trajectory, potentially reaching a magnitude that is twice that of the current rate. These observations underscore the profound impact of climate change on the planet and have implications for a range of fields, including precise timekeeping and space navigation.

The speed and range of West Antarctica ice loss are important for the expected sea level rise. In 2024, a group of researchers presented evidence that

> the prospect may be more remote than previously thought. As humans burn fossil fuels and heat the planet, West Antarctica's ice remains vulnerable to destruction in many forms. However, this form, in which ice cliffs collapse one after the other, looks less likely, according to the scientists' computer simulations.[21]

The multitude of these individual observations culled from the state of the art of climate science led some observers to arrive at rather pessimistic conclusions about the fate of humankind and the dark ages just ahead as efforts to cope with the consequences of climate change fall far short. David King,[22] the chair of the *Climate Crisis Advisory Group* and a former UK chief scientific adviser, for instance, dramatically concludes,

> On our current path, civilisation as we know it will disappear. If we meet current commitments only – net zero by 2050 – perhaps some form of humanity will survive, managing the challenges of continued extreme weather events, ice loss, and sea-level and temperature rises.

To avoid the future present sketched by King, enormous policy and behavior efforts are needed. This means, at least as King adds, "reducing emissions; removing the excess greenhouse gases (GHGs) already in the atmosphere; repairing ecosystems; and strengthening local and global resilience against inevitable climate impacts."

As our knowledge about climate change in general, weather extremes and their growing intensity (cf. Clarke et al., 2024) continues to accumulate, there is, at the same time, "currently no standardised way of, or effort towards, documenting climate change related harms systematically. As a result, there is no systematic basis to quantify the major contribution of human influence on extreme weather to the costs of climate change" (Clarke et al., 2022). Predictions of the rate and magnitude of climate change in different regions of the world, or in the world, are a major concern in climate change research. However, the uncertainties in current temperature scenarios for the next few decades and when critical thresholds might be reached are considerable. Estimating the timing of critical regional thresholds is an important lever for climate policy, particularly adaptation. Recently, researchers have tried to adapt an artificial intelligence (AI) approach to make more accurate predictions. The findings I describe in the next section on the state of the art—the future present are dramatic. As a climate scientist, commenting on the 2025 Los Angeles fires, noted in a column in the *New York Times*[23]: "One lesson climate change teaches us again is that bad things can happen ahead of schedule. Model predictions for climate impacts have tended to be optimistically biased."

Discussions about the future present of climate policy are dominated by technical topics like geoengineering, solar radiation management, and carbon removal—while innovations like solid-state batteries, which could be transformative, receive less attention.[24] However, climate change in a future present is far more than a technical problem—it is a "wicked problem." As we noted in our *Hartwell Paper* (Prins et al., 2010), wicked problems are complex, open-ended, and difficult to fully understand or solve. Originally described by C. West Churchman (1967)[25] and later by Horst Rittel and Melvin Webber (1973; Peters, 2017) in the context of urban planning, wicked problems are "issues that are often formulated as if they are susceptible to solutions when in fact they are not" (Prins et al., 2010: 15). Technical knowledge is not a sufficient basis to respond to wicked issues; wicked problems "demand profound understanding of their integration in social systems, their irreducibly complexity and intractable nature" (Prins et al., 2010: 15).[26] To address the wicked problem of climate in the future present, I will focus on the theme of "adaptation" as a more comprehensive strategy to respond to what is a societal rather than a technical problem of climate change.

The tendency to treat present and future as binary concepts is tempting, as a theoretical and empirical shortcut. I have already referred to Niklas Luhmann's helpful suggestion of a future present and a present future. More generally, in the public policy literature on intergenerational politics, the concept of "*discount rates*" refers to how present and future benefits and costs are weighed when making decisions that affect multiple generations. In the context of climate change policy discussions, the present future and the future present can be linked through the concept of "discount rates." The discount rate is essentially the rate at which society is willing to trade present value for future value. A higher discount rate gives greater weight to present benefits and costs relative to future ones, while a lower rate treats future impacts more equally to present ones.

Considerations of discount rates in climate policy, which relate to a future present from the perspective of the present future, involve several concerns: From an economic perspective, discount rates reflect market interest rates, opportunity costs of capital, or observed social time preferences. From an ethical perspective, discount rates reflect, for example, the argument that future generations will be wealthier and better able to cope with costs. Lower discount rates reflect greater moral consideration for future generations, while higher rates prioritize current welfare. Some observers argue that future generations should be treated equally (e.g. Stern Review on climate change), implying a discount rate close to zero for ethical reasons. A high rate might suggest that the welfare of future generations is less important than that of the present generation. From an uncertainty perspective, higher discount rates are due to uncertainty about future conditions, technological change, or even human existence. Last but not least, the linkage involves distributional concerns, since discount rates implicitly make assumptions about intergenerational equity and the distribution of welfare over time.

The actual determination of a particular linkage between the future present and the present future, should such considerations be of concern, is of course a political decision. In any case, the use of a particular discounting rate may, for example, lead to over- or underinvestment in adaptation measures. Countries or organizations that adopt lower rates are more likely to invest in long-term adaptation, while those that adopt higher rates may prioritize immediate environmental or economic gains from adaptation.

State of the art—the future present

An October 2024 report by the *U.N. Environment Programme* (UNEP) shows that median temperature projections through 2035—if current carbon-cutting pledges by countries for 2030 were met—would limit the temperature rise to a still catastrophic 2.6°C to 2.8°C globally. However, the current country's pledges to cut emissions by 2030 will not be met. An unprecedented global mobilization of renewable energy, forest protection, and other measures is therefore needed to keep the world off the current path to a catastrophic temperature rise of 3.1°C. The finance and technology to slash emissions exists, the head of the *U.N. Environment Programme*, Inger Andersen, said, but "political courage (and some strong leadership)

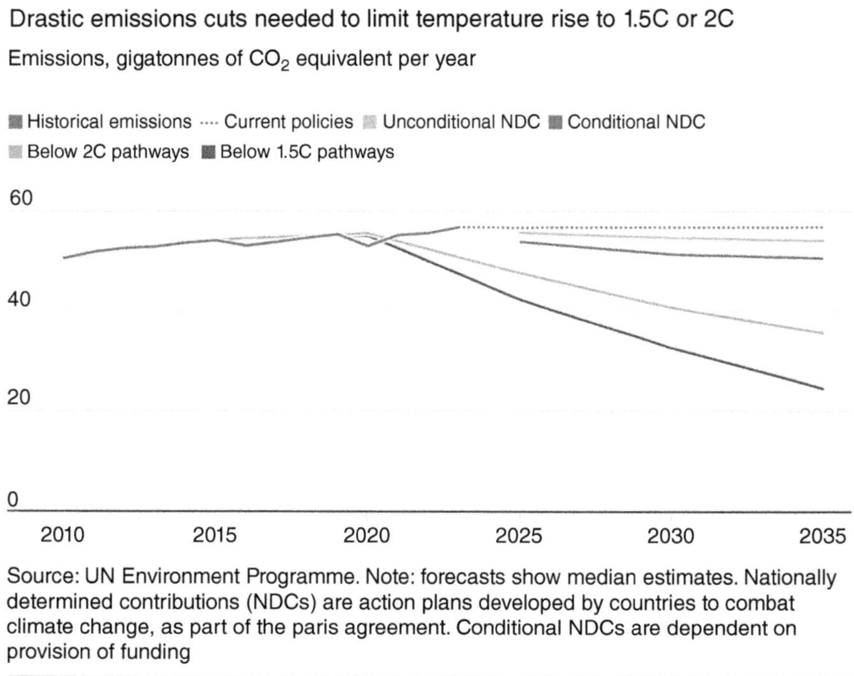

FIGURE 1.6 Annual drastic emission reductions that are necessary to limit the global temperature increase to 1.5°C or 2°C could be seen in GT CO_2.

was needed, particularly from the G20 nations (excluding the African Union) that cause 77% of global emissions" (Figure 1.6).[27]

Estimates of temperatures in the future present, especially temperatures in different regions of the world that are warming at varying rates, are important practical information. Hence, climate researchers are making many efforts to arrive at more precise forecasts. Recently, the emerging field of Artificial Intelligence offers tools to arrive at more refined estimates. For example, Elizabeth Barnes, Noah Diffenbaugh, and Sonia Seneviratne (2024) approach

> is to train a neural network on historical and future climate model simulations and then use historical observations as input to the trained network to make out-of-sample predictions based on the current state of the climate system. This method [… is] applied to make predictions of the time until global warming thresholds are reached based on the observed global map of annual temperature anomalies.

Their results indicate that the heating of the planet can practically no longer be limited to 1.5 degrees above pre-industrial levels. The research team, after the AI-supported analysis, suggested "a very high likelihood of 2.0°C of regional warming by 2040 for the majority of regions, along with a likelihood of 3°C by mid-century."

Aside from ongoing attempts to generate more accurate future present temperatures, the literature concerning the climate of the future present is vast. There are almost limitless predictions about the nature of the future climate and its societal consequences from all fields of the sciences. Predictions that foretell disaster or miraculous solutions, for example those that issue from the accelerated development of Artificial Intelligence (Kurzweil, 2024). More "conventional" is the account of the political scientist Ian Bremmer (2022: 6–7),

> if the pandemic was the biggest global challenge of our lifetime, climate change threatens our collective future on a larger and longer scale—by creating turmoil within the home all humans share, by making unprecedented economic demands on governments, by stoking political turmoil, by increasing inequality within and among countries, and by pushing more desperate people across borders than any war ever did.

Narratives about the climate of the future are pertinent to a discussion about the governance of climate change because of their influence on current political conduct, for example, in the sense that these scenarios may soften resistance against the abolition of liberties proactively. Peter Burnell (2014: 1233) considers such "calculations" reasonable:

> if at some future point there is a more universal conviction that immediate and drastic reductions in new greenhouse gas emissions are necessary, perhaps because of better understanding of tipping points, then it is presumptuous to say there would be no serious pressure to cede some liberal democratic values if that looked like a necessary price for taking requisite action.

Burnell's future orientation is, of course, speculative; an equally forceful case could be made that greater involvement and political participation in drastic future steps to adapt to considerable climate impacts could not only have greater political legitimacy but also much more successful outcomes. Calculating future risks against potential benefits of behavioral change could trigger assent without a loss of liberties.

I discuss a selected number of scenarios that I take to be close to the reality of the present in the future. In the coming decades, heat stress is projected to increase dramatically, and deaths are expected to increase, especially among citizens of low-income countries, who bear the least responsibility for global warming.[28] The population throughout the world is

> aging at the same time as heat exposures are increasing due to climate change. Age structure, and its biological and socio-economic drivers, determine populations' vulnerability to high temperatures [...] >23% of the global population aged 69+ will inhabit climates whose 95th percentile of daily maximum temperature exceeds the critical threshold of 37.5 °C, compared with 14% today, exposing an additional 177–246 million older adults to dangerous acute heat. The effects are most severe in Asia and Africa, which also have the lowest adaptive capacity.
>
> *(Falchetta et al., 2024)*

The first, more detailed sample concerns a counterfactual case of a major blackout in large urban areas. The strongest impact of a major blackout would be in areas that had finally become habitable, thanks to an invention that mitigates periods of excessive heat, that is, air-conditioning. The recent history of the city of Phoenix, Arizona, is a case in point. Before the 1950s, the population of Phoenix was relatively small because of the harsh local climate. Phoenixes experience extremely high temperatures, especially during the summer months, with average highs often exceeding 38°C. The advent of air conditioning in the mid-20th century marked a turning point. After air conditioning became widespread, Phoenix experienced rapid population growth. Air conditioning improved the overall quality of life for Phoenix. The city's population has grown from approximately 107,000 in 1950 to over 1.5 million today. Today, almost all buildings in Phoenix have air conditioning.

The city of Phoenix, Arizona, suffered a record 113 consecutive days with temperatures above 100°F (38°C) in 2024, leading to hundreds of heat-related deaths and more acres burned by wildfires across the state, officials said. The city of 1.5 million, the largest city in the Sonoran Desert, had its hottest summer on record, breaking the previous record set in 2023 by nearly two degrees, according to the U.S. National Weather Service.[29]

In brief, significant increases in population growth rates coincide with the adoption of air conditioning technology and an increased risk for urban populations due to the failure of the infrastructure, which makes living in these geographical areas comfortable. The ability to cool indoor spaces has facilitated, for example, economic growth in the city of Phoenix. Businesses, particularly those that require temperature control (e.g. data centers, tech companies, manufacturing), have been able to operate efficiently. This economic growth has created jobs and, in turn, attracted a workforce, contributing to its population growth.

However, the incidence of electric grid failure or blackouts across the United States has been steadily increasing. The majority of the failures occur during the summer months when demand is high and grids are stressed during extreme weather events in the form of tornado, heat wave, drought, or wildfire events:

> Since 2015, when the U.S. Energy Information Administration commenced monthly reporting on major blackout events (defined as power outages lasting more than 1 h and impacting more than 50,000 customers), the number of such events nationwide has more than doubled, increasing by 151% between 2015–16 and 2020–21.
>
> *(Stone et al., 2023: 8245)*

In 2023, the city of Phoenix broke records with 54 days above 43.3°C, including 31 days in a row reaching that mark. The number of days with extreme heat is expected to increase in Arizona and elsewhere.[30] Brian Stone and colleagues (2023) model the possible health consequences of a major, two-day city-wide blackout during a heat period in the cities of Phoenix, Atlanta, and Detroit. The electricity in their model was restored gradually over the following three days. In Detroit, Michigan, 53% of the buildings have air conditioning, and in Atlanta, Georgia, 94% of the buildings have air conditioning. Concurrent heat waves and blackouts can have serious health consequences. The temperatures during the heat wave were the highest in Phoenix, followed by those in Atlanta and Detroit.

Given a range of uncertainties, for example, the adaptive reaction of the population or the government response, the authors of the model conclude that in Phoenix, approximately 800,000 people—roughly half the population—would need emergency medical treatment for heat stroke and other illnesses. More than 13,000 people would die. Under the same scenario of the compound heat waves and blackouts of historical intensity in Atlanta, researchers reported that 12,540 visits to emergency rooms occurred, and six people died. In Detroit which has a higher percentage of older residents and a higher poverty rate than other cities do, 221 people die. The greater

> pronounced risk of heat mortality and morbidity in Phoenix relative to Atlanta and Detroit can be attributed both to the higher temperatures experienced during an extreme heat wave event and a disproportionate impact of blackout conditions for cities with a high AC prevalence.
>
> *(Stone et al., 2023: 8252)*

The impact of climate change on health in general could be significant, as current trends already indicate, for example, an accelerated spread of viruses.

Evidence suggests that climate change may facilitate the spread of viruses through geographic range expansion of vectors, extended transmission seasons, altered migration patterns of reservoir species, increased host-pathogen contact, thawing permafrost and increased extreme weather events.

In the following chapter, I address the growing momentum of authoritarianism with a focus on environmental authoritarianism, although the rise of authoritarian sentiment is hardly limited to environmentalism, for example, the ultraconservative and extremely wealthy corporate leaders in the United States despair about democratic governance and promote its displacement (cf. Reich, 2022).[31] The 2024 election could have potentially been the first in American history in which a majority of voters in the United States consciously choose the oligarchy.[32]

Perhaps the central set of overall scenarios concerning the climate of the future present involves discussion about expected *climate tipping points* (CTPs). CTPs refer to parts of the Earth system that are associated with a specific region (or collection of regions) of the globe and are at least subcontinental in scale. CTPs are defined as occurring when changes in large components of the climate system, known as tipping elements, become self-perpetuating beyond a specific warming threshold. In the first instance, the idea pf climate tipping points refers exclusively to natural or physical processes. Natural transformations inevitably have societal consequences. Whatever the societal consequences of specific climate tipping points may be, for example, the infrastructure of a region is a black box.

The concept of "climate tipping points" was first discussed and formalized by the American climatologists Wally Broecker (e.g. Broecker, 1975; Broecker and Hemming, 2001) and Timothy Lenton (2011) in the late 1970s and early 2000s. Broecker was a pioneer in climate science and is often credited with popularizing the term "global warming." He recognized that the climate system could undergo abrupt and irreversible changes once certain critical thresholds, some on a global scale, or "tipping points," were crossed. Tipping points represent specific levels or thresholds of certain climate variables (e.g. temperature, ice cover, and ocean circulation). Tipping points are assumed to represent an irreversible process. Once a tipping point is crossed, the changes that occur are difficult or impossible to reverse on a human timescale. Crossing a tipping point may lead to feedback loops that amplify the changes, making the situation worse. For example, the melting of polar ice reduces the Earth's albedo (reflectivity), leading to more heat absorption and further melting.

Some examples of potential climate tipping include the following[33]: the melting of the Greenland ice sheet, the collapse of West Antarctica, the mass death of coral reefs, the death of the Amazon rainforest, the shutdown of the Atlantic Meridional Overturning Circulation (AMOC), the sudden shift in the West African monsoon, abrupt permafrost thawing and methane release or enhanced warming over the long run. A recent study, combined data from studies of warm periods up to 3m years ago; observations of ice melting and sea level rise in recent decades; and climate models. It concluded: "Continued mass loss from ice sheets poses an existential threat to the world's coastal populations with an estimated one billion people inhabiting land less than 10m above sea level and around 230 million living within 1m" (Stokes et al., 2025: 1). The concept of tipping points is considered important in climate science because it highlights the potential for sudden, dramatic changes in the Earth's climate system, rather than gradual, linear changes. Understanding climate tipping points, if taken at face value, has significant implications for climate policy and the urgency of addressing global warming.[34]

Those who criticize the concept of tipping points (e.g. van der Sluijs et al., 2010; Hulme, 2011; Peters et al., 2013) argue that it oversimplifies the nonlinear nature of climate systems. The Earth's climate is influenced by a multitude of interrelated factors, and the notion of a singular tipping point may prove insufficient for capturing the intricacies of these interactions. Some scientists have highlighted that changes in the climate are frequently gradual and continuous in nature rather than abrupt. The tipping point model can create a misleading binary perspective, implying that climate change is either manageable or catastrophic with nothing in between. The precise thresholds at which tipping points might occur are often unknown and are difficult to predict accurately. The economist William Nordhaus (2015: 56) agreed and emphasized that,

> like financial crises, the exact timing and magnitude of such events are often impossible to predict. They may occur rapidly and unexpectedly or they may not happen at all. This uncertainty can make it challenging to use the concept for practical policy-making. Furthermore, a focus on tipping points might divert attention from at least equally important policy issues such as enhancing resilience and adaptation strategies, which tend toward core socio-political attributes.

As noted, the discussion of climate tipping points tends to be one-sided. The societal effects remain in darkness. The wide raging, perhaps even popular discussion of catastrophic tipping points and the associated media attention has prompted a concern with "relational tipping points," that is, a reference to broken treaties and unfulfilled promises of colonial nations and wealthy regions of the world (Fritsch, 2023: 1006).

Not surprisingly climate change and future scenarios related to the impact of climate change attracted intensive attention to *national* security interests, such as the *National Intelligence Council of the United States* (2021) and the *Department of the United States* (2021). In a report of the Council issued in 2021,[35] entitled it "Climate change and international responses. Increasing challenges to US National Security Through 2024," various probable geopolitical risks and tensions are enumerated. For example, geopolitical tensions are likely to grow as nations increasingly argue about how to accelerate the reductions in net greenhouse gas emissions and "the increasing physical effects of climate change are likely to exacerbate cross-border geopolitical flashpoints as states take steps to secure their interests." As a critic (Dumaine, 2022) however, notes, the national security perspective the Report generally choses means that it

> looks through the proverbial keyhole: the basic narrow picture is accurate, but vital information and perspectives are missing. To meet the [United States] administration's stated objectives, future national security approaches must go beyond a traditional state-centric national security lens to grapple with the larger ecological, social, political, and intergenerational dynamics that may arise in the wake of climatic changes and other complex, transnational challenges.

Climate change is a *global* environmental issue: "Intelligence analysis must encompass a wider range of disciplines, cultures, and perspectives, and it must establish connections between global and nation-centric analysis."

In principle, the same restrictions in geographical range apply to the *Department of Defense of the United State*s (DoD) report on the impact of climate change on security concerns although climate change hazards and potential impacts around the world are referenced:

> To keep the nation secure, we must tackle the existential threat of climate change. The unprecedented scales of wildfires, floods, droughts, typhoons, and other extreme weather events of recent months and years have damaged our installations and bases, constrained force readiness and operations, and contributed to instability around the world. Climate change touches most of what this Department does, and this threat will continue to have worsening implications for U.S. national security.
>
> *(Department of Defense, 2021: 5)*[36]

Climate risks will hence force, the DoD Report (2021: 15) claims, be incorporated into modeling, simulation, and wargaming, for example, with respect to the critical defense infrastructure: "climate-related delays, disruption, and/or degradation of DoD to produce, package, repair, and distribute materiel and ammunition and its effects on readiness and/or operations."

In addition to security concerns of a warming climate, a fundamental policy concern about the future present is the likely impact of a warming Earth on the supply of staple foods. A prospective study by a group of climate scientists led by Sara Heikonen (2025) of *Aalto University* in Helsinki,

Finland, examined the impact of future changes in temperature, precipitation and drought on the growing conditions of 30 major food crops.[37] Their results revealed a somewhat unsurprising pattern: regions at lower latitudes, which are closer to the equator, are expected to face far more severe consequences than those at middle or high latitudes. Depending on the extent of global warming, up to half of the food production in these areas could be lost. The climatic conditions there would simply become unsuitable for growing many crops, it says. According to this study, the most affected region would be in sub-Saharan Africa, where almost three-quarters of current food production would be at risk if global warming exceeds 3°C. On the other hand, the study finds "substantial increases in potential crop diversity in mid- to high-latitude regions, especially in North America, providing emerging climatic potential for transformational climate change adaptation and improving the climate resilience of crop production through diversifying production or selecting climate-change-resilient crops" (Heokonen, 2025). There is a significant need to mitigate the negative impacts of climate change on food crop production and future potential crop diversity, given these findings concerning the future present of food production around the world.

The intellectual and political support for authoritarian practices as a response to the dangers of climate change manifests itself in different forms and is based on different logics of urgency. Protest alliances such as Extinction Rebellion and the Last Generation—which advocate for a fundamental common interest, namely the prevention of the collapse of essential ecosystems and the safeguarding of the planet's habitability for humans—may, at least in the term run, exercise, and advocate support for a form of mild authoritarianism through their protest style. Responsible for the catastrophe on the immediate horizon is the industrial society and the lifestyle of every citizen of the Global North (cf. Freier and Schneider, 2022).

However, narratives about the climate of the future present are not always pessimistic. Upbeat accounts of the future present of the climate, especially the emergent capacity to cope with trends toward more damage to societies or the ability to arrest these trends, are now often linked to the promise of the rapidly augmented development of AI A report by the *World Economic Forum*[38] lists nine ways in which AI can assist climate change. The ideas range from "Icebergs are melting—AI knows where and how fast," "mapping deforestation," "AI is cleaning the oceans," "AI helps predict climate disasters," "AI can help industry to decarbonize," to "reforesting hills in Brazil using drones." All in all, these are rather slight expectations compared with the much more confident predications of Ray Kurzweil (2024), the proponent of the highly speculative so-called Singularity thesis.

Kurzweil argues that technological advancements, particularly in computing, follow a pattern of exponential growth rather than linear progress. This is exemplified by **Moore's Law**, which predicts the doubling of computing power approximately every two years. Kurzweil extends this idea to other technologies, such as AI, nanotechnology, biotechnology, and robotics, which he believes are all advancing at accelerating rates. Not surprisingly, Kurzweil extends the idea of the exponential growth of AI to environmental issues. Kurzweil's singularity thesis envisions a future where AI and other technologies evolve to the point where they surpass human intelligence, potentially leading to a profound transformation of humanity and society by the mid-21st century. Specifically, as Kurzweil maintains,

> three-dimensional printing will dematerialize manufacturing. Miniaturized solar cells will replace fossil fuels. Biocomputers will reduce the energy needed to power an infinitely expanding computational churn, allowing us to beat a hasty AI-powered retreat from the brinks of runaway climate change and ecosystem collapse.

It follows that

> once humanity has extremely cheap energy (largely from solar and, eventually, fusion) and AI robotics, many kinds of goods will be so easy to reproduce that the notion of people committing violence over them will seem just as silly as fighting over a PDF seems today. In this way, the millions-fold improvement in information technologies between now and the 2040s will power transformative improvement across countless other aspects of society.
>
> *(Kurzweil, 2024: 169)*

Engineering solutions

The World Meteorological Organization reported that CO_2 concentrations at Mauna Loa in Hawaii,[39] the benchmark observatory in WMO's Global Atmosphere Watch network, reached 420.13 ppm on 30 March 2022, up from 416.95 ppm on 30 March 2021 (Figure 1.7)[40]:

Unless a technical and scientific solution to climate change can be found, such as giant solar sails that can be used to control solar radiation,[41] global warming cannot be stopped. As an article in the *Economist* notes, sea levels have risen somewhere between 21 cm and 24 cm since 1880. Most sea level rise is a consequence of water physically expanding as it warms, but in recent decades, meltwater flowing off Greenland and Antarctica has significantly contributed.[42] As the contribution to sea level rise from melting polar ice becomes more important, suggestions to prevent melting embrace the notion that it could be slowed by technological means ranging from underwater curtains to ice-thickening pumps. Efforts to re-engineer the natural environment have moved from the fringe of science and technology

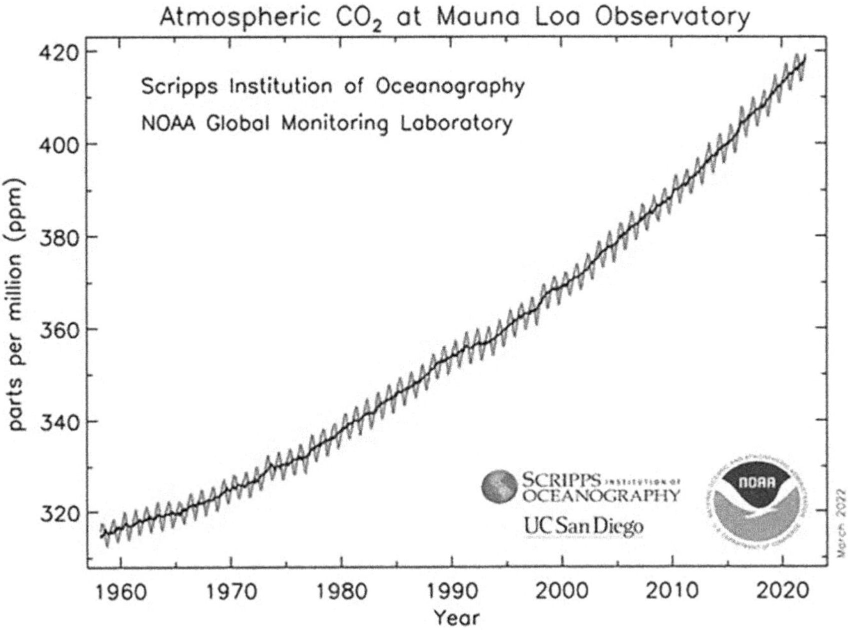

FIGURE 1.7 Atmospheric CO_2 at Mauna Loa Observatory, 1960–2020.
Source: https://scripps.ucsd.edu/

closer to undertakings that are taken seriously (cf. Keith and Dowlatabadi, 1992).[43] Tests have been carried out to ascertain whether the addition of iron to the ocean could add carbon dioxide to the floor of the ocean. Interest in sun shields, for example, once a far-flung idea, has grown. A similar comprehensive solution may be offered by artificial photosynthesis in plants, which may efficiently release CO_2 from the atmosphere.[44]

Experimental efforts

An Australian start-up is experimenting with the idea of employing fungi to extract carbon dioxide from the atmosphere and sequester it underground. It is part of a larger effort around the world to determine whether "dirt can clean up climate pollution." In the Australian case, farmers add a pulverized dust of fungal spores as they seed their crops:

> The fungus latches on to the crop roots, takes carbon that is absorbed by the plants from the air and locks it away in subterranean storage in a form that may keep it underground for much longer than the natural carbon cycle.

The appeal of the process is that it is simple and places few demands on farmers. It also has the side benefit of restoring soil health. The exact amount of carbon stored is still uncertain, but the potential is significant. The soil holds three times as much carbon as the atmosphere does. After oceans, soil is the second-largest carbon store in the world. However, the effort needed, in addition to emissions from agriculture itself, would be enormous and require vast areas of land.

Another recent engineering effort by University of Washington scientists in the United States involves a technology, the *Solar Radiation Modification* (SRM), which is designed to brighten clouds and bounce some of the sun's rays back into space. If it works on an experimental scale, then the "next stage would be to aim at the heavens and try to change the composition of clouds above the Earth's oceans."[45] The apparent downside of *Solar Radiation Modification* efforts, for example, in attempts to reduce high temperatures in California, could inadvertently intensify heatwaves in Europe. A study published in *Nature Climate Change* (Wan et al., 2024) models the consequences of regional tinkering with a changing climate. The model was designed to investigate the efficacy of regional SRMs in the North Pacific in mitigating extreme heat in the western United States. The authors find significant unintended consequences of the technique: it "reduces the relative risk of dangerous summer heat exposure by 55% and 16%, respectively. However, the same interventions under mid-century warming minimally reduce or even increase heat stress in the western United States and across the world."

On a broader scale, and perhaps the geoengineering solution with the greatest hope and greatest risk, both of which are difficult, if not impossible, to predict, is the proposal to release sulfur dioxide or other materials, for example, calcium carbonate or diamond dust, into the stratosphere to lower temperatures worldwide. David Suzuki, a Canadian environmentalist, is quoted in the *New York Times* as saying that the release of sulfur dioxide into the stratosphere has the following effects, the "whole notion of spraying sulphur compounds to reflect sunlight is arrogant and simplistic. There are unintended consequences of powerful technologies like these, and we have no idea what they will be."[46] Another attraction of the solar proposal is that it can be implemented without much delay and at relatively low cost. Proponents argue, of course, that the risks are small and manageable, and that they would help prevent millions of heat-related deaths. However, public opposition to early field trials has been strong and has prevented such trials.

In addition, increasing the ocean's capacity to absorb and store carbon is an emerging area of research within climate science. This concept is often discussed under the broader umbrella of "*ocean-based carbon dioxide removal*" (CDR) strategies. For example, a study by Andreas Oschlies et al. (2010) explored artificial upwelling as a method to increase the ocean's carbon uptake by bringing nutrient-rich deep waters to the surface, promoting phytoplankton growth. Froehlich and her colleagues (2019) explored the potential for large-scale seaweed farming to contribute to carbon sequestration and its role in climate change mitigation (also Gattuso et al., 2018).

A distinctive engineering approach to CDR was attempted by *Frontier*, a one billion dollar fund-backed tech company such as Stripe and Alphabet, to remove greenhouse gases from the atmosphere by transforming the earth's rivers and oceans into giant sponges that absorb these gases.[47] The plan is to employ a machine that grinds limestone. The pulverized limestone is released into the water:

> Rivers contain carbon dioxide that is constantly escaping into the air, where it traps heat and warms the planet. However, adding limestone converts some of that carbon dioxide into a stable molecule that instead stays underwater and washes into the sea, where it should remain trapped for thousands of years.

This process is known as alkalinity enhancement, which speeds up the process of carbon dioxide absorption by the water; it converts carbon dioxide into stable bicarbonate that sinks to the floor. The impact of altering the ocean's chemistry has, of course, unknown consequences for the ocean environment and could for example produce risks for aquatic life.

With respect to carbon capture and storage (CCS) research, a group of chemists at the University of California, Berkeley, is pursuing the idea of using a lab-created powder to absorb carbon from the air. It may not be the first material to absorb carbon, but as one scientist involved in the research claims, "it's a quantum leap (!) ahead [of other compounds] in terms of the durability of the material."[48] On an experimental basis, 250 grams of yellow powder can extract as much CO_2 as a tree is able to absorb. Once carbon is absorbed, it can be released for safe storage or industrial use. The powder was used multiple times. Removing carbon from the air remains a technical challenge because the high energy intensity of absorption and scaling up the process is difficult. A paper by Howard Herzog and colleagues (2024) cautions that studies on carbon capture from the air "rely on assumptions that are overly optimistic regarding the cost and scaling-up of DAC [direct air capture] systems."

Geoengineering experiments are being contemplated to reduce the amount of sunlight reaching the planet, or efforts are being considered to ensure that more heat escapes from the planet. For example, one area of research is sunlight reflection methods, which include stratospheric aerosol injection, where tiny particles are released into the stratosphere to reflect sunlight. Another potential solution is marine cloud brightening, which involves ships spraying sea salt particles into the sky to increase the reflectivity of low-lying clouds. Other geoengineering ideas include seeding cirrus clouds to allow more heat to escape into space. At present, the wispy high-altitude clouds act as a blanket, trapping heat.

Finally, to mention one additional experimental geoengineering effort underway, an international team of engineers is working on a machine designed to alter the composition of the atmosphere of the Earth. The machine, located on a windswept plateau in Iceland, is designed to extract vast quantities of air to extract the carbon dioxide.

All these technological ideas and experimental attempts to re-engineer the natural environment are highly contentious undertakings. Nicholas Stern (2015: 29) offers a sobering conclusion:

> There are also very worrying governance and democracy issues around the research, development, and deployment of geoengineering techniques. For example, do appropriate governance mechanisms exist, or can they be developed for geoengineering techniques with uncertain and cross-border impacts, such as the injection of aerosols into the atmosphere?
>
> *(see also Steffen et al., 2011: 859)*

A faculty member at Stanford University's Department of Civil and Environmental Engineering is warning that climate interventions are generally dangerous distractions from the much more urgent tasks to rapidly reduce the use of fossil fuels. There are no regulations governing regional geoengineering interventions that may have consequences for climate and weather patterns elsewhere in the world.[49] The model from the *Nature Climate Change* article only confirms that intervention in *complex* systems is fraught with immense problems.

Efforts to place a global moratorium on geo-engineering based on a provisional 2010 UN resolution on Biodiversity continue to be a demand by groups of climate scientists. In this context, the *Guardian* (September 14, 2023) newspaper cites Peter Kalmus, a climate scientist at NASA's Jet Propulsion Laboratory, who warned:

> Geoengineering, like direct air capture, is a deeply uncertain techno-solution that fossil fuel executives love to push to take pressure off their core business of selling oil, gas, and coal, which, as more and more people are realizing, is causing rapid and irreversible destruction of our planet's habitability. Fossil fuel elites will use geoengineering as an excuse to continue business-as-usual. As a climate scientist, my worst nightmare is continued fossil fuel expansion accompanied by solar geoengineering followed by termination shock. This would be game over for human civilization and much of life on Earth.

It is, of course, difficult to anticipate how scientific and public opinion may evolve under circumstances of grave environmental constraints in a future present. Opinions may change radically. However, it is evident that the pressure for the regulation of geoengineering ideas will increase in the coming years (cf. Victor, 2009; Gerrard and Hester, 2018),[50] responding, for example, not only to the dilemma that geoengineering can be carried out by a single country while impacting all other countries in the world. However, as the impact of climate change on societies accelerates, pressure to employ geoengineering technology will increase as well. In a cautionary description of the future present, David Victor and colleagues (Victor et al., 2009) anticipate, under the ominous heading "brave new world," even though "fiddling with the climate to fix the climate strikes most people as a shockingly bad idea" and

> any worry that research on geoengineering will make governments less willing to regulate emissions. However, serious studies are more likely to reveal the many dangerous side effects of geoengineering, revealing it as a true option of last resort. However, because the option exists, and might be used, it would be dangerous for scientists and policymakers to ignore it.

It is yet to be seen how the balance between promising and dangerous narratives about geoengineering options will evolve in the future.

What is already clear, however, given the current record of government efforts to address the urgency of climate change, is that climate scientists tend to be increasingly pessimistic: in a survey of leading climate scientists conducted by the *Guardian* newspaper in the spring of 2024, many respondents said, "they had been left feeling hopeless, infuriated and scared by the failure of governments to act despite the scientific evidence provided."

Mind the gap

However, for now, how is this possible under these circumstances, scientists ask, and their position is one of the main focuses of my chapter; Does sound scientific evidence motivate and inspire comprehensive policy action in all societies and changes in the behavior of members of civil society around the world? How is it possible that democracies have done so little to effectively address the risks of climate change? After all, it is obvious that without the sciences,

> the ecological mutations would have remained invisible. In some sense, scientists are the activists on behalf of this new "social" issue. They were the first to politicize the mutations [...] by becoming their representatives and

introducing them into the old question of democracy and representative government.

(Latour, [2015] 2017: 264)

Disappointment with the workings of democracy and the blaming of democracy for a variety of social, economic, and political ills is not a new complaint: Complaints about the shortcomings of democracy have been a constant feature of democratic life, persisting through crises of governance as well as successes (Runciman, 2013a). However, the reference point of climate change is a novel cause for fundamental concern about the fate and future of democracy.

The future present depends largely on decisions made now. The narrative of the future present is, of course, determined not least by the combination of two phenomena, namely, human existence and its environment. The main perspective is, of course, a fixed human nature inscribed in a static environment. As the historian Arnold Toynbee (1949: 72) noted, this linkage only makes sense if one assumes that the relationship is unchanging and permanent. If one accepts the narrative about the intimate nature of the link between the environment and human existence, one emerges as a proponent of climate determinism.

In Bernard Shaw's 1903 four-act drama *Man and Superman*, he includes a famous preface in which he discusses various philosophical ideas, including the theory that climate can influence culture and human characteristics. He mentions how people from colder climates, such as northern Europe, have developed a more industrious nature, whereas those from warmer climates, such as southern Europe, may have developed a more relaxed and artistic temperament due to their environment. This reflects Shaw's interest in how external factors, including climate, can shape the development of societies and their moral and social behavior.

The emergence of AI as a practical tool poses a dual problem for climate change. On the one hand, it can facilitate solutions, as I will discuss shortly, or the climate crisis may prove to be AI's greatest challenge. Aside from the growing energy consumption of e-mobility and cryptocurrencies, AI is expected to consume more energy than many countries.[51] These extrapolations into the future present of energy consumption have reignited the discussion of nuclear power generation.[52] For now, crypto mining and data centers account for 2% of global electricity use and nearly 1% of global emissions, but their footprint is growing.

Generative AI and the energy problem

Calculations of energy consumption and production in the future present suffer, of course, from the difficulty of anticipating the amount and nature of energy production, let alone energy consumption. There are several so-called elephants in the room. One of these rarely considered applications is the significant but highly contested energy consumption of the rapid development and deployment of generative AI as an energy-intensive technology. Generative AI requires more power than "older" AI models do. Older AI models extract patterns from data, but do not generate new text and images such as generative AI does. AI innovation is accelerating rapidly, perhaps at an unprecedented rate. The AI experiment is still at the starting point. The pioneer of ChatGPT alone, OpenAI, has a multi-trillion-dollar plan to build new computer chip factories and data centers around the world to provide computing power to create the next generation of AI.[53] From a global perspective, the future energy demand for AI operations is expected to be comparatively small. However, as with all projections, the future level of electricity demand is a matter of debate (Chen, 2025) (Figure 1.8).[54]

Data centers power the artificial intelligence boom. Data centers make cloud-computing services possible. With most industries shifting toward AI, a massive volume of data centers is needed, and with new data centers comes the demand for energy. For the time being, data centers in the United States account for a small percentage of total electricity consumption. This figure is bound to increase as the infrastructure needed for AI grows rapidly around the world. The same applies to data centers:

> As of late March 2024, online giant ***Amazon*.com, Inc.** appears to be going all in on AI-driven data centers, with a $150 billion investment to retain

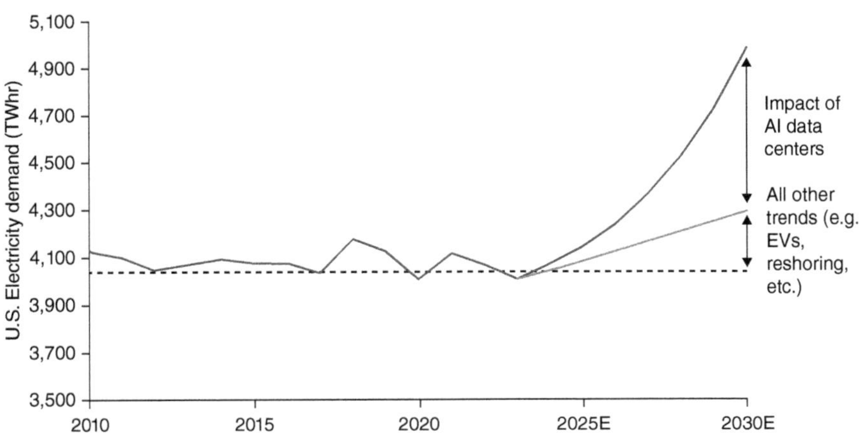

FIGURE 1.8 Estimated aggregate increase in power demand in the United States (up to 2030).

Source: Reed Cassady, Wells Fargo Securities: www.clearbridge.com/perspectives/institutional/2024/the-challenge-of-powering-ai.

> its cloud computing edge over competitors like ***Microsoft*** and ***Google***. Over the next several years, Amazon plans to build approximately 30 data centers in Indiana that will be packed with hundreds of thousands of specialized computer chips. With hundreds of thousands of miles of fiber connecting every chip and computer, the entire complex will function as one giant machine dedicated solely to artificial intelligence. The facility will consume 2.2 gigawatts of electricity, which is enough to power one million homes. It will use millions of gallons of water each year to prevent the chips from overheating. The biggest headline grabbing element of the giant investment is that one of the largest nuclear power plants in the USA[55] will directly power new ***Amazon Web Services*** (**AWS**) data center.

It is estimated that there are (in 2024) 5,000 data centers in the United States. Goldman Sachs estimates that AI will lead to a 160% increase in data center electricity demand.[56] In short, Big Tech companies are looking to secure electricity as U.S. power demand rises for the first time in two decades, driven by demand from artificial intelligence and data centers. In June of 2025, Meta Platforms struck an agreement with Constellation Energy to keep one of the utility's nuclear reactors in Illinois operating for 20 years, in the Big Tech company's first such deal with a nuclear power plant. The deal could serve as a model for other Big Tech companies to support existing nuclear plants while they plan to power data centers with new nuclear and other energy sources.

The electricity demand of data centers is linked not only to computation but also to cooling needs:

> Data centers use water in evaporative cooling systems to keep equipment from overheating. One non-peer-reviewed study, led by researchers at UC Riverside, estimates that training GPT3 in Microsoft's state-of-the-art US data centers could have consumed 700,000 liters (184,920.45 gallons) of freshwater.
>
> *(Li et al., 2023)*

Moreover, a search by "ChatGPT may consume ten times the electricity of googling."[57] Reed Cassady (2024), in an assessment of the energy needs of AI development, concludes the following: AI "like all forms of cognition, is power intensive and investors that are conditioned to think of digital innovation as purely deflationary may be in for a surprise."[58] Optimists, however, are unperturbed by such developments, convinced that technological solutions will be found and that it will be possible to deliberately alter the Earth's climate. To date, there is no evidence that AI will make a significant contribution to climate change. Corporate spokespeople also claim that the increase in energy consumption will be offset by AI, making other industries more efficient. The most prominent efforts are geoengineering solutions to climate change, which involve technologies that alter nature, preferably on a global scale.

To conclude the section on future present responses to climate change, I turn to adaptation as the most realistic and necessary way to respond to a warming planet in the years and decades ahead. Adaptation measures can, of course, be taken proactively to address both the risks of climate change and the damages of climate change impacts on societies.

Societal protection: adapting after the fact and/or proactively

In an editorial in the journal *Science*, the then-President of the British Royal Society (2005–2011) and Master of Trinity College (2004–2012) at the University of Cambridge, astronomer Martin Rees, noted in August 2006:

> The urgent challenge is to meet global energy demand [which is projected to rise by more than 50% over the next 25 years, according to the International *Energy Agency* (IEA)] while simultaneously reducing the impact of greenhouse gas emissions on climate change.

He reiterated these observations in an interview with the BBC on August 4, 2006:

> If we look at what is happening globally, there is an ever-increasing demand for energy, especially in developing countries, particularly in India and China, and at the same time, carbon dioxide levels are rising very rapidly. It is clear that if we cannot control carbon dioxide, we will reach a dangerous level of potential climate change within 50 years.

In the decades since, little has changed in the validity of these observations—except for one critical factor. Global emissions of various greenhouse gases have not decreased and are unlikely to decline significantly in the coming years. The world's appetite for energy remains equally insatiable. Meanwhile, the timeline for global warming approaching dangerous levels has shortened. Given these developments, the true political challenge of global climate change is not just *climate protection*—i.e. mitigating greenhouse gases—but above all *societal protection*, meaning adaptation.

Starting points for these considerations must be:

1. The catastrophic consequences of Hurricane Katrina for the city of New Orleans (in August 2005) and the surrounding U.S. states, or Hurricane Maria for the island of Puerto Rico (in September 2017), are prime examples of failed climate policy. However, the fault does not lie with the

Bush or Trump administrations because they refused to join the Kyoto Protocol or continue supporting the Paris Agreement.

2. Instead, climate scientists should be asked when we would see improvements if the United States, as well as China, Russia, and India, were to drastically reduce their greenhouse gas emissions. Would the impacts of hurricanes like Maria then be less severe, and how much smaller would the resulting damages be?
3. More important for our societies, however, is the question of how we can protect ourselves in the coming decades from *weather* phenomena like Hurricane Maria, heatwaves, floods, heavy snowfall, and other extreme weather events—and what climate research and policy would need to look like to achieve this. The alternative to the prevailing climate policy of mitigation is *adaptation*.
4. Nature is slow to respond. The moderate greenhouse gas reduction measures discussed so far and politically feasible have little influence on climate change, despite claims to the contrary. The reduction in global greenhouse gas emissions necessary to halt the worsening of climate change amounts to roughly 70% to 80% or more. How such a reduction could be achieved without ignoring the hopes and expectations of more than 80% of the world's population is *rarely discussed*. The resilience and security of vulnerable nations and populations will play an increasingly critical role as CO_2 levels continue to rise.

Economic, environmental, education, and research policies share several important commonalities. Their gains or losses are difficult to calculate; their successes and failures become apparent, if at all, only after decades; future generations reap their benefits or suffer from their mistakes. Voters, encouraged and reinforced by politics, have short memories and reward only what they experience directly. Environmental policy, like the other disciplines, is a matter that only yields results *in the long term*. Because of this, it becomes a pawn of current events. Extreme weather events bring climate policy to the forefront. Suddenly, the topic is on the agenda. The next day, however, news about the potential transfer of a national soccer player successfully pushes climate issues aside. But that's not all. The state's capacity for action—like that of other major societal institutions—is limited in matters of economics, research, education, and the environment. This is not due to a lack of reliable scientific knowledge. It is the loss of power by large societal institutions. The state can no longer enforce its will to the extent it could a few decades ago. In knowledge societies, the resistance capabilities of even small groups have grown disproportionately. Moreover, economic, environmental, education, and research policies have global dimensions. When it comes to climate change, this is immediately obvious to everyone. Here, politics seemingly can only act on an *international* scale. The same is fundamentally true for education and research policy. Bright minds are mobile, and the mobility of knowledge cannot rely on small-state opportunities and effects. There is another commonality: environmental, research, and education policies are *power-defining* policy areas. Those who succeed here secure a better position in the future rankings of economics and politics. This is a consequence of power, but one that feeds back into its cause. And that is what politics is interested in.

Finally, discourse in these policy areas thrives on binary extremes. Survival *or* collapse, or, as in the case of climate research and policy, mitigation *or* adaptation. How do we find solutions? Through sensible, science-backed adaptation policies, for example. As we emphasized in our *Hartwell Paper* (Prins et al., 2010) on climate policy, human history can be described as the story of humanity's emancipation from the natural variability and changes of the climate. Today, significant populations live in almost all of Earth's climate zones. Technological innovations (e.g. air conditioning, building designs, crop variants) and cultural innovations (e.g. patterns of social interaction, diet), which have developed at varying speeds, demonstrate humanity's ability to adapt to a wide range of climatic conditions. Cultural change has made larger parts of the world habitable for humans. However, adaptation owes its effectiveness throughout human history to the relative stability of climatic conditions, which, though not always, mostly varied within predictable limits. In historical times, the ability of European societies to cope with extreme climatic conditions was tested by strong deviations from experienced norms, such as the bitterly cold winters of the 16th to 18th centuries—evidenced, for example, by Bruegel's paintings of winter pastimes. The range of modern climatic conditions is both a promise and a challenge for adaptation. Yet, in the history of climate policy regulations, adaptation remains the mocked poor cousin of emissions reduction. Adaptation aims to avoid losses (and seize opportunities). It is thus an active response to recognized risks.

The prevailing consensus on climate change and the steps derived from it to address global warming lead to a political dead end. The alternative to this way of thinking is adaptation. It involves political measures that focus—not exclusively, but primarily—on adapting to climate change. Where is the crucial difference? In the consensus on the cause of climate change, the political outcome is always the same: reduce greenhouse gases, particularly carbon dioxide emissions. CO_2 is bad. This is endlessly emphasized. This mantra has little to do with the practical problem of environmental protection and preventing the dangerous consequences of environmental changes. It does, however, explain why previous measures have been so unsuccessful. They are mitigation strategies. But the issue is *adaptation*.

Adaptation measures are politically much easier to implement and justify than mitigation strategies. And they have a tremendous advantage over all mitigation strategies, whose success (or failure) lies in the distant future: adaptation processes operate on a relatively short time horizon. When it comes to finding solutions through innovations in science and technology, these are much easier to present when conceived as adaptation measures. The knowledge-based economy makes possible what was long unimaginable: reconciling ecological and economic goals. For example, if traditional objectives of business—maximizing profits—are to be maintained in the future, resources must be used more sparingly, efficiently, and productively. Adaptation occurs. The dynamics of societal transformation have grown, and with them, the chances of adapting to new realities.

Adaptation strategies also make it easier to achieve multiple goals simultaneously: improving quality of life, reducing social inequality, and increasing political participation are not mutually exclusive. The risks and dangers of dealing with uncertainties, such as new technologies, are lower in the case of adaptation measures. Adaptation processes can become the engine of what we now call sustainable development. Adaptation can lead to greenhouse gas reductions because adaptation and mitigation are not contradictory. Only: reduction does not necessarily lead to adaptation. All sustainability is local. We must rethink.

Resolving these contradictions against the backdrop of what is feasible quickly reveals that the main burden of politically realistic measures falls on adaptation strategies. Adaptation strategies describe what is possible. Consider the warnings that climate change will lead to famines and epidemics. Health is at stake. Yet personal behaviors are far more decisive than climatic conditions. And people can influence their behaviors more easily and sustainably than any attempt to deliberately alter the global climate. Adaptation means giving everyone the chance to respond to changes.

But: The fear of disasters triggered by extreme weather events is used to garner public support for mitigation efforts. This is an extremely dubious strategy. In politically relevant timeframes, the mitigation measures propagated by science and sanctioned by politics have no impact on the likelihood or intensity of extreme events or the long-term societal consequences of global warming. It is conceivable that the public will rebel against the burdens imposed on them. The dynamics of climate demand politically viable adaptation strategies that remain stable over much longer periods. This stability can hardly be achieved on the basis of fear of extreme events. Paradoxically, as our knowledge of the human-made share of global warming improves and expands, the chances of reaching a sustainable and planned reduction of greenhouse gases through negotiation in modern societies diminish—let alone the question of who should bear the costs and how the benefits should be distributed.

Adaptation, on the other hand, works. Proactive measures, for example, are effective in preventing heat-related deaths. While a tragedy unfolded in

Chicago in mid-July 1995, with more than 700 heat-related deaths, the "Hot Weather Health Warning Watch System" in Philadelphia saved about 300 lives that same summer. Experiences with extreme heat in Philadelphia in 1993 and 1994 triggered the development of an efficient warning system and social networks benefiting the elderly and other vulnerable individuals. What does this mean? In reality, it was the isolation of elderly people in Chicago who did not know how to help themselves, or the concentrated poverty (and thus helplessness) in these areas, that led to the high death toll.

This is also the issue on a global scale: Those who fight poverty create the foundation for ensuring that climate change does not bring about the disasters still invoked by politics to achieve mitigation. Adaptation means spreading knowledge and creating access to opportunities. Where people are exposed unprotected to changes, there will always be disasters—also caused by climate change. The necessary adaptation is not just a problem for the future; it is also a vital question for the present. Many populations are poorly equipped to adapt to their current climatic conditions, and we must strengthen the adaptive capacity of all communities to cope with the vagaries of—however caused—extreme weather. An environmental policy that understands this would be truly sustainable. And enforceable.

The following chapter examines the increasing advocacy of authoritarian and technocratic governance in response to climate change. Drawing on Pitirim Sorokin's theory that calamities often lead to increased governmental control and diminished individual freedoms. Environmental authoritarianism and populism are on the rise, often fueled by a lack of trust in experts, institutions, and climate science. While technocratic approaches such as 'follow the science' may offer efficiency, they risk undermining democratic legitimacy. The politicization of climate science, disinformation campaigns, and corporate influence all contribute to public skepticism. Populist and authoritarian leaders often exploit climate denial or delay to maintain power. The chapter argues that both authoritarian and technocratic responses can erode democracy, albeit in different ways. It concludes that climate change is a complex "wicked problem" requiring nuanced, democratic, and inclusive approaches, not rigid control.

Notes

1 *Sylva Sylvarum* (or *A Natural History in Ten Centuries*) is a posthumously published collection of Bacon's observations and experiments on natural philosophy. The work explores various phenomena, with heat being a recurring subject of interest. In full context (Modernized Spelling): *"It is certain, that of all powers in nature, heat is the chief; for it is the active principle of all things. And we see that in all generations and motions, heat is the great instrument of nature."*

2 The term "biota" is used to describe the collective of living organisms within a specific region, habitat, or time period. This encompasses all forms of life,

including flora (plants), fauna (animals), and microorganisms such as bacteria and fungi. In the field of ecology, "biota" is used to describe the complete range of life in an ecosystem, highlighting the interrelationships and diversity of species in a particular environment.

3 Dian Hitchcock in a letter to James Lovelock dated 13 December 1966 discussing their joint work on the Gaia hypothesis as cited in Jonathan Watts, "A cool flame: how Gaia theory was born out of a secret love affair," *Guardian*, September 5, 2024; www.theguardian.com/environment/article/2024/sep/05/gaia-theory-born-of-secret-love-affair-james-lovelock; cf. Hitchcock and Lovelock (1967).

4 Global sea ice fell to a record low in February of 2024 (see "Global; see ice hit record low in February, scientists say," *Guardian*, March 6, 2025).

5 According to the World Meteorological Organisation, the last five years have seen lower-than-average levels of rivers and reservoirs around the world; https://wmo.int/publication-series/state-of-global-water-resources-2023.

6 "A ride in a chemical-sniffing van shows how heat amps up pollution," *New York Times*, June 21, 2024; www.nytimes.com/2024/06/21/climate/heat-wave-air-pollution-chemicals.html, https://arxiv.org/pdf/2407.12447.

7 Air pollution; www.who.int/health-topics/air-pollution#tab=tab_1.

8 The average global temperature for the year 2024 surpassed that of the previous year: "global average temperatures passed the benchmark of 1.5 degrees Celsius above preindustrial times for the first time. Simultaneously, the growth rate of the world's energy demand rose sharply, nearly doubling over the previous 10-year average. As it turns out, the record heat and rapidly rising energy demand were closely connected, according to findings from a new report from the International Energy Agency" (Claire Brown, "The vicious cycle of extreme heat leading to more fossil fuel use," *New York Times*, March 24, 2025; www.nytimes.com/2025/03/24/climate/extreme-heat-emissions-energy-trends.html).

9 Piyu Ke et al., "Low latency carbon budget analysis reveals a large decline of the land carbon sink in 2023," 2024; https://arxiv.org/pdf/2407.12447; as a comment on the remarkable findings, the *Guardian* ("Trees and land absorbed almost no CO_2 last year. Is nature's carbon sink failing?," October 15, 2024) notes: "Even a modest weakening of nature's ability to absorb carbon would mean the world would have to make much deeper cuts to greenhouse gas emissions to achieve net zero. The weakening of land sinks – which has so far been regional – also has the effect of cancelling out nations' progress on decarbonisation and progress towards climate goals, something that is proving a struggle for many countries."

10 For example, Henley et al. (2014) report that the Great Barrier Reef (GBR) has experienced the warmest ocean temperatures in the past 400 years, primarily due to human-induced climate change. Extreme heat events, particularly between 2016 and 2024, have caused widespread coral bleaching. Using coral-based reconstructions and climate models, researchers have confirmed that the recent warming is unprecedented and poses an existential threat to the biodiversity of the GBR. Without urgent climate action, annual bleaching could become the norm, endangering the reef's ecological function and its World Heritage status. Also, more generally, *Woods Hole Oceanographic Institution* on ocean warming: www.whoi.edu/ocean-learning-hub/ocean-topics/climate-weather/ocean-warming/.

11 See Carl Pope, Director of the Sierra Club from 1992–2010, "What environmentalist like me got wrong about climate change," *New Yrok Times*,

June 23, 2025; www.nytimes.com/2025/06/23/opinion/climate-change-methane-natural-gas.html?searchResultPosition=1.

12 For an analysis of how to evaluate and adapt adaptation policy from the perspective of economists, see Carleton, Duflo, Jack, and Zappalà (2024).

13 Cold Nordic countries in the summer months of 2025 experienced "truly unprecedented" heat as hot weather, strengthened and lengthened by carbon pollution, continues to blaze across northern Europe. A weather station in the Norwegian Arctic Circle recorded temperatures above 30°C (86°F) on 13 days in July, while Finland experienced three consecutive weeks of 30°C heat. According to scientists, this is the longest streak in records dating back to 1961 and 50% longer than the previous record: www.theguardian.com/environment/2025/aug/02/nordic-countries-hit-by-truly-unprecedented-heatwave.

14 The extreme heat drives up enery demand, see *International Energy Agency*, "Growth in global energy demand surged in 2024 to almost twice its recent average," March 2025; www.iea.org/news/growth-in-global-energy-demand-surged-in-2024-to-almost-twice-its-recent-average : The International Energy Report finds "that global energy demand rose by 2.2% last year – lower than GDP growth of 3.2% but considerably faster than the average annual demand increase of 1.3% between 2013 and 2023. Emerging and developing economies accounted for over 80% of the increase in global energy demand in 2024."

15 In the middle of July 2025, the Spanish Environment Ministry reported that high temperatures caused 1,180 deaths in Spain in the past two months, a sharp increase from the same period last year: www.reuters.com/sustainability/climate-energy/heatwaves-spain-caused-1180-deaths-past-two-months-ministry-says-2025-07-14/.

16 A decision by the United States Supreme Court on January 13, 2025, as reported by the *Guardian*, indicates that the "supreme court has shot down the fossil fuel industry's attempts to kill a Hawaii lawsuit, which is seeking to hold the sector accountable for an alleged decades-long misinformation campaign [...] In March 2020, the city and county of Honolulu and the Honolulu board of water supply sued oil companies for violations of state law, including for creating a public nuisance and failing to warn the public of the risks posed by their products." (www.theguardian.com/us-news/2025/jan/13/supreme-court-hawaii-fossil-fuel-lawsuit). The decision of the Supreme Court allows the litigation to proceed toward a trial.

17 Stanley Reed, "BP to 'fundamentally Reset' strategy in shift back to oil and gas," *New York Times*, February 26, 2025; www.nytimes.com/2025/02/26/business/bp-oil-gas-renewables-strategy.html.

18 Source: Damien Gayle, "Private equity firms ploughing billions into fossil fuels, analysis reveals," *The Guardian*, October 1, 2024; www.theguardian.com/business/2024/oct/01/private-equity-firms-us-pensions-fossil-fuel-projects.

19 The symbol "Gt CO_2-eq" stands for "gigatons of CO_2 equivalent." It is a unit of measure used to compare the emissions from various greenhouse gases based upon their global warming potential (GWP), expressed in terms of the amount of CO_2 that would have the same GWP. Essentially, Gt CO_2-eq allows us to quantify and compare the impact of different greenhouse gases on global warming in terms of the amount of CO_2 that would cause the same amount of warming.

20 "Richest 1% account for more carbon emissions than poorest 66%, report says," *Guardian*, November 20, 2023; www.theguardian.com/environment/2023/nov/20/richest-1-account-for-more-carbon-emissions-than-poorest-66-report-says. To stay on track for 1.5C, the richest 1% would have to cut their 2015 emissions by 97% by 2030. But according to Oxfam's analysis, they are likely to reduce emissions by only 5%. A study by Schöngart, Nicholls, Hoffmann, Pelz, and Schleussner (2025) concludes that the richest ten percent of the planet, with an annual income of more than €43,000 in 2019, are responsible for two-thirds of global warming since 1990. The richest one percent, with an annual income of around €150,000 or more, contributed 26 times more than the global average to the century's heat waves. Regions such as the Amazon rainforest, Southeast Asia, and southern Africa, whose populations have historically contributed little to global greenhouse gas emissions, are particularly vulnerable to climate change and extreme weather events.

21 Raymond Zhong, "Worst-Case' Disaster for Antarctic Ice Looks Less Likely, Study Finds," *New York Times*, August 21, 2014; the article, Mathieu Morlighem et al., "The West Antarctic Ice Sheet may not be vulnerable to marine ice cliff instability during the 21st century," *Science Advances*, 2024; www.science.org/doi/10.1126/sciadv.ado7794, www.nytimes.com/2024/08/21/climate/antarctic-ice-cliff-collapse.html.

22 David King, "Humanity's survival is still within our grasp – Just. However, only if we take these radical steps," *Guardian*, May 27, 2024; www.theguardian.com/commentisfree/article/2024/may/27/humanity-survival-emissions-resilience-ecosystems-greenhouse-gases.

23 Peter Kalmus, "As a climate scientist, I knew it was time to leave Los Angeles," *New York Times*, January 10, 2025; www.nytimes.com/2025/01/10/opinion/la-fires-los-angeles-wildfires.html.

24 See Jack Ewing, "A decade-long search for a battery that can edn the gasoline era," *New York Times*, May 9, 2025; www.nytimes.com/2025/05/09/business/mercedes-factorial-solid-state-battery.html.

25 I am grateful to B. Guy Peters for the reference to C. West Churchman's initially published discussion of the issue of "wicked problems."

26 For the first application of the concept of wicked problems to climate issues, see Rayner (2006).

27 " 'Crunch time for real': UN says time for climate delays has run out," *Guardian*, October 24, 2024; www.theguardian.com/environment/2024/oct/24/crunch-time-for-real-un-says-time-for-climate-delays-has-run-out.

28 The *Guardian* reports on April 11, 2025 that "Northern India has been experiencing early extreme heat this week as temperatures topped 40°C (104°F), including in the capital, New Delhi. Hot weather across the north-west of the country peaked on Tuesday as Barmer, a city in the state of Rajasthan, reached 46.4°C – more than 6°C above the average maximum in April [...] The Indian meteorology bureau said last week that most parts of the country would experience an intense heatwave this summer, with two to four more such days than normal": www.theguardian.com/environment/2025/apr/11/weather-tracker-early-heatwave-sweeps-northern-india.

29 On August 6, 2025, the temperature in Phoenix reached 48° Celsius (118.4° Fahrenheit).

30 Adrienne Heinz, a clinical research psychologist at Stanford University, describes the *cognitive* consequences of extreme heat as a way of hijacking "our brain's ability to think clearly. Those executive functions like decision-making and inhibition and sequencing tasks, they all become harder. It becomes like walking through mud. It affects your ability to learn if you're a student or your ability to perform if you're an employee. It doesn't leave anything untouched, really" ("US faces another summer of extreme heat as fears rise over Trump cuts," *Guardian*, May 26, 2025; www.theguardian.com/us-news/2025/may/26/extreme-heat-summer-weather-forecast).

31 Robert Reich, "America's billionaire class is funding anti-democratic forces," *The Guardian*; 2022; www.theguardian.com/commentisfree/2022/may/23/americas-billionaire-class-is-funding-anti-democratic-forces

32 Timothy Noah, "Democracy is no bulwark against oligarchy," *The New Republic*, December 26, 2024; https://newrepublic.com/article/189230/democracy-trump-musk-oligarchy-inequality?utmsource=newsletter&utmmedium=email&utmcampaign=tnrdaily&vgoee=To4ju7zLklQO3wnxTUANVGdB%2B8XA3%2F%2FuO1CkBXX80ik3WawFItg%3D%3AkssMAAqBhY551xR4%2FSZPg4et%2FeTR4S4b.

33 "How close are the planet's climate tipping points?," *New York Times*, August 11, 2024; www.nytimes.com/interactive/2024/08/11/climate/earth-warming-climate-tipping-points.html.

34 A 2024 report by Femke Nijsse, Simon Sharpe, Rishi Sahastrabuddhe, and Timothy M. Lenton titled "A positive tipping cascade in power, transport and heating" advocates the concept of "*positive tipping points.*" Positive tipping points are "regulatory mandates" that are "the most powerful policies for bringing forward positive tipping points where clean technologies become cheaper than fossil fuels in power, heating, light road transport and heavy road transport … well-targeted clean technology subsidies and regulations could often be more effective and efficient" than carbon pricing; chrome-extension://efaidnbmnnnibpcajpcglclefindmkaj/https://global-tipping-points.org/.

35 chrome-extension://efaidnbmnnnibpcajpcglclefindmkaj/ www.osti.gov/servlets/purl/1836980.

36 The *New York Times* reports, "Almost the entire United States faced drought conditions during the last week of October [2024 …]. The past four months were consistently warmer than normal over a wide swath of the country, said Rich Tinker, a drought specialist with the National Weather Service. However, in June, while roughly a quarter of the country was dry to some degree, he said, now 87 percent of the nation is." Cf. "In a record, all but two U.S. states are in drought," *New York Times*, November 5, 2024; www.nytimes.com/2024/11/04/climate/united-states-record-drought.html.

37 A report from May 2025 by *Christian Aid*, as a significant example of the impact of climate change on food production, points out that bananas are the world's most consumed fruit and the fourth most important food crop globally, in terms of production volume after the usual suspects wheat, rice, and maize. In fact, more than 400 million people rely on them for 15%–27% of their daily calories. And yet this vital crop is under threat from climate change in some of its key growing regions, posing a risk to global supply. As well as growing conditions and extreme weather directly harming the growing conditions for bananas, they also

contribute to the spread of fungal diseases. www.christianaid.org.uk/sites/default/files/2025-05/going-bananas-2025.pdf.

38 Victoria Masterson, "9 AI is helping to tackle climate change," www.weforum.org/stories/2024/02/ai-combat-climate-change/#:~:text=The%20use%20of%20artificial%20intelligence,the%20World%20Economic%20Forum%20says.

39 It was the "fossil fuel industry [that] funded some of the world's most foundational climate science as early as 1954, newly unearthed documents have shown, including the early research of Charles Keeling, famous for the so-called "Keeling curve" that has charted the upward march of the Earth's carbon dioxide levels. A coalition of oil and car manufacturing interests provided $13,814 (about $158,000 in today's money) in December 1954 to fund Keeling's earliest work in measuring CO_2 levels across the western US, the documents reveal. Keeling would go on to establish the continuous measurement of global CO_2 at the Mauna Loa Observatory in Hawaii. This "Keeling curve" has tracked the steady increase of the atmospheric carbon that drives the climate crisis and has been hailed as one of the most important scientific works of modern times"; "'Smoking gun proof': fossil fuel industry knew of climate danger as early as 1954, documents show," *Guardian*, January 30, 2024; www.theguardian.com/us-news/2024/jan/30/fossil-fuel-industry-air-pollution-fund-research-caltech-climate-change-denial. As the New York Times in July 2025 reports, "President Trump's proposed 2026 budget would put an end to Mauna Loa, along with three other key observatories and almost all the climate research being done by the National Oceanic and Atmospheric Administration," www.nytimes.com/2025/07/17/climate/budget-cuts-climate-observatories.html?searchResultPosition=1.

40 The *New York Times* (April 22, 2024) reports, "In 2023, global levels of the greenhouse gas rose to 419 parts per million, approximately 50 percent more than before the Industrial Revolution. That means there are approximately 50 percent more carbon dioxide molecules in the air than there were in 1750"; www.nytimes.com/interactive/2024/04/20/upshot/carbon-dioxide-growth.html.

41 "Could a giant parasol in outer space help solve the climate crisis? Interest in sun shields, once a fringe idea, has grown. Now, a team of scientists says it could launch a prototype within a few years," *New York Times*, February 2, 2024.

42 "Zany ideas to slow polar melting are gathering momentum," *Economist*, June 6, 2024; www-economist-com.login.ezproxy.library.ualberta.ca/science-and-technology/2024/06/06/zany-ideas-to-slow-polar-melting-are-gathering-momentum.

43 Somini Sengupta et al., "Can dirt clean the climate?," *New York Times*, August 10, 2014; www.nytimes.com/2024/08/10/climate/climate-change-fungus-carbon-australia.html.

44 see "Better than trees," *Der Spiegel*, February 17, 2024; www.spiegel.de/wissenschaft/synthetische-fotosynthese-gegen-die-erderwaermung-besser-als-baeume-a-f3584b99-2582-4406-bb35-446f05458c3c.

45 Christopher Flavelle, "Warming is getting worse. So they just tested a way to deflect the sun," *New York Times*, April 3, 2024; www.nytimes.com/2024/04/02/climate/global-warming-clouds-solar-geoengineering.html.

46 *New York Times*, "This scientist has a risky plan to cool earth. There's growing interest," August 1, 2024; www.nytimes.com/2024/08/01/climate/david-keith-solar-geoengineering.html.

47 Brad Plumer and Raymond Zhong, "They've got a plan to fight global warming. It could alter the oceans," *New York Times*, September 23, 2004; www.nytimes.com/2024/09/23/climate/oceans-rivers-carbon-removal.html?searchResultPosition=1.

48 Katharine Gammon, "New powder that captures carbon could be 'quantum leap' for industry," *The Guardian*, November 29, 2024; www.theguardian.com/environment/2024/nov/29/covalent-organic-framework-carbon-capture-powder.

49 The *Guardian* (Jonathan Watts, "Climate engineering off US coast could increase heatwaves in Europe, study finds," June 21, 2024) notes in reporting about the *Nature Climate Change* article, "experts have said the paucity of controls means there is little to prevent individual countries, cities, companies or even wealthy individuals from trying to modify their local climates, even if it is to the detriment of people living elsewhere, potentially leading to competition and conflict over interventions."

50 Michael Gerrard and Tracy Hester's book *Climate Engineering and the Law* focuses on the legal aspects of these geo-engineering technologies: what government approvals would be needed; how liability and compensation would be assessed if something went wrong; and how a governance system could be structured and agreed internationally.

51 As far as the present state of carbon emissions in the tech sector is concerned, a report produced by the International Telecommunication Union (ITU) and the World Benchmarking Alliance (WBA) indicates that their emissions have continued to rise in recent years, fuelled by rapid advances in artificial intelligence (AI) and data infrastructure. The ITU and the WBA track the greenhouse gas (GHG) emissions, energy use, and climate commitments of 200 leading digital companies, beginning in 2023. The most recent report also indicates some progress. Worldwide, more companies have set emissions targets, sourced renewable energy, and aligned with science-based frameworks (cf. www.worldbenchmarkingalliance.org/news/wba-and-itu-publish-fourth-annual-joint-report-assessing-digital-companies-climate-commitments-and-progress/).

52 Cf. World Economic Forum, www.weforum.org/stories/2024/11/meeting-global-climate-goals-requires-a-step-change-in-nuclear-investment/.

53 Cade Metz and Tripp Mickle, "Behind OpenAI's audacious plan to make AI flow like electricity,"*New York Times*, September 25, 2024; www.nytimes.com/2024/09/25/business/openai-plan-electricity.html?searchResultPosition=3.

54 The demand for water of the new data centers also constitutes an unknown but citical develoment; see "Big tech's new datacentres will take water from the world's driest areas," *Guardian*, April 9, 2025: In parts of the world where water is plentiful, datacentres' high water usage is less problematic, but in 2023 Microsoft said that 42% of its water came from "areas with water stress," while Google said 15% of its water consumption was in areas with "high water scarcity." Amazon did not report a figure; www.theguardian.com/environment/2025/apr/09/big-tech-datacentres-water.

55 https://electrek.co/2024/03/05/amazon-just-bought-a-100-nuclear-powered-data-center/.

56 "The AI power grab," *New York Times*, October 23, 2024; www.nytimes.com/2024/10/22/climate/ai-big-tech-emissions.html.

57 "Generative AI has a clean-energy problem," *The Economist*, April 11, 2024; www.economist.com/business/2024/04/11/generative-ai-has-a-clean-energy-problem?utmmedium=cpc.adword.pd&utmsource=google&ppccampaignID=18151738051&ppcadID=&utmcampaign=a.22brandpmax&utmcontent=conversion.direct-response.anonymous&gadsource=1&gclid=CjwKCAjwwiwBhApEiwAuG6ccEIIf2RZ0o470rd3G0vJL5ThocGAeD2ckPPp7iSM8pnSInpASYzjdRoClQgQAvDBwE&gclsrc=aw.ds.

58 According to a Guardian analysis, dated September 15, 2004, "from 2020 to 2022 the real emissions from the 'in-house' or company-owned data centers of Google, Microsoft, Meta and Apple, are likely about 662% – or 7.62 times – higher than officially reported"; "Data center emissions probably 662% higher than big tech claims. Can it keep up the ruse?" www.theguardian.com/technology/2024/sep/15/data-center-gas-emissions-tech.

2

THE MOMENTUM OF AUTHORITARIANS OR THE OLIGARCHIC TEMPTATION

> [...] the main uniform effect of calamities upon the political and social structure of society is an expansion of governmental regulation, regimentation, and control of social relationships and a decrease in the regulation and management of social relationships by individuals and private groups.
>
> Pitirim A. Sorokin ([1942] 2010: 122)

> Unfortunately, [...] prominent Western democracies currently lack the leadership necessary to counter the rise of authoritarianism.
>
> Aryeh Neier (2024)[1]

Aryeh Neier fears that the future of democratic governance around the world is under serious threat. However, the threat comes not directly from climate change, but from the lack of political leadership to vigorously defend democracy. In contrast, in the middle of World War II, the now often forgot Russian-American sociologist Pitirim Sorokin who Lenin undeservedly called a "muddle head," who stressed that "we live in an age of great calamities" (Sorokin, [1942] 2010: 9). The term "calamities" even then did not often find its way into the sociological lexicon (Horowitz, 2010: xi). Sorokin could not foresee the calamity of climate change. However, what he foretold on the basis of a wide range of socio-historical examples of calamities was the almost universal "expansion of governmental control means a decrease of freedom, a curtailment of the autonomy of individuals and private groups in the regulation and management of their individual behavior and their social relationships, the decline of constitutional and democratic institutions"; if the calamity "is acute and of long duration, the totalitarian, autocratic,

DOI: 10.4324/9781003682066-3

or dictatorial trend assumes a more drastic and less ephemeral character" (Sorokin, [1942] 2010: 122).

Given that the term calamity or catastrophes, which Sorokin applies to pandemics, famine and pestilence, war and revolution, we easily recognize their effects as the societal conditions that are anticipated to accompany climate change in the future present. The effects of calamities are an unforgiving influence "on the mental processes, behavior, social organization, and cultural life" (Sorokin, [1942] 2010: 9) of the entire population.

In numerous contemporary democratic countries, there has been a discernible surge in the influence of political authoritarians, accompanied by a notable ascendance of the notion of environmental authoritarianism. Without delving deeply into the intricacies of political theory and the contested concept of populism and authoritarianism (cf. Elçi, 2024), for heuristic purposes, *authoritarianism* can be defined as a political system where power is concentrated in the hands of a single leader or a small elite, with limited political pluralism, suppression of dissent, and control over various aspects of society, including the media, civil rights, and opposition parties. The link between political authoritarianism and authoritarian capitalism is also noteworthy.[2] The label of "free-market authoritarianism" (Darian-Smith, 2022: 67) signals, moreover, the affinity of neoliberal capitalism and antidemocratic political platforms. Political scientist Ian Bremmer (2010; Keane, 2020) introduced the concept of "state capitalism" to describe countries such as China and Russia, where governments wield significant control over key sectors of the economy while still allowing capitalist enterprises. Bremmer views authoritarian capitalism as a competitor to liberal democratic capitalism in global politics. Given that preferences for an authoritarian political regime are expressed, a link to populism is not uncommon. *Populism* represents a supplementary political program that asserts the representation of "the people" against a corrupt or distant elite (see Norris and Inglehart, 2019: 66). As Kurt Weyland (2024: 2) generally notes, "while each version of populism did remain a temporary phenomenon, populism as a political strategy proved irrepressible and recurring; and through its rapid mutations, it demonstrated its adaptability, which boosted its chances of electoral success."

Pippa Norris and Ronald Inglehart (2019: 66) describe the typical dichotomous populist discourse:

> The authority of the voice of the people is valued even when at odds with professional specialists ('Britain has had enough of experts'), legal authorities ('so-called judges'), mainstream media commentators ('fake news'), scientists ('climate change is a hoax'), and elected politicians ('just out for themselves'). Thus, in a black-and-white moral vision, most hardworking people are portrayed as homogeneous and inherently 'decent,' betrayed by dishonest pols and corrupt elites ('Crooked' Hillary/ 'Lying' Ted).

Another excellent example of fake news is the linking of weather phenomena (hurricanes) to abortion, as claimed by "journalist" Tucker Carlson in the run-up to the presidential election in the Fall of 2024 in the United States: "And I have to say, I'm sure I'll be attacked for saying this, but I really believe it," he said.

> People are like, 'Oh, well, we had another hurricane. Must be global warming.' No, it's probably abortion, actually. Just being honest. Like, you can't do that. You can't kill children on purpose, knowing that you're doing that in exchange for power or freedom or happiness or whatever you think you're getting in return. You can't participate in human sacrifice without consequences.[3]

Populism advocates hostility toward the educated strata and the distrust of institutions. Populist leaders often portray politics as a conflict between virtuous, ordinary citizens and an immoral, powerful elite. Populism is more ideologically flexible than authoritarianism. Populists, whether on the left or the right of the political spectrum, share a political style which claims to be more democratic than that of their opponents.

Populism sees itself in a positive light, describing aspects of a direct democracy or, if its proponents view their political agenda as exemplifying "real democracy." More often, it is now seen as serving as a pathway to authoritarianism, with populist leaders using their popularity as a justification for authoritarian measures. The populist movement may be right-wing or left-wing, depending on the ideological orientation. For Pippa Norris (2019: 983), "authoritarian populism" movements in more general terms

> a) advocate conformity with conventional moral norms and traditions within a group, expressing intolerance of out-groups perceived to threaten accepted group norms, b) expect deference and loyalty to the group and its leaders, being intolerant of dissent, and c) seek to strengthen collective security against perceived group threats.

Contemporary populism and authoritarianism are driven by political and economic developments in addition to environmental challenges, such as fear of mass migration, increased outbreak of terrorism, and/or austere governmental responses to the global financial crisis in 2008 and the more recent coronavirus pandemic. Vlad Surdea-Hernea (2025) explores the enduring impact of authoritarianism on contemporary climate policy preferences, using post-World War II Germany as a natural experiment. Using a geographic regression discontinuity design, the study shows that districts from former East Germany today exhibit significantly lower support for ambitious climate policies compared to the West. This divergence persists independently of other political attitudes, suggesting that authoritarian

FIGURE 2.1 Political sentiments toward the contemporary political order in the United States.

Source: New York Times, January 17, 2025.

Notes:

Question: Which of the following comes closest to your opinion about our nation's political system, even if none are exactly right?
It has been broken for decades
It has been broken only for the last few years
It is not broken
Democrats and Republicans include those who identified with or leaned toward each party. Based on a poll by The New York Times and Ipsos of 2,128 U.S. adults conducted from Jan. 2 to 10. Graphic by June Kim.

suppression of pro-environmental organizations has a lasting effect on environmental engagement.[4] In a representative poll commissioned by the *New York Times* and conducted in January 2025, most respondents—across race, gender, and party—said that *the political system was broken* and that *the economy is working against them* (Figure 2.1).

An interpretation of the poll's results could be that most of the American public expresses a preference for a strong political leader, confirming the trend of democratic backsliding. As previously stated, the concurrence of political and environmental authoritarianism represents an increasingly prevalent association of political beliefs that collectively espouse "new eco-political pessimism" (Hulme, 2009: 309). This phenomenon jointly advocates for the concentration of political power. The political position in question is exemplified by the natural scientists David Shearman and Joseph Wayne-Smith (2007). They diagnose a fundamental conflict between the goals of individual liberties and ecological survival. The conflict between civil rights and the realization of the goal of ecological survival leaves but one choice in their view: the curtailment of individual liberties. This leads Shearman and Wayne-Smith to the conclusion that

> it may well be non-Western states, including China, [who] will find ways to deliver [on climate change], while the West continues to display its extreme

> liberty with ineffectual debate and a surrender to powerful interests in its grinding democratic institutions.[5]

It is perhaps obvious that the conflation of political and environmental authoritarianism hardly represents a constructive position in a world of massive societal challenges, while at the same time, it is not surprising that the same conditions call forth extreme political voices. In any case, the ascendance of authoritarian beliefs represents a symptom of a crisis in governance and is incompatible with the institutions of a constitutional democracy, including the rule of law, checks and balances, and minority rights.

The broad general momentum of the authoritarians and the decline in the belief that democracies are the most effective and capable response to a turbulent era may generally be a reaction to exposure to the structural and material societal changes underway and the already felt as well as expected that follow from such secular societal changes, for example, anticipated damage to one's economic interests, massive social inequalities of power and privilege, and transformations in desirable consumption patterns. In addition, what is more, ideology-driven rejection of the status quo manifests in an aversion to globalism, liberalism, and the loss of national sovereignty (Lockwood, 2018).

More than half a century ago, Norbert Elias (1977: 358) observed the common-sense interpretation of momentous and threatening social change which is still valid today

> the increasing stream of innovations makes the affected people insecure; the growing tempo of change strengthens their desire for enclaves of peace and for symbols of changelessness. Above all, however, people search for deliverance from the unremitting conflicts of human groups – whether they delude themselves that everything could be peaceful and harmonious if only the others, the disturbers of the peace, the agitators, did not rebelliously threaten the good life, or whether they see the means of salvation in an overthrow of existing power relations and the establishment of another order in which they hope for greater peace, harmony and freedom from conflict.

Resilience and adaptation to major social changes appear to affect one's life and the foreseeable future is often limited. This is a response to societal transformations that are secured and strengthened by the belief in a good *existing* life.

In many countries, including democratic countries, governments attempt to directly influence the development of scientific research. For example, funding priorities are set for specific medical conditions such as Alzheimer's disease or social problems such as migration across international borders. What is at issue here is that the reverse is also true. The scientific community or specific fields within science attempt to persuade or direct the government to follow science. The government, in response to the interest and defense of basic constitutional rights, insists that science is only able to offer advice. Politics follows politics. The Federal German Minister of Justice at the time, Marco Buschmann, in an essay in the *Frankfurter Allgemeine Zeitung* on the 75th anniversary of the Basic Law (*Grundgesetz*), offers the same defense of the autonomy of politics, especially with reference to the diversity of scientific advice:

> Deriving a universal 'ought' from the various scientific descriptions of existence is always a process of political deliberation. Every scientific discussion result can only ever be a piece of advice that has to face up to a democratic debate and cannot replace it.[6]

However, the role of scientific knowledge in the present political context can have the dual but paradoxical effect of supporting essentially authoritarian political agendas. Each side favors or fears a direct line from science to politics. On the one hand, there is resistance to scientific knowledge. Efforts to push back, i.e. to "not follow climate science"—not least because climate science knowledge is labeled political—are best achieved in a non-democratic political "climate" in the eyes of opponents of climate policies that, for example, seek to advance decarbonization. On the other hand, those social movements, scientists, and businesses that support the implementation of decarbonization policies are convinced that a technocratic policy approach is best suited to achieve the desired outcomes. A positivist self-understanding of the "nomological sciences lends countenance to the substitution of technology for enlightened action. It directs the utilization of scientific information from an illusory viewpoint, namely that the practical mastery of history can be reduced to technical control of objectified processes" (Habermas, [1968] 1971: 316). Political opposition is labeled illegitimate and contrary to the best available knowledge and the best interests of the people. Climate science lends itself to these interpretations of support for authoritarian political ambitions because, as I have noted, climate change is a wicked problem. Climate change is more than an environmental issue.

Pushing back

> Acknowledging manmade global warming [...] undermines the belief of neoliberalism in unregulated economic growth as the standard for evaluating economies, societies and political leaders and as a mean for addressing a wide variety of problems.
>
> *Ulrich Beck and Joost van Loon (2011: 116)*

Anthropogenic climate change is not only an essentially human catastrophe but also a perceived threat to the main beneficiaries of the modern economic order, and who have disproportionately prospered from the burning of fossil fuels. On a smaller scale, like other industries that fear threats to their business model, such as pharmaceutical companies, tobacco industries, banks, fossil fuel interests, and "the godfathers of climate chaos,"[7] have used a variety of tactics, such as "greenwashing,"[8] lobbying, donating money to sympathetic politicians and political parties, supporting reassuring research, relenting to the pressure brought to hear on the media to reduce the volume of the coverage,[9] and setting up "grassroots organizations," to convince the public and politicians that their products are not harmful but are even beneficial, for example, that their fuels are fuels of the future. As these efforts are often quite transparent and have often been exposed, more tacit efforts to support their interests are likely to be more successful.[10]

Efforts designed to push-back the message of climate change by diverse tribes opposed to measures to address climate change, take on all sorts of subtle and perhaps at times less subtle strategies, spreading "disinformation" or "misinformation," for instance, in the hope that climate discourse will simply vanish (see Treen et al., 2020). In the United States,

> disinformation about climate change is being strategically disseminated to the American public through well-funded networks. Over 90% of environmentally sceptical books published in the U.S. have been linked to conservative think tanks. The fossil fuel industry is providing donations to members of Congress in direct proportion to their propensity to vote against environmental policies.
>
> *(Lewandowsky et al., 2023)*

The government of Florida, the hottest state in the mainland United States introduced a new law on May 16, 2024, that makes climate change a lesser priority and largely *removes* the phrase from statutes in the state. The legislation,

> which was signed Wednesday by Republican Governor Ron DeSantis and comes into effect in July [of 2024], also bans power-generating wind turbines offshore or near the coast while reducing regulation on gas pipelines [… the Governor added] We're restoring sanity in our approach to energy and rejecting the agenda of the radical green zealots.

In the state of Texas, fossil fuel interests have been successful in persuading the Texas Board of Education to reject textbooks that describe and explain the impact of climate change and extreme weather events accurately.[11]

In the run-up to the November 2024 election and in search of campaign funds, Donald Trump held a private dinner, some observers called a "policies-for-money transaction" deal for oil executives in Mar-a-Lago in April. Trump

> told a group of oil executives and lobbyists [including executives from ExxonMobil, EQT Corporation and the American Petroleum Institute] gathered [... at the] resort [...] that they should donate $1 billion to his presidential campaign because, if elected, he would roll back environmental rules that he said hampered their industry. [...] Mr. Trump repeated his public promises to delete Mr. Biden's pollution controls, telling the attendees that they should donate heavily to help him beat Mr. Biden because his policies would help their industries.[12]

The Mar-a-Lago dinner in April 2024 for fossil-fuel industry representatives was among several such dinners during the presidential campaign of Donald Trump.[13] Paradoxically, as the *New York Times* noted during this period, the

> oil and gas production have set records under the Biden administration and the United States is the world's leading exporter of liquefied natural gas. Even with the pause on permits for new export terminals, the United States is still on track to nearly double its export capacity by 2027 because of projects already permitted and under construction.[14]

As the *Guardian* reported,[15] the president of Cop 28, in the fall of 2023, UAE Sultan Al Jaber, who is also the chief executive of the United Arab Emirates' state oil company, *Adnoc*, claimed in November of 2023 that there is "no science" indicating that a phase-out of fossil fuels is needed to restrict global heating to 1.5°C.

Denying a problem is not limited to efforts to counter scientific evidence; at times, it becomes violent, as a 2024 report "For Internews's Earth Journalism Network" and Deakin University demonstrate. This study reports on the state of climate and environmental journalism *globally*, assessing the challenges and trends faced by journalists in reporting on these topics. The study does not analyze media content but rather emphasizes the experiences and responses of journalists. The study is based on 744 survey respondents and 74 in-depth interviews, for a total of 818 responses. Journalists from 102 countries, with the highest participation rates from Tanzania, India, Kenya, and the United States, participated in the study.

Approximately one-third of the respondents identified themselves as specialists in climate change and environmental reporting. Almost 40% of the journalists covering climate and environment issues have been threatened as a result of their work, with 11% being subjected to physical violence. This study also investigated

> whether safety concerns affected news that is produced, for example, by resulting in self-censorship [...] a slim majority of our respondents had not experienced threats or danger as a result of their work; however, many said they needed to self-censor at times.[16]

The study also noted that

> In the digital media landscape where anyone with access to the internet can be a producer and broadcaster of information, mis/disinformation are increasingly complicating the job of journalism. We heard from journalists that erroneous information was widespread, especially on social media, and that the speed with which it proliferated could 'drown out' carefully researched, factual journalism on climate and environment.[17]

Part of the push-back against climate policies originates with political parties and their strategies to appeal to voters who are disenchanted with secular developments in a variety of political domains, especially in the field of the economy, be it inflation, mounting social inequality, a lack of mobility prospects, or legally binding mitigation targets in the home-building sector or mobility. A strategy that does not carry the designation of "greenlash." Voters agree that action on the climate front is needed, but they are less willing to bear the costs of the green transition. Political parties will attempt to deflect the backlash against climate policies by changing the political narrative. The war in Ukraine for example becomes a justification for climate policies. The policies lower the dependence on Putin's Russia. Climate policies can further be framed to constitute ways of maintaining the competitive edge of the continent in its economic struggle with China or the United States.

Part of the media fight against climate change involves what is variously called "misinformation," which is defined as false information, "disinformation," or misleading information. Another set of pushbacks attempts to use the legal system. The legal system can indeed be a formidable ally of social movements, conservative activists, political parties, a coalition of corporations, and other entities who find a sympathetic court impairing environmental legislation. Allies of fossil fuel companies, for example, are pressuring the United States Supreme Court to ensure that the companies are protected from litigation which could cost fossil fuel companies billions of dollars.[18] In recent years, the conservative majority of United Supreme Court judges have issued rulings that impact the "[U.S] Environmental Protection Agency's authority to limit pollution in the air and water, regulate the use of toxic chemicals and reduce the greenhouse gases that are heating the planet."[19] The legal framing brought forward by conservative activists against governmental regulation resonated with the opinions of most of the court. Most significantly, the doctrine of administrative law that the courts should defer to government agencies to interpret unclear laws was struck down. More remarkably, the *New York Times* report stresses that there "have been several decisions by the court to intervene to stop environmental regulations before they were decided by lower courts or even before they were implemented by the executive branch." This kind of intervention by the activist court is virtually without precedent. In sum, these rulings by the court endanger many environmental rulings of government agencies and impose a kind of anticipatory hesitation to formulate new rules, immobilizing environmental protection efforts.

The courts (see the section on climate change and the law) can also be a major obstacle to environmental activism. In the United States, for example, Greenpeace has been involved in protracted legal battles over the Dakota Access Pipeline (DAPL), particularly over liability claims made by the pipeline's developer, the Texas-based pipeline company Energy Transfer LP. The Dakota Access Pipeline is a crude oil pipeline that runs from North Dakota to Illinois. It became the focus of intense protests in 2016–2017, particularly by the Standing Rock Sioux Tribe, with opponents of the pipeline raising concerns about: Water contamination (particularly of the Missouri River), violation of indigenous rights and treaties, and climate impacts of fossil fuel infrastructure. Energy Transfer filed a $900 million lawsuit against Greenpeace and other organizations under the Racketeer Influenced and Corrupt Organizations Act (RICO): a law traditionally used against organized crime. A North Dakota jury awarded Energy Transfer more than $660 million in damages on March 19, 2025. The case is making its way through the courts.

Social media posts before and after hurricanes Helene and Milton exemplify the volume of misinformation about extreme weather events and climate change. The pushback instrumentalizes extremes for political gains:

> Infowars' Alex Jones, who claimed that Hurricanes Milton and Helene were 'weather weapons' unleashed on the East Coast by the U.S. government, and 'truth seeker' accounts on X that posted photos of condensation trails in the sky to baselessly allege that the government was 'spraying Florida ahead of Hurricane Milton' in order to ensure maximum rainfall, 'just like they did over Asheville!'[20]

Michael Caulfield[21] has argued, summarizing the primary function of spreading misinformation on social media in particular, that misinformation "is not to change the beliefs of other people at all. Instead, most of the misinformation is offered as a service for people to *maintain* their beliefs in face of overwhelming evidence to the contrary."

The sum of local, regional, sectional, or global pushback feeds into the narrative that climate change is not real, or if it is real, it is not significant enough to warrant major political, economic, or cognitive changes in societies. If climate change is real and significant, the answer "follow the science" is an often-heard advice and an often-branded solution: the technocratic response. Theodore Roszak, the inventor of the term counterculture and, as the term perhaps implies, the author of an anti-technocratic treatise in the late 1960s, had no sympathy for blindly following the "follow the science" advice. The counterculture he described and favored rejected claims to expertise by any kind of expert. Roszak (1969: 6, 9) insisted,

> in the technocracy, nothing is any longer small or simple or readily apparent to the non-technical man. Instead, the scale and intricacy of all human activities – political, economic, cultural – transcends the competence of the amateurish citizen and inexorably demands the attention of specially trained experts.

Technocrats might consider themselves benevolent forces, offering working solutions to complex public problems, but they exercised "totalitarian control."

Follow the science

> Science-based decision-making is what will ultimately help the world to resolve the [climate] crises it faces.
>
> *Nature* **636**, *521–522 (2024) December*

> Where power is based on science and scientifically recognized necessities, there is no room for political decision-making.
>
> *André Gorz ([2008] 2009: 33)*

For those who prefer what can only be called a technocratic approach to climate policy, the rewards of strictly "following the science," as an editorial in *Nature* expresses it, are too obvious. The center of power shifts. The political elite becomes the subordinate of the scientific elite. It is the reversal of the disadvantages of following a course of action that involves democratic accountability. In this case, adherence to scientific principles requires the possession of the necessary expertise, which in turn requires the implementation of climate policies that are free from the influence of vested interests, competing narratives, uninformed and irrational voters, or social movements that prioritize political participation. As Jürgen Habermas

([1969] 1971: 63–64) has made evident, the politician becomes "the mere agent of a scientific intelligentsia, which, in concrete circumstances, elaborates the objective implications and requirements of available techniques as well as of optimal strategies and rules of control."

Jacob Hale Russell and Dennis Patterson (2025) explore the idea that technocratic reasoning, or the demand that politics follow the science, can support *populism*. Russell and Patterson touch on a somewhat paradoxical dynamic: the reliance on expert authority in liberal democratic systems can fuel populist backlashes. The strict adherence to expertise, when elevated above democratic deliberation, can undermine the legitimacy of democratic decision-making. This perceived technocratic elitism provides fertile ground for populist movements that claim to be returning power to 'the people'. When experts are perceived to fail in their advice, make demands that are perceived as excessive (e.g. financial crises, pandemic missteps), or appear biased or captured by elite interests, public trust declines. Populists capitalize on this by portraying elites as out of touch or self-serving. Claiming to restore sovereignty and voice to ordinary citizens, populists often oppose expert-led institutions such as courts, regulators, or supranational bodies (e.g. the EU). They argue that these institutions are undemocratic and represent elite control.

If one is not prepared to describe the straightforward demand to follow climate science as a technocratic worldview, then one would at least have to point out that such a demand assumes that the political application of an imperative of climate science advice coincides with the political "desirability" of its implementation. Opponents of deferring to climate science in the political arena are not prepared to make such an assumption. Politics as politics is not about to disappear.

On the other hand, it could be argued that the maxim "follow the science" is embedded in a general secular trend in modern society described, for example, by Helmut Schelsky (1961: 5; emphasis added), as the "fundamental fact" of the enormous expansion and effectiveness of science in modern life: the "*scientization of our world* and life in it." The wide-ranging scientization of our existence, as Schelsky (1961: 5–6) further notes, "requires a new classification of science in relation to man and the world, and that this scientization of our lives gives rise to a new relationship between man and the world itself." With the technocratic approach, the idea of democracy

> loses [...] its classical substance: the political will of the people is replaced by the objectivity that man produces as science and labor. This fact changes the foundations of our state rule in general; it transforms the foundations of legitimacy, of government as rule, of state reason, of the relationship between states.
>
> *(Schelsky, 1961: 22)*

The downside of this approach, and the sacrifices it imposes, are all in the name of the overarching agenda. At the same time, the advantage, in the eyes of its advocates, is that following science has the benefit of not only acting promptly, depending on the circumstances but also quickly. The paradigmatic reference would be to an equally urgent situation such as a pandemic or a climate disaster, where in control of the necessary means of power, mitigating measures are promptly imposed, even in the face of open resistance.

The success of advocating a "follow the science" approach depends in part on the public's level of confidence in the role of science in policy. A recent PEW Research Center survey of 9,593 U.S. adults conducted October 21–27, 2024, in the United States, revealed that trust or confidence in science in policy matters is substantial. It varies by respondent's political affiliation and has improved since the results of the COVID-19 pandemic survey. Overall, 76% of Americans express a great deal or fair amount of confidence in scientists to act in the public interest. With respect to the political role scientists should play, 51% say scientists should take an active role in public policy debates about scientific issues, whereas 48% say they should instead focus on establishing sound scientific facts and staying out of political debates. Public support for scientists taking an active role in politics remains lower than it was in surveys conducted in 2019 and early 2020. In both of those surveys, 60% of Americans said that scientists should take an active role in policy debates about scientific issues. That share dropped to 48% in September 2022, well into the second year of the COVID-19 pandemic. The current survey marks a slight uptick (+3 points) in the share supporting an active political role for scientists (Figure 2.2).

Die Politik des Unpolitischen

The discourse in favor of a technocratic approach is also recommended as a means of overcoming mere political interests, conflicts, and struggles; in fact, the approach that advocates "scientific decision-making" obscures what remains political (Habermas, [1969] 1971: 65). A decision-making approach to climate policy that refers only to climate issues is a utopian idea. Therefore, following the scientific approach amounts to a social process of politics of the supposedly apolitical. As Héléne Landemore (2023: 39) illustrates:

> A niche of independent decision-making that would be strictly related to climate is implausible. The sheer complexity of the issue would introduce all kinds of uncertainties [...] In the end, climate change would have to recognized as merely one of many issues that bear on humanity's future. In this broader, fuzzier context, it would not be clear that experts have more authority than laypeople.

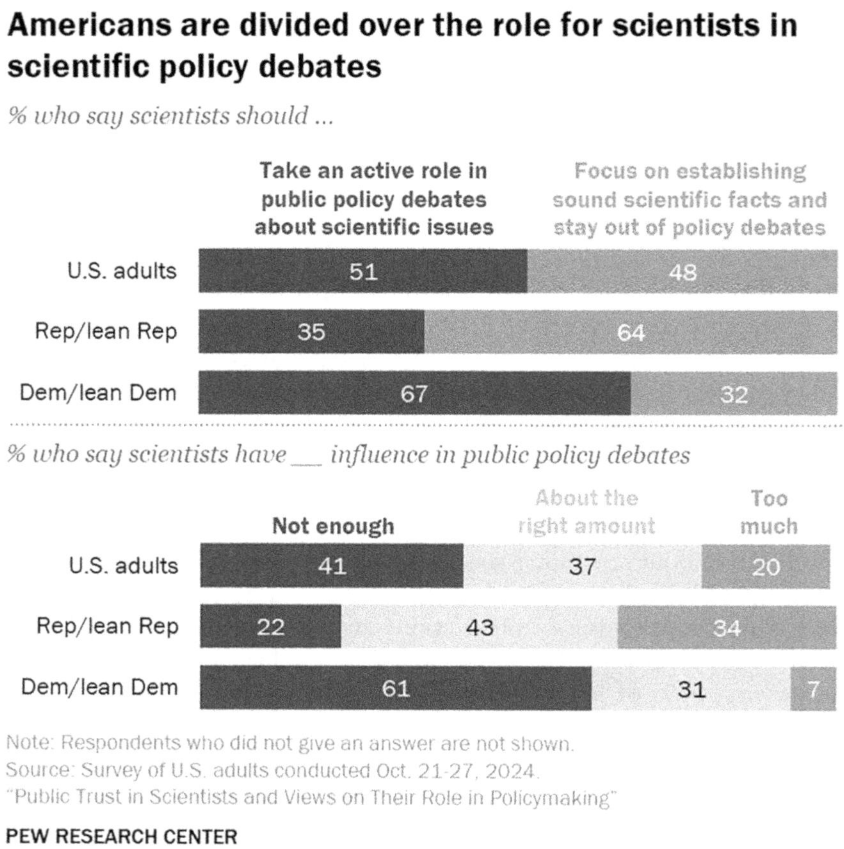

FIGURE 2.2 Americans are divided over the role of scientists in scientific policy debates.

Source: Pew Research Center, Report, November 14, 2024; www.pewresearch.org/science/2024/11/14/public-trust-in-scientists-and-views-on-their-role-in-policymaking/?utmsource=Pew+Research+Center&utmcampaign=441d4a5e6a-Internet-Science20241121&utmmedium=email&utmterm=0-441d4a5e6a-401353848#views-on-scientists-judgment-and-role-in-policymaking.

Even more specifically and more importantly, as I explain in a later section in more detail, climate change is a *wicked problem.* Climate change is not the only problem that needs to be solved. The climate problem is part of a larger complex of conditions, which include population, technology, wealth disparities, public values, and resource use. Therefore, it is not just an "environmental" problem.

With respect to the technocratic model, its general weaknesses are evident:

> On the one hand, it assumes an immanent necessity of technical progress, which owes its appearance of being an independent, self-regulating process only to the way in which social interests operate in it—namely through continuity with unreflected, unplanned, passively adaptive natural history. On the other hand, this model presupposes a continuum of rationality in the treatment of technical and practical problems, which cannot in fact exist.
>
> *(Habermas, [1969] 1971: 64)*

A recent "apolitical" or agnostic approach to the governance of climate policy rests on the suggestion that algorithms—instructions for action provided by data and executed by code—could or should take the place of policy, based more directly on "social" instructions for appropriate policy tools; for example, algorithms are mobilized to generate policy outcomes, assess mitigation and adaptation strategies, and design projects to deal with the consequences of climate change. For example, Ruth Machen and Eric Nost (2021) argue that as algorithms become central to climate policy tools, they carry with them built-in epistemic commitments that shape what counts as valid knowledge. They identify three key algorithmic "ways of thinking": (1) induction, where data-driven inference claims objectivity by 'letting the numbers speak' but obscures underlying choices; (2) abstraction, which renders complex socio-environmental contexts into universal, comparable categories but erases situated nuance; and (3) optimization, which presents "best possible" solutions as necessary while embedding existing constraints (e.g. cost, property value) into decision frames. Drawing on Laclau and Mouffe's theory of hegemony, Machan and Nost show how these commitments promote knowledge regimes that assert objectivity, universality and necessity—and marginalize alternative forms of place-based knowledge or transformative perspectives. They call for more empirical research into the limits and political implications of 'algorithmic thinking' in climate governance, and for spaces of agonistic debate in which algorithmic models become provocations rather than arbiters of decision-making and therefore do not see algorithmic thinking [and the erosion of the political] as a paradigmatic shift in governance (Machan and Nost, 2021: 565). The technocratic approach remains free of any references to democratic governance. The model acquires legitimacy by dispensing with democracy. Political rule as political power can be transcended for the sake of an efficient execution of scientific exigencies.

In contrast, the conviction not to follow climate science can amount to either a *simulated climate change opinion* or a *strict negation of climate science*. The simulated persuasion of climate change involves recognizing and accepting the results of climate science and the reality of climate change, but failing to follow its implications in one's life (e.g. Norgaard, 2011). The strict rejection of climate science findings, i.e., not adhering to science, is linked to active political opposition to science.

Do not follow the science

> What doesn't manage to get through to people bombarded by bad news about the ecological mutation is the activity, the autonomy the sensitivity to our actions, of the materials that make up the critical zones in which we all reside. These people seem incapable of responding to the agency of these materials.
>
> *Bruno Latour ([2015] 2017: 207)*

The lack of attribution of agency to non-human factors, according to Bruno Latour's narrative of climate change, is at the root of an essentially religious basis for climate denialism. Segments of society, such as populist social movements, are making a concerted effort not to follow science. However, the core of the political strategy of climate denialism in the 1990s, questioning the evidence accumulated by the scientific community, has lost much of its appeal for opponents of climate change. Today, the emphasis is on broad-based movements that aggressively defend the fossil fuel way of life. Even with considerable funding from business, the big lie became hard to uphold; Exxon and BP now acknowledge the existence of climate change (Tooze, 2021). Today, the emphasis has shifted to backing fossil fuels and "drill, baby drill." As the alleged blueprint for the new Trump administration demands:

> USAID [*United States Agency for International Development*] should cease its war on fossil fuels in the developing world and support the responsible management of oil and gas reserves as the quickest way to end wrenching poverty and the need for open-ended foreign aid.
>
> *(Heritage Foundation, 2023: 257)*

In many countries of the Global North, there is a clear ideological divide in public opinion on climate change. Typically, the extent of the ideological divide varies from country to country. What is consistent, however, is that respondents on the right of the political spectrum tend to be less committed to the reality of climate change and less likely to demand government and private action on climate change than those on the left of the political divide. Examples of this are the German public and the Austrian Freedom

Party. A climate of political polarization, in general, helps extreme political movements and parties. "Mobilising passions," as Robert Paxton (2004: 41) calls them, helps to enter polarization.

Hilde Coffé, Sam Crawley and Josh Givens (2025) study the growing ideological polarization around attitudes to climate change in Germany between 2016 and 2022. Using panel data from the German Longitudinal Election Study (GLES), the authors analyze how political ideology—divided into economic and cultural left-right dimensions—influences public support for climate action. They find a significant increase in polarization over time, driven in particular by the cultural right's decreasing support for climate policy, while the cultural left's support remains constant. The research identifies cultural ideology, rather than economic ideology, as the primary driver of divergence in climate opinion. While both dimensions show widening gaps in support for climate action, the gap is significantly more pronounced along the cultural spectrum. In particular, people with culturally conservative views—often associated with traditional lifestyles, scepticism about social change, and opposition to immigration—are increasingly resistant to climate policy. By contrast, economic views (such as support for free markets) are less strongly correlated with shifts in climate opinion.[22]

A key factor in this polarization is the role of political parties, in particular the right-wing populist Alternative for Germany (AfD) and the Greens. AfD supporters have become increasingly sceptical about climate change, while Green voters have consistently supported climate action. The study finds that polarization is not symmetrical; it is primarily driven by declining climate concern among AfD supporters, rather than changes on both sides. This asymmetry is further confirmed by a cross-lagged panel model showing bidirectional causality: AfD support predicts lower climate concern and vice versa.

The authors argue that elite cues, media framing, and the politicization of climate change by parties such as the AfD have reinforced polarization. The findings raise concerns about the implications for climate policy, suggesting that right-wing populist rhetoric could significantly hinder collective climate action by fostering ideological division. The study concludes that Germany's climate opinion landscape mirrors trends observed in the US, with growing resistance to environmental policies among culturally right-leaning voters, driven by party influence and broader political polarization.

Thomas Schwartz (2024) presents empirical data on the United States which suggests that the country's dual economy is a significant factor in the current political polarization surrounding climate change policy. The dual economy model refers to carbon-intensive industries in the interior of the United States and knowledge-intensive industries along the coasts. Oatley also indicates that, independent of the economic division, exposure to severe weather related to climate varies significantly across space. Communities

> that face a high risk of climate-related extreme weather will support policies that mitigate the crisis. In contrast, communities that face a low risk of climate-related events will not prioritize climate change as a concern and will be reluctant to support costly measures intended to mitigate the crisis.
>
> *(Oatley, 2024: 1360)*

Communities in the carbon economy are insulated from the climate crisis, while those in the knowledge economy are directly exposed. Using county-level aggregate data on economic geography, as well as individual political attitudes towards climate change policy and levels of support for Trump in the 2016 and 2020 presidential elections, the analysis indicates that residents of carbon economy communities oppose climate change policy, whereas knowledge economy residents support it and oppose Trump. The author is cautious about the potential *ecological fallacy* of his analysis and does not overstate the causal interpretation of county-level correlations. County-level aggregates may obscure significant variation within counties. For instance, a knowledge economy county may still have carbon-dependent workers, and vice versa. Assuming that counties are homogeneous in political and economic terms is problematic. Oatley's analysis is static. We do not know whether the economic and political cleavages widen or narrow over time. The role of other factors and processes in reinforcing or mitigating the carbon-climate divide, such as the influence of political elites, interest groups and the media, remains under-explored.[23]

The populist FPÖ (Freedom Party of Austria), to name another typical example, takes a skeptical and critical stance on climate change and climate policy. Here are the key positions from their political platform[24]: (1) The FPÖ questions the extent to which climate change is caused by human activity. They often argue that climate change could also be partly due to natural cycles and that the role of human activity is overemphasized in the debate. (2) The FPÖ opposes many of the stringent climate policies that are currently being proposed, especially if they are seen as economically harmful or as a burden on citizens. The specific policies they oppose include carbon taxes (CO_2 taxes) and other climate-related levies, which the FPÖ views as punitive measures for businesses and individuals as well as expensive government investments in renewable energy, especially if they lead to higher costs for consumers, such as increased electricity or fuel prices. (3) The FPÖ often refers to the climate debate as being driven by "climate hysteria" and accuses other political parties and environmental movements (e.g. Fridays for the Future) of exaggerating the issue. The FPÖ tends to downplay the urgency of action on climate change. (4) Rather than pushing for radical decarbonization and a green energy transition, the FPÖ emphasizes a pragmatic approach that prioritizes job security, economic stability, and affordable energy. They believe that climate policy should not harm economic growth or industrial competitiveness. The FPÖ advocates the continued use of fossil fuels, including coal, oil, and gas, for as long as necessary. They argue that these energy sources are still essential to the Austrian economy. In addition, the FPÖ supports the continued operation of nuclear power plants in Europe as part of the broader energy mix. (5) The FPÖ is critical of some EU climate policies, especially those that impose strict environmental targets on member states. They see these measures as top-down and potentially harmful to Austria's sovereignty and economic interests. (6) The FPÖ has tied **immigration** into its climate policy, suggesting that **mass immigration** contributes to environmental degradation and resource strain. Their platform suggests that controlling immigration would also benefit the environment by reducing population pressure. The catastrophic flooding in the Eastern provinces of Austria in the late summer of 2024 did not put a dent in the position of the FPÖ.

In many cases, the scientific community is the political opponent. The populist movement is predicated on the notion of a dichotomy between the "us" (the common people) and the "them" (the elites, experts, or institutions). The phenomenon of climate science denial can be situated within this framework, as it is often presented by international organizations, academic institutions, and expert groups that populists frequently label as part of the elite. This creates a tension between long-term, science-based climate solutions and the short-term, populist priorities of national economic and political interests tied to their anti-global sentiments.

By rejecting climate science and favoring fossil fuels, populists can appeal to constituents who distrust these elites and feel that their concerns are being ignored. Populist leaders often portray climate scientists as part of a global, disconnected elite that is out of touch with the realities of everyday people. This distrust in expert opinion leads to skepticism toward climate change evidence, even when it is overwhelmingly supported by the scientific community.

For populists, climate change policy can be framed as less urgent or as a drain on resources that could address more immediate concerns, particularly for economically vulnerable or marginalized groups. Climate change policy can also be portrayed as an elite luxury, especially when it involves lifestyle changes or higher costs (e.g. carbon taxes, energy transitions). For example, former and current leaders such as Donald Trump in the United States and Jair Bolsonaro in Brazil have openly opposed climate action, portraying environmental protection as detrimental to local economies or incompatible with economic growth. Trump's withdrawal from the Paris climate agreement was then, as it is now, partly justified by his administration on the grounds that it would hurt American industry and workers. Populist denial and active political opposition find their complementary counterparts in the "organized denial" of big business (McCright and Dunlap, 2011; MacDuffie and Allen, 2024).

Notably, large fossil-fuel companies have repeatedly failed to follow the science of climate in recent years. The largest global oil companies are "way off the track" on even their own emission goals. Large fossil fuel companies have made publicly bold climate pledges. However, as a report of the thinktank *Carbon Tracker* demonstrates,[25] none of the actual activities of the company's production plans are aligned with the goals of the Paris Agreement. In fact, most companies have plans to expand fossil-fuel production and, at the same time, meet their earlier emission targets. On the basis of the available evidence, none of the companies has plans to reduce fossil-fuel exploration or halt the approval of new exploration activities. In short, there is limited evidence, if any, indicating that the enormous fossil-fuel industry has designs to follow science. American oil companies, Chevron, ConocoPhillips, and Exxon Mobil, tend to lead the charge against the science and for the growth of profits. The collective efforts of the largest fossil-fuel corporations are expected to lead to a global temperature of 2.4°C.

The rule of law and climate change

Around the world, as of 2024, at least 88 lawsuits have been filed against the largest oil, gas, and coal producers. The vast majority have been filed since the signing of the 2015 Paris Agreement. Such lawsuits are spreading. Almost half of the 88 lawsuits sought compensation for climate-induced damage to the environment or communities (see also Chapter 8 on climate litigation). Other lawsuits contested misleading advertising by fossil fuel producers, claimed that the companies had failed to reduce emissions or lawsuits in the United States filed by state or local governments that claim fossil fuel companies have misled its citizens about global warming, covering up what they knew about climate change for decades in the quest for profits and that the corporations therefore should pay the cost of adapting to climate change. Oil companies consider the lawsuits about the failure to inform the public a major threat. Nearly 40 lawsuits have been filed since 2017. Oil companies are employing an unusual legal strategy in the case of these court battles. The companies are alleging that the opposing sides are infringing on their constitutional right to free speech.[26] However, to date, not a single corporation has had to pay liability claims.[27] Likely, the growing legal claims for damage will have an impact on the future business policies of the fossil fuel companies, especially if the lawsuits should be successful.

Eskander, Fankhauser and Setzer (2024: 44) offer a useful up-to-date guide the volume of legislation and litigation around the world involving climate change[28]:

> There is no country in the world that does not have at least one law or policy dealing with climate change. The most prolific countries have well over 20, and globally there are 1,800 such laws. Some of them are executive orders or policies issued by governments, others are legislative acts passed by parliament. The judiciary has been involved in 1,500 court cases that concern climate change (more than 1,100 of which were in the United States).

The political orientation of the parties or the government in power does not seem to have had an impact on the level of effort to legislate climate change. Efforts to introduce or pass climate-related legislation peaked in 2009–2014. The frequency of legislative activity, executive orders, or government policies related to climate change appears to have declined in line with economic downturns. In terms of judicial activity, judges outside the United States were more likely to rule in favor of climate protection. In the United States, rulings were more likely to be against climate action.

Importantly, climate change litigation is not just about damage caused by non-compliance with environmental regulations. Important issues concern "climate justice." For example, on April 29, 2021, Germany's Federal Constitutional Court issued a landmark ruling on the country's climate change law. The case was brought to court by a group of nine young climate activists, many of whom were supported by environmental organizations such as Fridays for Future.[29] The court's decision revealed that current climate policies disproportionately burden future generations with the responsibility of reducing greenhouse gas emissions while failing to impose sufficient restrictions in the short term.[30] The ruling was seen as an important step in the global climate justice movement, which addresses intergenerational equity in the fight against climate change. The court ruled that Germany's 2019 climate protection law was partially unconstitutional because it failed to set a clear and binding path for reducing emissions after 2030. The plaintiffs argued that by delaying significant climate action, the burden of drastic emission cuts would fall on future generations, violating their fundamental rights to life, health, and property. The court agreed, stating that current policies risk "irreversible burdens" on young people by delaying action. The ruling emphasized that climate action cannot be delayed to the point where future generations are forced to bear the brunt of drastic emissions reductions. The Court emphasized the need for intergenerational fairness, requiring that the responsibility for reducing emissions be shared by both current and future generations. Germany's Federal Constitutional Court opens jurisprudence for a new dimension of human rights: the intertemporal guarantee of freedom (Krämer-Hoppe, 2021). As a result of this ruling, the German government quickly adjusted its climate policy. By May 2021, the government had proposed updated legislation that aims to reduce greenhouse gas emissions by 65% by 2030 (from 1990 levels), 85% by 2040, and achieve net-zero emissions. However, these policy goals are far from uncontested, as different political actors desire to revoke them.

Climate change litigation is part of a wider trend in which stakeholders are increasingly scrutinizing companies for their sustainability practices.

For companies, the growing risk of climate change litigation may exacerbate broader sustainability risks and affect the value attributed to the company in the stock market. To demonstrate the response of financial markets to climate litigation, Misato Sato and her colleagues (2024: 1461) constructed a comprehensive database of filings and decisions relating to 108 climate lawsuits against US and European listed companies between 2005 and 2021. The authors show that "firms experience, on average, a 0.41% fall in stock returns following a climate-related filing or an unfavourable court decision."

The strong state

On the one hand, the discrepancy between our knowledge about climate change and the readiness of citizens to change their behavior within their immediate environment (not in my backyard; NIMBY)[31] is diagnosed as "inconvenient reason"; on the other hand, the inertia of politicians in implementing advances in climate science knowledge leads to the diagnosis of "inconvenient institutions." Hans Jonas (1984: 147) regretfully admits, "a well-intentional, well-informed tyranny [...] appears superior, for our uncomfortable purposes, to the capabilities of the capitalist-liberal-democratic complex." The sense of political inefficiency, which is particularly strong among some climate scientists, leads to robust disenchantment with democratic governance. Ulrich Beck (2010a: 175) refers to one of the conditions for disenchantment among scientists: "Climate researchers are great realists, but socially and politically idealists, because they see all people as small-scale climate researchers and consequently cannot understand why their apocalyptic model calculations do not immediately trigger the urgently needed countermeasures."

Therefore, it is suggested that political action based on the principles of democratic governance should be abandoned. The response to the adverse effects of climate change is too slow. On the other hand, climate deniers hope that authoritarian political structures can successfully lead the fight against rapid decarbonization in response to climate change. Both perspectives favor the strong state. A strong state, as Peter Drahos (2021: 2) specifies, for example, must

> organize an exogenous shock to change the direction, speed, and scale of innovation in world capitalism. Trillions of dollars of financing will have to be directed away from the fossil fuel industry and into circular and green economy projects. Big climate treaty meetings, filled with calculated bargaining and conditionality, deliver too little, too late. This exogenous shock has to target the energy system in the first instance.

The position favoring a stronger role of government in climate policies may not be only based on the perceived inefficacy of democratic governments or the desire to roll back existing environmental legislation that curbs, for example, the search for additional fossil fuel sources or regulations that restrict pollution. It can also be based on the consideration that *market-driven* solutions to environmental problems are inherently weak and ineffective. Naomi Oreskes and Erik Conway (2014: 44), in a most general sense, argued that "when scientist discovered the limits of planetary sinks, they also discovered market failure."[32] The toxic effects of DDT, acid rain, ozone layer depletion, and climate change are serious problems for which markets do not provide a spontaneous remedy. Rather government intervention was needed.

Theda Skocpol has advanced arguments in her work emphasizing the need for state capacity and robust institutions to facilitate the enactment and enforcement of transformative policies. In her 2013 paper on U.S. climate policy, "Naming the Problem: What It Will Take to Counter Extremism and Engage Americans in the Fight against Global Warming," Skocpol (2013) offers a critique of past efforts at climate policy that relied heavily on market-based solutions such as cap-and-trade. She posits that a more robust governmental role is necessary for the establishment of institutions and political coalitions that can effectively enforce climate policies and mobilize public support. Skocpol underscores the importance of state intervention not only for the regulation of emissions but also for the creation of a political environment in which climate policies can be democratically supported and sustained.

Moreover, the position that emphasizes the role of a strong state is also based on the observation that *scientific knowledge* alone is insufficient without **strong state action (Haas, 2004)**. This makes governments essential actors in the process of addressing environmental problems.[33] Haas highlights the importance of **governments as essential actors** that are mobilized by scientific expertise to create and enforce environmental regulations. Governments must adopt and institutionalize policies, and they need political will and the capacity to enforce environmental regimes effectively. Peter Haas's work underscores that **effective governance** in both the national and international context is central to tackling global environmental challenges such as climate change, ozone depletion, and biodiversity loss. Ultimately, however, the demand for a robust state apparatus gives rise to a call for a more authoritarian form of governance that eschews any pretense of democratic legitimacy.

On the global front, as the political scientist Daniel A. Bell and colleague (Bell and Pei, 2020: 109) stressed that international relations theorists underestimate the virtues of strong states in global relations in general and in response to global warming in particular:

> If it's just a matter of strong states bullying weaker ones to get what they want, normative theorists can just step aside. But strong states do good things for the global order as a whole. However much we worry about "rogue" leaders in strong states that sabotage global agreements, it would be much harder to forge agreements for dealing with global challenges such as climate change in an international system characterized by states with equal power to shape and withdraw from global accords [...] Moreover, hierarchical arrangements can actually benefit weaker states because this sense of dominance means that states have extra responsibilities. Security hierarchies, for example, reduce the level of defense expenditure in subordinate states. Unequal economic power can also benefit weaker states.

In general, the history of social science discourse on the impact of non-human entities on social relations can be described as varying between a strong interest, which will be examined in the next chapter under the heading of "climate determinism," and a complete lack of interest in much of modern social science (cf. Livingstone, 2024). Indeed, the rejection in social theory of any influence of non-human factors on social relations can be seen as an emphatic warning, for good reasons, to allow for the intrusion of non-human attributes into social science discourse.

Aside from the energy implications of running artificial intelligence programs, which I have already discussed, there are of course claims that AI will have a significant impact on dealing with climate change. Why algorithms are attractive for climate governance is perhaps obvious: algorithms offer accessibility, integration, timeliness, and volume of data, as well as the potential for new insights by bringing previously unknown correlations to the fore.[34] Promises to range from more accurate predictions of future temperatures, more exact weather forecasts, and improved climate models to the more ambitious expectation that AI will be the ultimate driver in dealing with the consequences of climate change. AI could become a tool to help build a greener and more sustainable future world (e.g. Fischer et al., 2021). Artificial intelligence might be better placed in the future present than "humans at finding and enacting the best policies in certain areas concerning science, engineering and complex societal and macroeconomic issues" (Sætra, 2020), leading to the conclusion that AI should be given the power to discover and enact climate policies. (Coeckelbergh and Sætra, 2023).

Addressing climate change through AI

On a purely "prosaic" level, AI, as a broadly applicable engineering tool, could play a significant role in addressing the climate crisis by providing advanced tools to optimize resource use or model complex climate patterns. In a most general sense, AI could facilitate climate change mitigation and adaptation. More specifically, AI can improve climate models by analyzing large datasets from satellites, weather stations, and oceanic systems, potentially producing more accurate predictions. AI could be helpful in addressing the operational efficiency of industrial heating and cooling systems or reducing waste in food systems (Kaack et al., 2020). Climate modelers have begun to make bold claims that they will soon be able to overcome some of the resolution and data problems of conventional climate models and produce results more quickly.[35]

Machine learning algorithms can **simulate climate impacts**: by providing detailed simulations of how specific policies or changes will affect global or regional climates over time. Alternatively, AI may be able to assist industries in reducing energy consumption by improving the efficiency of processes, machinery, and supply chains. However, the claims go far beyond what is efficient "research support" or AI as a technological aid to political decision-making, augmenting democratic governance. The path from being a mere tool to influencing decision making may be small (cf. Coeckelbergh, 2021). Claims already extend to AI-driven full technocracy:

> Technocracy is attractive when dealing with climate change because ordinary citizens might not have the expertise necessary to address it, and even human scientists might not understand everything. AI can discover patterns in big data and make more accurate predictions—thus fueling/making better decisions.
>
> *(Coeckelbergh and Sætra, 2023)*

Technocracy obliviously transcends democracy. This may be a greater appeal than authoritarian governance if it only appears "neutral" in the sense of being non-political.

However, the authors caution against "techno-solutionism"—an over-reliance on technology to solve inherently political and ethical problems—at the potential intersection of AI and climate policymaking. They highlight the risks of unintended consequences of proposed pathways to AI deployment, and advocate for a reflective approach that prioritizes ethical considerations, global governance challenges, and human expertise. Rather than accepting a deterministic trajectory, they urge society to critically shape the role of AI in politics, aiming for responsible integration that respects democratic values while addressing global environmental threats. Even more cautious claims for the effective solution potential of AI in the face of the growing dangers of climate change cannot neglect the often-exaggerated claims and faith in the potential of more technological fixes to climate change.

In the following chapter, I discuss several broad narratives about the interaction of society and its natural environment. The core of the discussion focuses on the ancient tradition of climate determinism, the belief that climate shapes human behavior, culture, and societal development, because the perspective of climate determinism can be traced back to classical Greek thought. In short, that "the history of like on earth is a story about the climate" (Kemp, 2025: 328).

As Fernand Braudel ([1979] 1985: 51) dryly observes, humans "of former times would not have been put out by this climatic explanation, implicating as it does the heavens." In stark contrast, however, as can be illustrated with reference to the well-known French historian Le Roy Ladurie ([1967] 1971: 119), for example, in his treatise Times of Feast, Times of Famine, a book published in 1967 and subtitled "A History of Climate since the Year 1000," Ladurie concludes—in line with the narrative that transcends recent climate disasters—that "in the long run" the human consequences of climate "appear to be slight, perhaps negligible, and certainly difficult to detect."

The chapter explores the evolving relationship between social science and climate (see also Grundmann und Stehr, 2010), focusing on the intellectual history of climate determinism. It contrasts early thinkers like Montesquieu and Huntington, who emphasized climate's direct influence on political and social structures, with modern skepticism from social scientists like Durkheim, who viewed societal phenomena as largely independent from environmental causes. I note how environmental factors were marginalized in post-WWII social science but experienced a resurgence in recent decades amid growing climate concerns. Modern climate determinism reinterprets these ideas, viewing climate as both influenced by and influential upon human societies. The chapter also critiques capitalism and democracy's responses to climate change, debating whether market forces or state intervention are more effective. It calls for integrating environmental factors into social science and policy, arguing that addressing climate change requires rethinking long-standing academic, economic, and political assumptions.

Notes

1 Aryeh Neier is the former executive director of *Human Rights Watch*. www.nybooks.com/articles/2024/12/19/rise-of-authoritarianism-aryeh-neier/.

2 Compare the analysis of Donald Trump's anticipated economic policies, in a second presidential term: Peter Coy, "Trump's economics, too, are sounding more authoritarian," *New York Times*, September 13, 2004; www.nytimes.com/2024/09/13/opinion/trump-authoritarian-economics.html?searchResultPosition=2.
3 Cf. www.mediaite.com/tv/tucker-carlson-wildly-claims-abortion-causes-hurricanes-you-cant-participate-in-human-sacrifice-without-consequences/.
4 Stephen Schwartz (2024) claims that it is the "fear in humanity around the world [...] a kind of unconscious gut feeling" triggered by climate change that is responsible for the decline of democratic governance. Political and economic changes are merely manifestations of the backsliding, not the cause of moving away from democracy. The cause is the fear of climate change.
5 As cited in Mike Hulme (2009: m308–309).
6 Marco Buschmann, "Paternalismus kann sich nicht auf das Grundgesetz berufen," ["Paternalism cannot invoke the Basic Law"] *Frankfurter Allgemeine Zeitung*, May 16, 2024; www.faz.net/einspruch/grundgesetz-paternalismus-ist-nicht-geschuetzt-19721566.html.
7 A characterization employed by António Guterres, Secretary General of the United Nations in a major speech on June 5, 2024, delivering dire new scientific warnings of global heating; "In his speech, Guterres announced new data from the World Meteorological Organization (WMO) showing there is an 80% chance the planet will breach 1.5°C (2.7°F) in warming above pre-industrial times in at least one of the next five calendar years"; cf. www.theguardian.com/environment/article/2024/jun/05/antonio-guterres-un-chief-fossil-fuels-advertising.
8 See Jonathan Watts, "Five of the world's biggest banks are 'greenwashing' their role in the destruction of the Amazon [...] The banks say they follow ethical policies that help to protect intact forests, biodiversity hotspots, indigenous territories and nature reserves," *Guardian*, June 11, 2024; cf. https://exitamazonoilandgas.org/reports/#greenwashing; Jonathan Watts, "World's top banks 'greenwashing their role in destruction of the Amazon'; www.theguardian.com/environment/article/2024/jun/11/worlds-top-banks-greenwashing-their-role-in-destruction-of-the-amazon.
9 The *New York Times* reports that in the United States "in 2023, the hottest year in recorded history, research conducted by the nonprofit Media Matters found that corporate broadcast television scaled back on its climate coverage by 25 percent"; www.nytimes.com/2024/06/20/climate/weather-forecaster-iowa-climate-change.html.
10 As the *Guardian* (May 20, 2025) reports, the fossil fuel industry funds as part of their lobbying efforts "cultural institutions such as museums, youth organizations and athletic groups in recent years, new research shows, in what appears to be a public relations effort to boost their image amid growing public awareness of the climate crisis"; www.theguardian.com/us-news/2025/may/20/oil-industry-reputation-girl-scouts-british-museum.
11 Andrew Dressler, "Oil and gas companies are trying to rig the marketplace," *New York Times*, June 1, 2024; www.nytimes.com/2024/06/01/opinion/clean-energy-solar-wind.html?searchResultPosition=1.
12 "At a dinner, Trump assailed climate rules and asked $1 billion from big oil," *New York Times*, May 9, 2024; www.nytimes.com/2024/05/09/climate/trump-oil-gas-mar-a-lago.html.

13 The 2025 Trump administration implements, as expected, the radical conservative blueprint, as outlined in its "Project 2025," for guidance on its national and international climate policies, demanding that the "USAID [United States Agency for International Development] should cease its war on fossil fuels in the developing world and support the responsible management of oil and gas reserves as the quickest way to end wrenching poverty and the need for open-ended foreign aid. The next conservative administration should rescind all climate policies from its foreign aid programs (specifically USAID's *Climate Strategy 2022–2030*); shut down the agency's offices, programs, and directives designed to advance the Paris Climate Agreement; and narrowly limit funding to traditional climate mitigation efforts. USAID resources are best deployed to strengthen the resilience of countries that are most vulnerable to climatic shifts. The agency should cease collaborating with and funding progressive foundations, corporations, international institutions, and NGOs that advocate on behalf of climate fanaticism."

14 Lisa Friedman, "Senate democrats open inquiry into Trump's $1 billion request of oil industry," *New York Times*, May 23, 2024; www.nytimes.com/2024/05/23/climate/senate-democrats-open-inquiry-into-trumps-1-billion-request-of-oil-industry.html.

15 "Cop28 president says there is 'no science' behind demands for phase-out of fossil fuels," *Guardian*, December 3, 2023; www.theguardian.com/environment/2023/dec/03/back-into-caves-cop28-president-dismisses-phase-out-of-fossil-fuels.

16 P. 20 of the report.

17 P. 90 of the report.

18 Dharna Noor, "Far-right fossil fuel company allies pressure US supreme court to shield firms in unprecedented campaign," *Guardian*, June 9, 2024; www.theguardian.com/us-news/article/2024/jun/09/fossil-fuel-allies-pressuring-supreme-court.

19 Coral Davenport, "A string of Supreme Court decisions hits hard at environmental rules," *New York Times*, June 29, 2024; www.nytimes.com/2024/06/29/climate/supreme-court-epa.html?searchResultPosition=3.

20 Charlie Warzel, "I'm running out of was to explain how bad it is," *Atlantic*, October 10, 2024; www.theatlantic.com/technology/archive/2024/10/hurricane-milton-conspiracies-misinformation/680221/?utmsource=www.status.news&utmmedium=referral&utmcampaign=the-atlantic-editor-in-chief-jeffrey-goldberg-warns-newsroom-decay-is-how-democracy-decomposes.

21 Michael Caulfield, "Copium addicts: What 'misinformation' is actually usually about," https://mikecaulfield.substack.com/p/copium-addicts-what-misinformation.

22 The critical limitation of the study by Hilde Coffé and her colleagues is that the dependent variable—support for climate action—is based on a *single* question comparing climate action with economic growth. This is unlikely to capture the complexity of climate attitudes, such as views on specific policies (e.g. carbon taxes, renewable energy) or the perceived urgency of the problem. In addition, the cultural and economic left-right dimensions are based on *one* question each (immigration and taxes/welfare). Such a measure oversimplifies the likely multifaceted nature of ideology and excludes other relevant dimensions (e.g. nationalism, religious values, environmental concern as a stand-alone ideology).

23 Also see my cautionary observations about extreme weather events and attitudes towards climate in the "Excursus" section of this study.

24 For example: www.fpoe.at/artikel/umfassender-umweltschutz-sinnvoller-als-aktionistischer-klimaschutz/ or www.fpoe-ooe.at/klima/.

25 "World's largest oil companies 'way of the track' on emission goals, report finds," *The Guardian*, March 22, 2024; www.theguardian.com/environment/2023/oct/30/climate-crisis-carbon-emissions-budget#:~:text=The%20analysis%20found%20the%20carbon,year%20of%20about%2040bn%20tonnes.

26 Karen Zraick and Sachi Kitajima Mulkey, "Oil companies fight climate lawsuits by citing free speech." *New York Times*, June 22, 2025; www.nytimes.com/2025/06/22/climate/oil-industry-anti-slapp-climate-lawsuits.html?searchResultPosition=1.

27 Karen Zraick, "Climate lawsuits are exploding. Are homicide charges next?," *New York Times*, September 12, 2024; www.nytimes.com/2024/09/12/climate/climate-lawsuits-major-cases.html?searchResultPosition=1.

28 The authors used the *Climate Change Laws of the World* (CCLW), a searchable, publicly accessible database created and maintained by the Grantham Research Institute on Climate Change and the Environment at the London School of Economics: https://climate-laws.org. By the end of 2019, the database contains 1,800 climate change laws and policies of similar status around the world. This compares to 35 laws in 1990 and 145 laws in 1999. About 40% of the laws in 2019 are acts passed by the respective parliaments; Accounting for government effectiveness and the length of time a law has been in effect, Spain, the United Kingdom, and South Korea are the most comprehensive legislators on climate change (Eskander et al., 2024: 69).

29 For an analysis of climate activism in Germany (and the United States), see Cameron Abadi's (2024) *Climate Radicals*. Abadi examines the political and moral dilemmas facing climate activists as they seek to rally public support and pressure on overly cautious political leaders.

30 For a discussion of the dilemma of "short-termism" or "presentism" of climate change policies in democracies, that is, shifting costs to future cohorts, see Fritsch (2023: 1007–1011).

31 The "not in my backyard" phenomenon refers to (individual) common public behavior where people support climate policy in principle but oppose specific local projects that would affect them directly. The "not in my backyard" phenomenon also manifests the frequent contradiction between attitude and behavior (e.g. Benninghaus, 1976). Examples of the phenomenon under consideration are in Germany, the *SüdLink* transmission project (bringing wind power from the north to the south) was delayed due to protests over property values and health concerns. In California, affluent neighborhoods such as Palo Alto have resisted state-mandated densification to reduce urban sprawl and emissions. A major global survey shows that people around the world are united in wanting action on the climate crisis, but remain a silent majority, mistakenly believing that only a minority share their views (cf. Andre et al., 2025). Angrist, Winseck, Patrinos, and Graff Zivin (2024) analysed "effects on pro-climate beliefs and behaviors, as well as novel data on policy preferences and voting for green parties. Results show that a year of education substantially increases pro-climate beliefs, behaviors, and policy preferences."

32 A perhaps counteruntuive perspective would maintain that climate change is not a market failure: "It's a market success" (Delay, 2024: 16).

33 The state or government should be conceptualized not as single, unitary actor. The state defined as "an amorphous complex of agencies with ill-defined boundaries, performing a variety of not very distinctive functions" (Schmitter, 1985: 33) is not only of importance in the field of environmental politics.

34 For a critical analysis of the political implications of algorithms in climate change research and policies cf. Machen and Nost (2021). Ruth Machan and Eric Nost stress that algorithms have the potential of strengthening hegemonic power through its attributes of objectivity, universality, and necessity.

35 "Artificial intelligence is helping improve climate models," *The Economist*, November 13, 2024; www-economist-com.login.ezproxy.library.ualberta.ca/science-and-technology/2024/11/13/artificial-intelligence-is-helping-improve-climate-models.

3

SOCIAL SCIENCE AND THE ENVIRONMENTAL DILEMMA

> Political power [...] embraces two radically different relations: control of nature, and control of man. Power over nature is mere intellectual power.
>
> *Franz Neumann (1957: 3)*

The political theorist Franz Neumann's account of the nature of political power, human power, and power over nature exposes the fundamental dilemma that can be found in much of the history of the social sciences when it comes to locating the role of nature in social affairs. Neumann explains,

> power over nature is mere intellectual power. It consists in man's understanding of the lawfulness of external nature for the ultimate purpose of subjecting external nature to man's needs. It is this accumulated knowledge which is the basis of the productivity of any given society. This power is powerless. It does not involve control of other men.

Nature is a distinct, benevolent servant, that is, objectively accessible and static but without power. Society's relationship with nature is a one-sided relationship. The one-sided relationship with nature marks social science off from natural science. Not all social scientists and humanists agreed. Climate determinists disagreed.

Those who have read Èmile Durkheim's *Suicide* ([1897] 1973: 101–102) are likely to be familiar with his seminal methodological argument, which has become the paradigm of modern sociology. This argument posits that seemingly idiosyncratic individual actions are social phenomena or that their distribution cannot be attributed to physical or even cosmic causes. In contrast, numerous contemporaries of Durkheim were convinced of a

DOI: 10.4324/9781003682066-4

causal relationship between climate or weather and the number of suicides. Durkheim's assessment of this phenomenon was severe: "The facts must be in unusual agreement. Now, far from there being a relation between suicide and a given climate, we know suicide to have flourished in all climates. Italy is today relatively exempt, but it was very frequent there at the time of the Empire when Rome was the capital of civilized Europe. [...] We must therefore seek the cause of the unequal inclination of peoples for suicide, not in the mysterious effects of climate nature of this civilization, in the manner of its distribution among the different countries." The field of environmental determinism gives way to the social sciences, in critical sense, to "sociocentrism" (Connolly, 2019: 5, 2025). Sociocentrism is the idea that socio-structural and socio-cultural phenomena and processes can only be explained or interpreted in terms of other, perhaps more fundamental, social processes.

In this respect, Durkheim's contributions to the social sciences have left an indelible mark on the history of ideas in the field. The distinction between the social and natural sciences is reinforced and celebrated in these works. Conversely, Durkheim's endeavors to overcome the fallacies of environmental determinism encountered remarkably limited intellectual resonance in other social science disciplines at the time. The genuine scientific blossoming of climate determinism and the somewhat analogous, yet also competitive, science of race occurred only at the advent of the 20th century (Weinstein and Stehr, 1999). In the post-World War II era, the notion of climate determinism or geo-determinism, which had previously been a dominant paradigm, came to be regarded as a simplistic and reductionist worldview. Among the scientific community, there was minimal impetus to further develop these paradigms, and the same was true for decision-makers in business and politics.

In recent years, however, these research questions have experienced a resurgence. For example, the American economic historian David Landes (1998) posits that it is time to reintroduce climatic and geographic factors into the context of comparative studies of the prosperity and poverty of nations as a dimension that aligns with empirical reality. He laments that climate determinism has long discredited this approach and adds,

> geography also bothers many people because it is obviously and intrinsically unequal. There are places with better climates and worse climates from the point of view of comfort and health, and a lot of social scientists are reluctant to accept this. It really bothers them to see evidence of nature's 'unfairness'.
>
> *(Landes, 1998: 14–15)*

Discussions of a crisis of democracy rarely focus on issues of the environmental crisis. These concerns are exacerbated by the dangers of a changing global climate. The notion of "climate" in the sense of a property of the earth as a whole, [...] as "the concept of the global climate entered popular discourse in central Europe in the 1870s, in the midst of a heated debate over anthropogenic climate change" (Coen, 2018: 7). Not much later, the earlier discussion of anthropogenic climate change in the 19th and early 20th centuries gave way to virtual silence about the dangers of global climate change (Stehr and Storch, 2000). Economic discourse followed a similar trajectory. Prices, for example, were once seen as closely linked to local and natural conditions of production. However, by the end of the 19th century, prices were largely tied to financial causes. Stock market values became self-sufficient, feeding on themselves. Economic discourse "unburdened itself from nature" (cf. Bonneuil and Fressoz, 2017: 211–215). Perhaps paradoxically, economic growth is seen as a process that is increasingly less harmful to the environment, or government-imposed measures, such as taxes, could more deliberately overcome what was once "market failures."

The dangers to democracy posed by climate change were not a topic of discussion in the social sciences before the 1980s. For example, Herbert Marcuse's 1979 lecture to students ([1989/1990] 2009) on "Ecology and Social Criticism," which makes no mention of climate. The renowned historian Emmanuel Le Roy Ladurie ([1967] 1971: 119), in his seminal study of society and climate, concludes: "In the long term, the effects of climate on humans appear to be small, perhaps even negligible and difficult to recognize." Sociologists in the post-World War II era, not surprisingly, were often convinced that modern societies lived increasingly outside of nature. In fact, they did. Even in the relevant political science literature on democratic forms of government, there have been restrained references to or discussions of the significance of climate change for democracy and democratization in recent years (e.g. Alonso et al., 2011; Haerpfer et al., 2019).

Three additional minor tests confirm the general picture of the almost complete lack of interest among social scientists in climate conditions during the decades after World War II: The very successful *The Affluent Society* of the *economist* John Kenneth Galbraith (1958: 387) mentions climate but once only to dismiss the idea that poverty is related to climatic conditions: "It is an explanation that is nearly devoid of empirical application." In short, climate is an irrelevant condition, and climate change does not appear at all. Similarly, the equally successful monograph *The End of History and the Last Man* by the *political scientist* Francis Fukuyama (1992) did not mention the natural climate once. At best, there will be a reference to the "political climate" in the relevant political science literature of the time. In his widely read volume *The Consequences of Modernity*, the *sociologist* Anthony Giddens (1990), does not use the term "climate change" in his discussion of modernity. Finally,

> in no discussion of freedom in the period since the Enlightenment was there ever any awareness of the *geological agency* that human beings were acquiring at the same time as and through processes closely linked to their acquisition of freedom. Philosophers of freedom were mainly, and understandably, concerned with how humans would escape the injustice, oppression, inequality, or even uniformity foisted on them by other humans or human-made systems. Geological time and the chronology of human histories remained unrelated.
>
> *(Chakrabarty, 2009: 208; emphasis added)*

Although the question cannot be answered with certainty, William Conolley (2019: 3–4) posed the counterfactual question—given the broad consensus in the humanities and social sciences—of what would (or could) have happened to the geosciences, humanities, and social sciences in Euro-America if—instead of being transfixed by themes that treat the earth as a set of glacial regularities that change very slowly—they had been inspired by thinkers such as Sophocles and Shelley? How much earlier might the galloping processes of planetary climate change have been diagnosed? As Conolley notes, observing nature in different times and places, Sophocles and Shelley advanced "overlapping insights that, had they been widely taken up in major Eurocentric theories, may well have advanced understandings sooner about the unruliness of this planet." For example,

> Sophocles—through the vehicle of the gods—appreciated how periodic eruptions of plagues, earthquakes, volcanoes, and raging seas bounce into the fabric of social life and civic spirituality [..] Mary Shelley, writing in the early 1800s after experiencing a mysterious year without a summer in Europe, populates her great novel *Frankenstein: Prometheus Unbound* with radical shifts in terrain, changing weather systems, thoughts about evolution, the dangers of scientific hubris, and hints about the partially self-organizing trajectories of planetary systems.

Climate determinism[1]

> It is amusing to think that the men of former times would not have been put out by [...] climatic explanation, implicating as it does the heavens. They found it all too tempting to explain the course of everything terrestrial, including individual or collective destinies and disease, by the stars.
>
> *Ferdinand Braudel ([1979] 1985: 51)*

Reflections on climate determinism begin best with a description of the term "climate": In his classical physical description of the world, *Cosmos*, Alexander von Humboldt ([1845]: 317–318) provides such a useful definition of climate that straddles, at least in part, the boundaries of a natural and social science framework of nature. Humboldt defines climate "in the most general sense," as

> all the changes in the atmosphere which sensibly affect our organs, as temperature, humidity, variations in the barometrical pressure, the calm state of the air or the action of opposite winds, the amount of electric tension, the purity of the atmosphere or its admixture with more or less noxious gaseous exhalations, and, finally, the degree of ordinary transparency and clearness of the sky, which is not only important with respect to the increased radiation from the Earth, the organic development of plants, and the ripening of fruits, but also with reference to its influence on the feelings and mental condition of men.[2]

Humboldt's conception of the influence of climatic conditions extends to the social, or at least socio-psychological, realm of humanity. Working at the same time, however, humanities scholars were certain that the influence of climate on society was much broader and more consequential.

At present, the perspective of climate determinism leads to a strange double existence: On the one hand, if I am not mistaken, it is a widely accepted view among laypeople and natural scientists, based on traditions of common sense and obvious "facts," and on the other hand, among social scientists, it is considered a long and deservedly discredited intellectual perspective. More than a century ago, perhaps the most famous climate determinist, the geographer Ellsworth Huntington, was confident that man could live in any region of the world but that his physical and mental energy and moral character would reach their highest development only in a few strictly limited areas. For Ellsworth Huntington (1876–1947), the link between climate and social change was central to this analysis. He never wavered in his views, even in the face of devastating criticism (see Olmstead, 1912; Sorokin, 1928). His prolific writings, his considerable fame, and his wide influence both in the scientific community and in early 20th-century American society were remarkable.

Huntington was generally driven by a concern for concrete ways to improve the human condition, and he rarely hesitated to offer practical suggestions and policy advice, including on climate issues. The practical advice he offered came directly from his research. One example was his suggestion that the seat of the United Nations should be in Newport, Rhode Island, because it has the most suitable climate for humans. His concern for the optimum climate also included a close association with the American Society of Heating and Plumbing Engineers (see Martin, 1973: xiv). Huntington's main work concerns the reasons for the "progress" of human civilizations.[3] These ideas first took shape approximately 1914–1915 and were worked into a definitive thesis that changed little over the next three decades. As the historian David Arnold (1996: 31) noted, "like so many environmentalists before him, he looked first to the East, seeking in climate and climate change an explanation for the differences between Western dynamism and Eastern stagnation." His early writings on post-glacial climate change were stimulated by a worldwide interest in the phenomenon of climate change in historical times: this peaked at the turn of the century and quickly gave way to the conviction that climate was essentially a static phenomenon (see Lamb, 1959). Whether the long-term changes observed were merely recurrent oscillations, fluctuations, or "pulsations," or whether they indicated progressive changes toward different climatic states, such as aridity, was one of the contentious issues in the discussions among scientists who shared the conviction that significant changes could be observed (see Brückner, 1890).

In his early work, there was a clear emphasis on climate change, variability, and fluctuations. In his later work on climate and weather, especially in his synthesis books published in the 1940s, the time horizon changed, and the emphasis moved away from climatic variability and fluctuations and was replaced by an emphasis on the essentially fixed nature of climate.

In approximately 1914 and 1915, Huntington began to collect a great deal of empirical information about the effects of changes in weather patterns on daily "nervous activity," productivity, "feelings, and energy." On the basis of these data, he formulated the essence of his thesis on the link between climate and human activity. The initial findings were published in 1914 in a series of three articles commissioned by Harper's Magazine. The essays, entitled "Labor and Weather," "Climate and Civilization," and "Is Civilization Determined by Climate?" "brought angry letters from Southern gentlemen, inquiries from physicians and psychologists, [and] a proposal for a civic celebration in Seattle" (Martin, 1973: 114).

According to his biographer, the idea that climate stimulates or inhibits human energy, health, and progress seems to have been first suggested to Huntington by Charles J. Kullmer, a professor of German at Syracuse University, who claimed to have found a close correlation between storm tracks and civilization, and that shifts in storm tracks account for shifts in the locations of civilizations.

Climate determinism may be a discredited and virtually abandoned line of research and perspective in the social sciences, but it is by no means entirely absent from contemporary social science discourse; one could point, for example, to Leonard Jeffries's climate-based theories of racial identity.[4] Jeffries "taught a climate-based theory of racial identity in which Africans were cast as 'sun people' who were 'communal, cooperative and collective,' while Europeans were scorned as 'ice people' who thrive on brutality and destruction."[5]

Arnold J. Toynbee (1973), writing in the early seventies of the last century in an introduction to the biography of the geographer Ellsworth Huntington, presents a somewhat different view of the process of intellectual influence. Toynbee is convinced that intellectual stimulation is capable of progressing in an almost unconscious manner, or in which amnesia (see also Gouldner, 1980), with respect to the work of an individual scholar, is not a serious obstacle to his cognitive authority: "Huntington is influencing present-day thinkers even if they are not aware of this, and also even if they are aware of it but dissent from Huntington's ideas." Ironically, Huntington himself displays a remarkable degree of his own amnesia regarding the great ancestry of climate determinists; for example, in what evidently constitutes the synthesis of his life's scholarly work, in his book *The Mainsprings of Civilization* (1945), one does not find a single reference to Montesquieu, Herder, or Virchow. One can only surmise that he did not see himself standing on the shoulders of giants but rather assumed that his scientific approach obliterated his intellectual predecessors.

Both views coexist in a mostly dogmatic and dramatically separated manner. In the decades after World War II, the idea of "climate determinism" appeared as a sometimes naive, simplistic world view, and little intellectual energy was invested in it by serious scientists, let alone by political and economic decision-makers. In the 1940s, Ellsworth Huntington was considered the gravedigger of geography. The famous historian and economist David Landes (1998: 4) did not mince his words: "Huntington gave geography a bad name." However, Landes (1998: 14–15) concludes his own venture into climate determinism by claiming that.

> life in poor climes [...] is precarious, depressed, brutish. The mistakes of man, however well intentioned, aggravate the cruelties of nature.
>
> Even the good ideas do not go unpunished. No wonder that these zones remain poor; that many of them have been growing poorer; that numerous widely heralded projects for development have failed abysmally [...]; that gains in health peter out in new maladies and give way to counterattacks by old.

Only in recent years has the idea enjoyed something of a renaissance, mostly in the form of a reinvention of an old idea. From a theoretical point of view, much of modern climate impact research is pure climate determinism, but the reinventors of this line of research are mostly unaware of their intellectual predecessors (Stehr and von Storch, 2000).

On the other hand, the potentially dramatic social, economic, and political consequences of the forces of climate change, such as significant sea level rise (cf. Tollefson, 2016) in the not-too-distant future, may rehabilitate climate determinism as a credible perspective on the "nature" of the climate-society relationship. The extent to which societal development is driven by the impacts of climate change may rehabilitate climate determinism as an appropriate perspective.

As much as the current concern about the impact of climate on democratic governance may appear to be a new concern, prompted of course by the progress of global warming, the reflections of scholars, especially philosophers, on the effects of prevailing climatic conditions on society are not entirely new. The impact of climate on political institutions was most prominently addressed by the philosopher Charles de Secondat, Baron de Montesquieu (1689–1755). His famous *The Spirit of the Laws* ([1786]) contains an extensive discussion of the impact of climatic conditions on political institutions. In short, many contemporary observers of the power of climate in its impact on human existence would have no hesitation in agreeing with Montesquieu ([1786]) who, in *The Spirit of the Laws*, pointedly asserted that "the empire of the climate is the first, the post powerful of all empires." Considering renewed thinking about the relationship between climate and governance, it is worthwhile to consider Montesquieu's reflections as well as the tradition of "climate determinism" in the social sciences, which was more recently promoted by natural scientists.

To avoid misinterpretations and misconceptions of earlier climate and environmental reductionism, I suggest that it is imperative to review the classical concepts of climate determinism; it is also imperative to examine the pitfalls of climate determinism at a time when calls for a revision of the deep intellectual divisions between the natural and social sciences, brought about not least by urgent environmental problems, seem to be gaining credibility and urgency.

The mid-1990s report of the *Gulbenkian Commission on the Restructuring of the Social Sciences* (Wallerstein et al., 1996: 76) does not, in my view, pay sufficient or satisfactory attention to this issue, although it is mentioned as a question for the future present "how to reinsert time and space as internal variables constitutive of our analyses and not merely unchanging physical realities within which the social universe exists." The problem of reformulating the place of natural processes in social science discourse goes beyond reflecting on the meaning of time and place, as does the question of the role of the social in natural science discourse, where it now occupies a slum dwelling. It seems that an examination of the legacy of climate determinism is even more important, considering the need to rethink the status of "nature" in social science discourse.

In my discussion of classical climate determinism, I focus on the views of the political theorist Montesquieu and the perspective of the political economist

Werner Sombart as representatives of "modern" climate determinism. As noted above, climate determinists are convinced that climatic conditions are the main explanans of the nature of human existence (see Livingstone, 2020).

Montesquieu: the mind, political institutions, and climate

Although the first discussion of climate determinism has an ancient heritage, traditional climate determinism dates to the 18th century. In the middle of the 18th century, the natural climate appeared to be a new topic of political theory. In the 18th century, the century in which, according to some social historians, modern sociological discourse emerged, the educated classes of France, England, and Germany invested enormous intellectual energy in a debate about the climatic influence on the particularities of specific civilizational phenomena of entire nations. As one contemporary observer put it, there were a myriad of writers at the time who attached great importance to climate (Kames, [1788] 2007). Although interest in this question did not completely die out, it was ultimately discredited and disappeared from the scene as a completely compromised field of research until very recently.

In social science theory, movement between the poles of nature and society can be observed. The literature of the 18th and 19th centuries, which advocated one or more versions of climate determinism, can be roughly divided into two currents. One current view sees a direct influence of climatic conditions on human behavior (Bodin, Montaigne, Montesquieu), and the other assumes a specific and ultimately organic adaptation to regional or local climatic zones of the earth, which is different in the various climatic regions of the earth (Wilhelm Wundt, Herbert Spencer). The first current sees a rather passive reaction to the climate; the second, which became dominant in the 19th century, equips the human organism (often with reference to Darwin and Lamarck) with the ability to adapt actively to climatic conditions.

Regardless of active adaptation or passive reaction, both versions of monocausal climate theory consider certain climatic conditions as the conditions for the possibility of civilization, science, art, technology, and literature, and even the readiness for war. In a type of counterfactual reasoning, the absence of certain phenomena while they are present in other societies is attributed to climatic conditions that have produced something specific in one case but not in another. For Gottfried Hegel, for example, it was self-evident that culture can only develop in a temperate climate. Karl Marx, who goes far beyond Darwin in his active view of man's ability to change nature (Grundmann, 1991), also explicitly states that different natural conditions, including climate and soil, determine different social forms of division of labor (cf. Grundmann and Stehr, 1997).

The central figure in this emergent climate discussion in the 18th century was Montesquieu. His 1748 *Spirit of the Laws* developed the central doctrine. Montesquieu ([1748] 1989: 2310) asks

> if it is true that the character [that is, a habitual way of acting and feeling] and the passions of the heart are extremely different in the various climates, *laws* should be relative to the differences on these passions and to the differences in these characters.

More elementary, social phenomena acquire a strong imprint of their environmental context. How exactly does this happen? Why and how do humans differ in various climates? Climate, according to Montesquieu, explained why different populations behaved as they did: by arranging nerve fibers, muscle tissues, and blood composition into uniform patterns within a region, climate assembled people through the body into collective practices and mass dispositions. Temperature, terrain, and atmospheric rates of fluid evaporation act on the physiological springs of personhood to standardize the way the inhabitants of a region take in and respond to sensations.

Montesquieu dispenses with questions of consent altogether by making the involuntary impulses of the flesh the source of mass behavior. To belong to a population, then, was to be a certain kind of body, where different modes of government accreted around the sensibility climate installed in a regionally shared body. By shaping the somatic apparatus underlying consciousness, climate guided individuals into forms of mass behavior antecedent to any conscious decision to participate. Montesquieu translated mass behavior into biological terms, thus encompassing the exteriorized features of a population beneath the skin of the individual.

Montesquieu has long been recognized as the most influential writer to link climate to questions of why populations behave as they do, but comparatively little has been said about his detailed interest in the body as the place where climate creates that effect. Montesquieu argues that environmental forces condition nerves, muscles, and blood such that people react in the same way to similar stimuli. The free exercise of consciousness falls in line with mass behavior because the climate determines the somatic capacity for thought and affect in a region. The sections of *Spirit of the Laws* devoted to the analysis of climate constitute the linchpin for the entire work, in which Montesquieu reframes the task of government as the regulation of whatever tendencies to vice climate installs in a regionally shared body. In this context, government and social organizations take shape through the automatic behavior of the body long before any nominal moment of conscious consent arises and seemingly without the need for deliberate decisions about the government. When Montesquieu makes "the general spirit of a nation" a central concept of his work, he is affirming that every polity is a mixture of ideas (religion, mores, laws, maxims of government), practices (history), and natural context (climate, geography).

In not only everyday life but also fields of social science or among medical practitioners, Montesquieu's idea that climates shape the minds, emotions, and beliefs of people turns out to be of continuing fascination:

> the conviction that the empire of climate extends its influence into the deepest depths of the human psyche remains strong. Indeed, there have been continuing efforts to test Montesquieu's eighteenth-century speculations about the climatic determination of human psychology. In an era of unprecedented anthropogenic climate change, that fascination shows no sign of abating and may even be intensifying.
>
> *(Livingstone, 2024: 219)*

The French social theorist, Gabriel Tarde ([1890] 1903: 326–327), in his classic treatise, *The Laws of Imitation*, shows curiosity about the impact of climatic conditions on social life. However, his judgment is cautious. In fact, he rejects references to climatic conditions as a *sociological* (or extra-logical) explanation:

> If one wished to explain the architectural style and fashion of a given locality merely on the ground of the exigencies of its climate, one would be greatly handicapped. [... It] is a question of fashion or custom, or, rather, of an ancient fashion that has become a custom. From one end to the other of the United States, from top to bottom, throughout all classes, even among good-looking women (and there is certainly no more striking example of the power of imitation than this) we find the repugnant habit of tobacco chewing, a fact that explains the universal presence of the spittoon, the most indispensable piece of furniture in America. Is this a habit that is made necessary by the exigencies of race and climate?

The French sociologist Èmile Durkheim, in one of the earliest and most definitive assessments of Montesquieu's perspective, expresses open skepticism about the role of geography, especially topography, in the life of societies. Durkheim (1960: 39–40), in his volume on Montesquieu and Rousseau as "forerunners of sociology," comments with notable reservation,

> although Montesquieu does not put topography and climate in the same rank as size of population and although he recognizes that they are dominant only among savage peoples, it must be admitted that their influence has nowhere been so great as he thought. Domestic, political, and private virtue are found in countries that differ completely in climate and

soil. However, even this exaggeration shows how strongly Montesquieu felt that social phenomena are subject to definite laws.

Werner Sombart: the circulation model

> Soil and climate together determine the natural fertility of a country and largely determine the nature of the people, which they either induce to indolence or to activity.
>
> *Werner Sombart (1938: 361)*

In his introduction to *Vom Menschen. Versuch einer geisteswissenschaftlichen Anthropologie*, published by Werner Sombart (1938), only three years before his death, he characterizes his monograph as an attempt to explain the natural foundations of human existence. Despite the subtitle, his "philosophical anthropology" is written primarily from the perspective of nature or the natural environment rather than from that of social life. However, in a self-critical mood, Sombart (1938: 406) emphasized that "I consider it unlikely that it will ever be possible to determine in principle which aspects of the human soul can be influenced by the environment and which cannot." The cautionary note Sombart injects in his treatise is often violated by the author himself and contributes to a strong sense of deliberate opacity and ambiguity in his perspectives.

Overall, *Vom Menschen* contains detailed observations on the influence of environmental factors on the social and intellectual conditions of society. Of particular interest to Sombart is the influence of *climate* on nations and their peoples. He postulated a possible optimum between mind and nature, which then developed a state of harmony with each other (Sombart, 1938: 338). Such high points of culture can be seen in a few periods of history, such as the European Middle Ages, "perhaps again in the 17th century and in some hidden places in the 18th century" (Sombart, 1938: 339).

In the end, he concludes that one will have to be satisfied with raising questions about the relationship between nature and society. For example, Sombart (1938: 429) is tempted to respond to any of the following questions in the affirmative:

> Is the strong development of fantasy in tropical peoples a result of their environment? Is the observation correct that all highland peoples have a greater wealth of fantasy than lowland peoples? Are belligerence, energy, and strong willpower characteristics of peoples who have grown up in Nordic climates, as the milieu theorists of earlier times claimed with preference?

However, despite his own cautions about the centrality of climatic conditions for social manifestations, Sombart (1938: 430) becomes a proponent of climate determinism without much resistance:

> Thus we can state quite generally that in Europe everything is directed towards increasing human dignity, whereas in the tropics it is diminished. It is only in Europe that the fine arts and architecture achieve the high degree of freedom and harmony that is based on the rule of the spirit over matter. And within Europe, in this area, which are apparently due to the climate and have nothing to do with any kind of "racial" peculiarity: namely, the climate determines whether the human form appears to the artist naked or clothed: the south represents the human form, the north geometric decoration. A Dürer "Melancholia" would probably never have been created in a southern country, for purely external reasons.

In *Vom Menschen*, Sombart also emphasized the extent to which the natural environment has been transformed by society. Any form of direct experience of nature has become rare in modern societies. Nevertheless, in social science discourse, talk of the "spiritual" environment retains the same justification as that of the natural environment. One of the main elements of the natural environment, along with landscape, retains its analytical significance despite the phenomenon of "climate creation (!)," is the natural climate.

Sombart developed a circulation model that describes the relationships among individual, mental, and natural environmental factors (1938: 400). He cautiously noted that it is not possible to work out exactly how the interdependence between environmental factors and individual factors works in specific cases: one can emphasize the contingency of these relationships and the conditions on which different outcomes depend. That is, people are confronted with external phenomena, including environmental conditions, which they either passively (or almost unconsciously) or actively cope with, facilitating or impeding action. The degree to which active or passive effects influence individuals or people depends on the "weakness" or "strength" of the individuals or people concerned, their level of civilizational development, their location, and, finally, the possibility that different effects may cancel each other out. Strong individuals and collectives are likely to respond more effectively to external pressures and opportunities. These statements quickly led Sombart to make rather suspicious observations about weak people who are condemned to react to their environment in a predetermined way (Sombart 1938: 401).

Although Sombart is convinced that modern social science's knowledge of the relationship between man and the environment has not made much progress and has probably not progressed further than what Jean Bodin

(1529–1586) formulated in the 16th century he does not hesitate to offer rather deterministic observations about the effects of the natural environment on the development of certain mental faculties and human capacities, i.e. to make statements about what he calls the nature of the "objective" culture found in different societies. His self-critical comments, which are partial criticism and partial confirmation of what various authors have said in the past about this relationship, are characterized by strong ambiguity. He conveys both his fascination with the thesis that climate determines human life and his skepticism about uniform or causal explanations for the diversity of human nature. Perhaps it is appropriate to assess Werner Sombart's deliberate shifts between social and natural causality as an attempt to arrive at an epistemological position of "causal circularity," that is, a conception of social and natural processes that could simultaneously be the cause of another.[6]

Even the origins of the natural sciences can be linked to environmental conditions: according to Sombart, these are only possible in a context in which the natural milieu is no longer perceived as a threat or an overpowering force. Although Sombart poses these questions, he is aware that there are many prominent examples of knowledge development and cultural achievements that thoroughly refute his idea that certain mental traits and behavioral patterns can be confined to specific climatic zones. Nevertheless, he ends up endorsing the thesis, supported by Hegel and many others, that significant cultural achievements thrive only in temperate climates.

In contrast to Èmile Durkheim, Max Weber, and Georg Simmel, Sombart does not limit himself to analyzing social factors in isolation from all other factors that may have a determining influence on human beings. Nor does he take sides in the various attempts to prove scientifically that there are unchanging environmental determinants of human behavior. His approach insists on the analytical distinction between natural and mental factors, limiting the scope of social science to the latter. Although this brings him closer to the program of sociological classics, it still leaves open the question of whether (and if so, in what way) natural variables are decisive. There is no doubt that much of his effort in *Vom Menschen* is characterized by criticism of such naturalistic explanations as speculative and without an empirical foundation. However, he does not want to completely reject such approaches, which is why he considers the influence of race and climate on individuals and people possible—with all due caution. However, his statements can hardly be separated from his general normative and political orientation, which encountered both consonances and dissonances in the intellectual environment of the context of the Third Reich in which it was made.

As the effects of climate change, especially its visibility, become much more apparent in societies, it is not entirely surprising to observe a revival of climate determinism in science and society. In physical anthropology, for example,

> it "was the resort to the role of climate or climate change in the process of hominization in general and in both the enlargement of the brain and the expansion of mental powers in particular" (Livingstone, 2024: 171). These views are not necessarily widely shared. While the emergence of modern climate determinism does not represent a return to classical climate determinism, there is a renewed interest in understanding the ways in which climate influences economic and social outcomes, and as is the case for traditional climate determinism, modern climate determinism potentially leads to simplistic or deterministic interpretations.
>
> *(e.g. Wolf, 2023: 157)*

Modern climate determinism

> Humans, unlike Animals, do not allow themselves simply to be absorbed by the naturally given order of the world. Instead, he tears himself loose from it, places himself in opposition to it, making demands of it, overpowering it, then overpowered by it.[7]
>
> *Simmel. Georg ([1911] 1981)*

Georg Simmer offers a valuable observation about the tragedy of human emancipation from and servitude to nature. The latter is captured by modern climate determination. What distinguishes traditional climate determinism, especially as represented in the work of Montesquieu, from modern climate determinism or as Mike Hulme (2011: 249) has chosen to call it "neoenvironmental determinism," which represents the insistence on the utter dependence of social phenomena on climate. Societies, depending on regional environmental conditions, develop, perhaps are forced to develop, social arrangements prescribed by their climate. The regional climate is fixed and an independent source of constraints. At times, traditional climate determinism amounted to blatant racism. The assumption put forward was that "native" populations were irradicable dominated by their surrounding climate, whereas the Europeans could "control" their natural milieu (Geroulanos, 2024: 131–132).

Modern climate determinism is based on very different assumptions. Climate is not a fixed phenomenon. It is a dependent source of constraints. The dynamics of climate in historical times are dependent on society. Ultimately, however, the result is the same: sociopolitical and socioeconomic conditions are at the mercy of the climate. However, in both cases, and this justifies the notion of climate determinism from both perspectives, the nature of society must respond to its climatic conditions.

A recent example of what should be interpreted as a case from the perspective of modern climate determinism is a study by Philip Riris (2024) and his colleagues examining the resistance and resilience of human populations across a broad archeological sample. The authors claim that "frequent disturbances enhanced the resilience of past human populations." Based on this assertion, the authors infer that understanding past responses will support the development of solutions to future crises. The research design is based on

> a global comparative approach to long-term resilience to identify the factors that structure the response of prehistoric populations to disturbances through time. The approach measures population capacity to withstand changes, as well as the rate of recovery following a disturbance through the common proxy of radiocarbon time–frequency data.
>
> *(Riris, 2024: 837)*

Disturbances are the driver, on the one hand, of reductions in populations but, on the other hand, of recovery after the downturn. The modelling exercise suggests "that the frequency of downturns is likely to have an important effect on resistance and recovery" (Riris, 2024: 840), which tends to unfold at multidecadal to centennial timescales. The authors are convinced that the existence of a common mechanism among human populations that confers resilience to disturbances (Riris, 2024: 840).

It is therefore permissible to conclude that human societies evolve in significant ways in a direct response to climatic conditions. The response over time is almost assured, guaranteed, and sustainable: "humanity's overall constant long-term population growth may have been sustained owing, in part, to the emergent positive feedback between vulnerability, resistance and recovery" documented in their study (Riris, 2024: 841). Economists such as Jeffrey Sachs and colleagues (2004) favor a simple determinism, or at least a strong correlation, linking economic prosperity to temperate northern climates and poverty to southern tropical climates. Most industrialized countries are in temperate zones and most poor countries in the tropics. The pattern Sachs described is a modern development that came about only after the Industrial Revolution and prior to anthropogenic climate change (Fukuyama, 2014: 235). The "reversal of fortune" that occurred after 1500 and has lasted into the present era is due to differences in the early and persistent institutional structures found in these regions of the world (e.g. Acemoglu et al., 2001). For much of human history, the prosperous societies were in southern regions of the world.

The perspective on the interaction between climate disturbances and the development of human societies has considerable affinity with the shorter-term hypothesis advanced by Ellsworth Huntington. Huntington ([1915]

1924: 91) ascribes the *fluctuations* he observes in factory productivity to the extraneous, in this case invigorating forces of the climate. In his earlier study, *Pulse of Asia,* a book judged at the time as one of the best in geography of the year, Huntington (1907: 382) offers the following conclusion of his longer-term study of parts of Chinese Turkestan and neighbouring regions in Central Asia:

> The evidence of *climatic changes* which we have found in the Old World seems to render it probable that these conditions have prevailed in each of the great countries of history at the time when it has risen to the highest degree of civilization and power. Therefore we may conclude that long-continuing changes of climate have been one of the great controlling causes of the rise and fall of the great nations.

A plausible conclusion is that the interpretation of the evidence produced by Riris and colleagues in 2024 closely resembles Huntington's (1907: 14) perspective which has a greater degree of specificity. In dry regions, increased aridity

> is a dire calamity, giving rise to famine and distress. These, in turn, are fruitful causes of wars and migrations, which engender the fall of dynasties and empires, the rise of new nations, and the growth of civilizations. If, on the contrary, a country becomes steadily less arid, and the conditions of life improve, prosperity and contentment are the rule. There is less temptation to war, and men's attention is left more free for greater arts and sciences which make for higher civilization.

A liberal interpretation of modern climate determinism as advanced and represented by Ellsworth Huntington and Philip Rise and his colleagues would be to suggest that a changing climate itself takes care of the "disturbances" produced, in the first instance.

At the center of politics

For a significant number of observers, climate change is the greatest threat to liberal democracy (Gaskill, 2022). However, it could also represent a historic opportunity for democracy. Apparently, it is not difficult for many observers to argue that the future consequences of climate change can best be dealt with effectively against democracy. The argument that democracy is a burden and a brake on a promising climate policy is finding an increasing number of partisans in science and society. On the other hand, climate deniers are convinced that the best way to fight climate change is to fight democracy. A widespread dislike of "elites" goes hand in hand with the opposition to parliamentary democracy. In both views, I argue that only democratic societies are better able to address the dangers of climate change in a sustainable way. However, countries of all political stripes are currently struggling to cope with a warming planet and generate consequential political action against climate change.

However, there is not only a growing debate about the presumed failure of democracies to cope with the present and anticipated consequences of climate change; other major social institutions are also seen as ineffective in responding to climate change. Foremost among such discussions are concerns about the role of the economy, for example, in what sense "capitalism" itself may serve as the solution or to what extent specific economic policies, for example, how tax credits or monetary policies could offer answers to climate change. In the final analysis, all debates, whether they focus on the political economy of capitalism or the state-crafted economies of "socialist" nations, emphasize the reliance on the role of the state and its functions, even when it comes to engineering solutions toward a decarbonization of everyday life, for example, heating, transportation, or the economy generally. The course of the pandemic has also impacted the debate over the role of global shocks and the response of the state to such risks.

Regardless of whether critics of climate policy attribute responsibility for the slow response to the adverse effects of climate change to the political system, everyday business practices, or lack of technological progress, the consensus is that the response is simply not fast enough. Thus, alternative socio-political and socio-economic arrangements have come into view. Alternative perspectives include technological solutions to climate change, which I will discuss in the following section, and the idea that capitalism should be seen as the savior from the worst effects of global warming. In addition, the world of finance, for example, central banks everywhere, independent of political regimes, is sometimes seen as an efficient tool against the effects of climate change (e.g. Monnet, [2021] 2024: 97).

The question of whether capitalism can be a "savior" in addressing climate change is complicated not only because there are numerous forms of rising and declining capitalism (Stehr, 2025) but also since capitalism is both blamed for accelerating the climate crisis and hailed for innovating potential solutions. There are several arguments on both sides of this debate, each with implications for how climate change can be mitigated within or despite the current economic framework. Of course, there is also the (remote) possibility that the ecological crisis could "solve" the crisis of capitalism by generating significant shifts in the economic order of capitalism, such as a greening of capitalism (Stehr, 2008).

The spirit of capitalism has a dual face: capitalism is considered the major contributor to the possibility of anthropogenic global warming resulting from the capitalist mode of accumulation and the pursuit of profit (e.g. Schneider et al., 2023: 112); and, as I will discuss in the next section, capitalism as the savior from the worst impacts of climate change. Peter Newell and Matthew Paterson (2010: 1) coined the term "climate capitalism" to capture the idea that decarbonizing the economy, while amounting to a radical response to climate change on a global scale, cannot be achieved by transcending the capitalist economic order. In fact, Newell and Peterson argue that the wholesale transformation of how the economy and daily life are organized is best realized under the continuing umbrella of capitalism. Replacing oil and gas *without* abandoning the prospect of continued economic growth is not possible by abandoning capitalism.

Capitalism as the savior[8]

In addition to democracy as a key social institution, the economic system is the other major target of critics of ineffective climate policy. Slavoj Žižek (2010: 333–334) makes the unmistakable point that "[t]he key to the ecological crisis does not reside in ecology as such" but in a transformation or abolition of capitalism. The very success of global capitalism proves to be a threat to our freedom:

> The limitation of our freedom that becomes palpable with global warming is the paradoxical outcome of the very exponential growth of our freedom and power, that is, of our growing ability to transform nature around us, up to and including destabilizing the very framework for life.
>
> *(Žižek, 2010: 333)*

Similarly, Naomi Klein has promoted the idea that *neoliberalism* has played a major role in perpetuating global warming. Klein's (2014: 72) book *This Changes Everything: Capitalism* vs. *the Climate* argues that "free market orthodoxies [...] threaten our capacity to respond boldly to this crisis."

In more detail, Anders Fremstad and Mark Paul (2022) argue that the myth of the free market has prevented effective climate policy in the United States. The authors argue that three core tenets of neoliberalism—the decentralization of democracy, the dismantling of public investment, and the deregulation of the economy—have collectively stymied efforts to reduce greenhouse gas emissions. By promoting federalism, neoliberalism has led to fragmented climate policies and a 'race to the bottom' among states. Fiscal austerity has slashed public investment in green technologies and infrastructure, while deregulation has weakened environmental protection. The authors criticize carbon pricing as insufficient and advocate the Green New Deal as an alternative framework that emphasizes large-scale public investment, federal leadership, and binding regulations to achieve rapid decarbonization. It follows that overcoming neoliberal ideology is essential to addressing the climate crisis and implementing transformative policies. The difficulty with Framstad and Paul's position is, on the one hand, that neoliberalism is only one of a number of possible societal antecedents of anthropogenic climate change and, on the other hand, that an overarching issue is the attachment to economic growth, regardless of socio-political ideology. At the same time, the critique of neoliberalism underestimates the inventiveness of neoliberal responses to climate change, for example, carbon pricing (Akerlof et al., 2019),

Capitalism, however, can be seen not only as the generator of the warming of the world but also as the key lever for tackling climate change. In particular, the best way to decarbonize the economy is through the pursuit of low-cost renewables. Indeed, in developed economies, the price of energy produced from fossil fuels and from renewables has converged in recent years. The widely held assumption for the widespread economic success of renewables has been that fuel-based electricity is more expensive than electricity generated by solar and wind. This state of price competition was reached at the end of the 2010s. However, the competitive success of renewables, i.e. their lower costs, is based on government subsidies, such as tax credits in the United States. As renewables become more successful in the market, government subsidies tend to be withdrawn.

If, as reported by the *International Energy Agency* in 2015, the cost of fossil fuels is higher than the cost of electricity generated by solar and wind, then the law of the market should dictate that the success of renewables will become self-sustaining, and that fossil fuel-generated electricity will soon be squeezed out of the market. However, this did not appear to have occurred in the decade since 2015. As Brett Christophers (2024) stressed, the *profitability* of renewables is significantly lower than the profits from fossil fuel-produced energy. Large energy corporations will therefore prefer to pursue the more profitable business of fossil fuels, as they are open to proclaim. In a 2023 statement, Shell's CEO Wael Sawan responded during an "earnings call" to a question about whether he considered renewables' lower returns acceptable for his company:

> I think on low carbon, let me be, I think, categorical in this. We will drive for strong returns in any business we go into. We cannot justify going for a low return. Our shareholders deserve to see us going after strong returns. If we cannot achieve the double-digit returns in a business, we need to question very hard whether we should continue in that business. Absolutely, we want to continue to go for lower and lower and lower carbon, but it has to be profitable.
>
> *(cited by Christophers, 2024: 600)*

Companies such as Shell expect double-digit financial returns on investments in fossil fuel energy production. British Petroleum (BP) announced in early 2015 that it would sharply increase spending on oil and gas, while reducing investment in various forms of renewable energy. The *New York Times* comments, "the move appears to be a response to a combination of investor pressure for higher returns and a realization that the so-called energy transition to cleaner fuels is not paying off as once expected."[9] The return on investment in renewables is in the single digits, in the 5%–8% range. Corporate investment policies are driven not so much by the cost of renewables but by the profit of fossil fuels.[10] Given present market conditions, the profitability of the investment for the investors is not the choice between environmentally friendly investment and dirty energy generation that drives corporate investment decisions.

What is to be done? If market conditions, for example, the risks private investors are willing to take on board, are insufficient to advance decarbonization which lacks sufficient pace, regulations imposed, or incentives proposed by the state would have to "heal" the deficient market conditions.[11] For example, taxing profits from fossil fuel electricity at a high rate moves profitability below investments from renewables, thereby accelerating a shift in investment decisions. As William Davies (2024) stresses, such state policy changes amount to "something akin to wartime mobilization might then take place, in which the state stretches its financial credibility to the absolute limit to invest in renewables at the pace that the climate emergency demands." Such policies put democratic governance in considerable jeopardy.

In addition, another related set of arguments emphasizes that the private sector, such as insurers, banks, and other corporations, could be instrumental in decarbonizing the economy: capital from these companies could help finance the energy transition. Hundreds of private companies have pledged $130 trillion in the form of the *Glasgow Financial Alliance for Net Zero*. The financial firms voluntarily committed to ensuring that the companies in which they invest reduce their emissions. They also pledged to align their lending policies with the goal of limiting global temperature increases to 1.5°C above pre-industrial levels.[12] The Alliance currently consists of 675 financial institutions from 50 countries.

Clearly, trillions of dollars will need to be invested in clean energy to achieve the goal of a decarbonized world. If decarbonization is to be achieved, it will have to be on many fronts. A study by the *European Central Bank*, as an obvious proponent of the virtue of competition and market pricing, is sceptical about the power of the marketplace to achieve a transition toward a carbon-free economy. The 2024 study titled "Business as usual," notes the results thus far cast "doubt on the efficacy of voluntary climate commitments for reducing financed emissions, whether through divestment or engagement." In fact, banks that are part of the Alliance have reduced their lending commitments to fossil fuel projects they had targeted in their Glasgow commitment by 20%. However, banks that are not part of the COP26 Alliance also reduced their lending by the same amount. The European

Central Bank study leaves no doubt and broadly concludes that voluntary private-sector initiatives may have relatively little impact on decarbonization and, more generally,[13] the markets of capitalism are "unlikely to yield the needed transition" to a carbon-neutral economy.

The implication is that democracy may have to give way—if we are to accelerate the decarbonization of energy production and not rely on ineffective market voluntarism.

A related question that needs to be posed concerns innovative ways of consuming energy. For example, will the world reach its net-zero value while discovering new ways of using energy? The new form of energy consumption is linked to the rapidly growing deployment of AI or the exchange of Bitcoins.[14] According to an announcement of the *White House Office of Science and Technology*, Bitcoin mining uses almost as much energy in the United States as all home computers combined. However, the statistics quoted are far from reliable. It is not necessarily in the interest of the corporations that advance Bitcoin mining or AI to be transparent about their electricity needs. Moreover, it was developed early in the development of large language models. Beyond dispute is that AI has an enormous appetite for electricity. However, Bitcoin and AI have been largely invisible in calculations about efforts to reach net zero worldwide.

One of the still underrated issues in any combat against the repercussions of climate change is the role of the world of finance everywhere, for instance, the role of central banks. The central banks of the world are increasingly becoming aware that they ought to pay attention to the issue of climate change. As Éric Monnet ([2021] 2024: 125) points out,

> one of the crucial questions for central banks today concerns whether they may grant loans with privileged conditions to finance activities that fight climate change. This question is particularly pressing for central banks whose mandates feature overarching objectives of economic development or support for public policies; others have seen objectives concerning the climate written into their mandates by parliaments.

In practice, the central banks of China and Japan have already extended loans at favorable conditions for banks that pledge to finance activities with small carbon imprints.

The modern knowledge-based economy exists not only in the virtual world but also requires massive physical support. The data centers needed to store and crunch the information of the knowledge economy are energy-hungry infrastructures. The *Guardian*[15] reported that

> Ireland's energy-hungry datacentres consumed more electricity last year [2023] than all its urban homes combined, according to official figures. The country's growing fleet of datacenters used 21% of its electricity, an

increase of a fifth on 2022, according to the *Central Statistics Office*. It was the first year that datacenters supporting the Irish tech hub surpassed the electricity used by homes in its towns and cities, which consumed 18% of the grid's total power last year.

The infrastructure of the knowledge economy, and more recently with the advent of Generative AI, consumes massive amounts of energy and therefore leaves a major footprint and has a possibly growing impact on the warming world. The accelerating power demands of AI are underappreciated in climate change development and represent a new challenge to climate policies. The surge in power demand by datacenters could derail climate targets in Ireland and all of Europe. Measuring the energy impact of AI could be a first step toward its regulation. However, in addition to the contribution of AI to an increase in emissions, it also appears to be possible, at least in the eyes of some observers, that its development may also contribute to its containment, for example, through an acceleration of inventions reducing carbon emissions and lifting efficient technologies.[16]

Climate change is often seen as a global example of the tragedy of the commons. The Earth's atmosphere is a common resource. Collective emissions of greenhouse gases and other activities are responsible for warming the Earth. Countries, companies, or individuals may benefit disproportionately, economically or socially, from these activities, but the negative consequences are shared by all. However, the analogy with the classic tragedy of the commons perspective may be limited.

Tragedy of the commons

The tragedy of the commons, a concept popularized by ecologist Garrett Hardin (1968), is obviously relevant to the issue of climate change. The tragedy of the commons describes a situation in which individuals, that is, farmers, let their cows graze on the communally owned village green, act independently and rationally according to their self-interest, overuse and deplete a common resource, leading to its eventual degradation, despite their collective dependence on it. In principle, such a relationship applies to the common resources of the Earth's atmosphere. The atmosphere is used by all, but no single entity owns or controls it. The sum total of human behavior causes degradation of the atmosphere. Some nations, regions, institutions, or acronyms avoid taking meaningful action on climate change, assuming that they recognize the problem and perhaps hope that others will address it. This results in insufficient collective efforts to reduce emissions. Decision-makers often prioritize immediate benefits over the long-term risks of climate inaction, compounding the tragedy of the commons. However, the analogy of the classical perspective/example explicated by Garrett Hardin is limited.

Climate change is the result of a wide range of activities, not a single resource such as the common pasture described by Hardin. The impacts of climate change span ecosystems, economies, and societies, making the issue far more complex than the relatively simple overuse scenarios described by the classic tragedy of the commons perspective. In addition, technological advances (e.g. renewable energy, carbon capture) can mitigate climate impacts in ways that Hardin's tragedy framework does not take into account because it assumes limited technological solutions. Climate change is driven by externalities, where the costs of emissions are not borne directly by emitters. This problem goes beyond the common resource framework of the tragedy of the commons and involves correcting market failures. Climate change raises issues of intergenerational equity, ethical responsibility, and ecological stewardship that are not addressed from the Hardinian perspective. These dimensions involve moral (cf. Hannigan, 2025) and legal obligations that go beyond rational self-interest. For example, Stephen Gardiner (2002) noted that the real tragedy of the commons should not be understood as consisting of a dilemma for social collectivities living today but as a dilemma occurring between generations. The tragedy of the commons provides a useful lens to understand the overuse of the atmosphere as a carbon sink, but its applicability to climate change is limited by the global scale, complexity, and unequal dynamics of the problem.

But addressing climate change requires moving beyond the tragedy of the commons narrative to consider the mix of governance, innovation, inequality, law, participation, ethics, and the potential for cooperative policy solutions. Most importantly, the embeddedness of climate change and climate policy in the governance structure of societies requires a closer examination of the state of democratic governance in the face of climate change risks and hazards. The social science literature on the interplay between climate change and the capacity of democracies to cope with it, especially in the future, is largely shifting between encouraging outlooks and pessimistic narratives. At the same time, the diagnosis of the capacity of democracies to respond adequately to the challenges of climate change is invariably embedded in much broader and more spirited discussions about the evolution of democracy in the face of numerous challenges, including, of course, active and at times already successful efforts to transcend democracy that are now underway or anticipated for the future present.

In the following chapter, I examine the challenges that democracies face when it comes to addressing climate change. I contrast democratic systems with authoritarian regimes, noting that, although democracies may struggle with short-term electoral cycles and public resistance, they have historically fostered stronger environmental policies thanks to civic engagement and institutional capacity. The chapter also examines societal barriers, such as public skepticism, economic interests, and political polarization, which hinder climate action. The discussion then moves on to consider the influence of technology, the media, and elites on climate discourse. Ultimately, the chapter argues for the strengthening of democratic institutions to tackle climate change rather than resorting to authoritarian solutions. It emphasizes the need for systemic innovation and public participation to bridge the gap between scientific knowledge and political action.

Notes

1 I am referring to an account of the history of climate determinism by Nico Stehr and Hans von Storch (1999) were we developed a more detailed analysis of climate determinism and in which we refer to the difficulties scientists have in advancing the idea that climate determines the evolution of human civilizations.
2 Alexander Humboldt's closing part of the definition of the term climate in the original *Kosmus* version of 1845 reads as follows: "all changes [... that are important] for people's feelings and overall state of mind" (Humboldt, 1845: 340).
3 On the contentious concepts of "civilization and culture," cf. Grundmann and Stehr (2015).
4 See *New York Times*, Late Edition, East Coast, January 4, 1997, Section 1, Editorial Desk, p. 22.
5 For an explicit example and proponent of climate determinism from today's natural sciences, see Beck (1993).
6 Compare the remarks by Evgeny Morozov on the epistemology of AI: "AI and the techno-utopian path not taken," *LeMonde Diplomatique*, August 2024; https://mondediplo.com/2024/08/07ai-cold-war.
7 Slightly amended translation.
8 Peter Newall and Matthew Paterson (2010) have attempted to put forward a balanced blueprint of the contours of a narrative of "climate capitalism."
9 Stanley Reed, "BP to 'fundamentally Reset' strategy in shift back to oil and gas," *New York Times*, February 26, 2025; www.nytimes.com/2025/02/26/business/bp-oil-gas-renewables-strategy.html.
10 In the Spring of 2024, British Petroleum reports that it expects a strong quarter due to higher oil and gas production boosts (cf. www.wallstreet-online.de/nachricht/17964293-kurs-legt-zu-bp-rechnet-starkem-quartal-hoehere-oel-gas-produktion-befluegelt-gewinn). This announcement by BP follows the recent announcement by BP's largest competitor, Shell, which also announced an increase in its production expectations for integrated gas and liquefied natural gas for the first quarter of 2024.
11 A "powerful" solution often advocated is *setting* a *global carbon price*. A global carbon price would, it is maintained, correct market failures and set a process in motion whereby households and corporations would reduce their energy use. However, setting a global carbon price would itself be evidence of a market failure and evidence of a version of state capitalism.
12 "The internationally agreed goal to keep the world's temperature rise below 1.5°C is now 'deader than a doornail', with 2024 almost certain to be the first individual year above this threshold, climate scientists have gloomily concluded – even as world leaders gather for climate talks on how to remain within this boundary"

(*Guardian*, November 18, 2024; www.theguardian.com/environment/2024/nov/18/climate-crisis-world-temperature-target.

13 Quoting from a speech by Isabel Schnabel, Member of the Executive Board of the ECB, at the European Sustainable Finance Summit, Frankfurt am Main, 28 September 2020; www.ecb.europa.eu/home/html/index.en.html.

14 As Azeem Azhar notes in a guest comment ("The 19th Century Technology That Threatens AI") in the *New York Times*. "The development of advanced AI systems requires vast amounts of energy. At the heart of training these systems are large numbers of specialized computer chips. One estimate suggested that training GPT-4, the latest ChatGPT AI system, consumed roughly the same amount of electricity as several thousand U.S. households use in a year. These extraordinary demands are already pushing up against real-world constraints. America's power grid, hindered by decades of underinvestment and regulatory log jams, isn't equipped for the rapid growth in AI's electricity needs;" www.nytimes.com/2024/12/28/opinion/ai-electricity-power-plants.html.

15 "Ireland's datacentres overtake electricity use of all urban homes combined statistics raise concerns that rise in demand for data processing driven by AI could derail climate targets," *Guardian*, July 23, 2024; www.nytimes.com/2024/12/28/opinion/ai-electricity-power-plants.html, www.theguardian.com/world/article/2024/jul/23/ireland-datacenters-overtake-electricity-use-of-all-homes-combined-figures-show.

16 Steve Lohr, "Will AI ruin the planet or save the planet," *New York Times*, August 26, 2024; www.nytimes.com/2024/08/26/climate/ai-planet-climate-change.html.

4

CLIMATE POLITICS

Overburdened democracy?

> Certain kinds of states, driven by utopian plans and an authoritarian disregard for the values, desires and objections of their subjects, are indeed a mortal threat to human well-being.
>
> *James C. Scott (1998: 7)*

There is a broad consensus among most observers that the activities of democracies at the national and global decision-making levels need to shift from limited climate policy action, from weak policy commitments to real and binding political action (e.g. Held and Hervey, 2009). Barber (2012: 168) minced no words,

> if democracy means only the rule of majority as measured by an aggregate of private opinions, there is no question that it will be unable to deal with climate change [...] Prudent dictators will have to do what imprudent consumers refuse to do.

However, it is rather easy, perhaps even self-evident, to offer such an observation about the apparent inertia of democratic governance in the face of growing risks and dangers from climate change in contemporary societies. Are effective climate policies within the range of democratic government action?

In the 1980s, the ultimate threat to humankind resulting from the proliferation of nuclear weapons was widely discussed. Even a mere accident might lead to widespread destruction. Analogous to the present dilemma, half a century ago, we faced "the ultimate problem of governability: it appears that we have created technical possibilities of destruction which no

DOI: 10.4324/9781003682066-5

conceivable human governments can contain. We have reached limits of governability" (Hennis et al., 1977; Dahrendorf, 1980: 394). Can we today assume that democratic governance has and may have a superb record in drawing up and implementing effective climate policies on mitigation and adaptation? Without a doubt, the state, corporations,[1] and their citizens must shoulder the responsibility to take on climate change and deliver.

The effects of democracy on climate policies: the historical record

Much of the effort to determine how democracy relates to climate change policies tends to have a present-day or even forward glance, but the past record is likely not insignificant in offering insights into the effects of democracy on environmental policies. Per Frederiksson and Eric Neumayer (2013) carried out a relevant historical analysis. This study focuses on the relationship between democracy and climate change policies, specifically the effects of countries' long-term history of democratic experience on environmental policies. The focus is on climate change policies rather than carbon emissions, as policies take time to translate into noticeable changes.

The empirical approach utilized is the "Climate Laws, Institutions and Measures Index (CLIMI) developed by Franklin Steves and Alexander Teytelboym" (2023; emphasis added). The CLIMI composite index, which is based on information contained in the *United Nations Framework Convention on Climate Change* (UNFCCC),

> focuses on *policy inputs* climate laws, institutions, and measures rather than policy outcomes, viz., emissions. This is the first index that considers all major sectoral and cross-sectoral policies and measures as well as all government institutions focused on climate change for 95 countries. A country that scores highly on CLIMI is not only committed to reducing emissions today but is also building the institutional capacity to reduce emissions in the future.

The index classified the 95 countries as democracies or autocracies, going back to the year 1800, and ranked their national legislative, fiscal, and institutional frameworks. There are considerable difficulties in designing models for climate change policymaking. Policies mitigating emissions can vary vastly from those that attack emissions directly, short-term, or long-term, indirectly or even tacitly.

Per Frederiksson and Eric Neumayer (2013: 11; emphasis added) maintain that "democratic capital," that is, a "country's *accumulated* stock of civic and social assets built by historical experience with democracy [...] has a robust positive effect on national and multi-lateral policies addressing climate change." The main dependent variable in Fredriksson and Neumayer's empirical study involves the years 2005–2010 in the CLIMI Index: Alternative definitions of the long-term historical experience with democracy consistently show a positive and highly significant effect on climate change policies (defined as mitigation policies). Per capita income has a positive effect on climate change policies, and economic freedom has a positive but statistically insignificant effect (Frederiksson and Neumayer (2013: 14). Domestic policy making depends on the number and nature of key actors involved and the political context within which they operate.

Overall, the study's robustness tests provide further evidence that democracy, particularly the historical presence of executive constraints, is associated with stronger climate change policies. The evidence allows for the general observation that the consolidation of democracies not in the short run allows for the expectation that climate change policies are more likely to be generated and implemented. While the evidence of effective climate-related policies in democratic countries is promising, the current achievements cannot be simply generalized to the future. Among the major barriers is the political power of societal change, which could weaken the future present of democracy. I will examine some of these developments, including those that issue from the growing digitalization of social life. Initially, the impact of a warming planet itself should be examined.

Stressed democracies

> The mansion of modern freedoms stands on an ever-expanding base of fossil-fuel use. Most of our freedoms so far have been energy-intensive.
>
> *Dipesh Chakrabarty (2009: 208)*

The historian Charles Maier (2023: 390) summarized the essential dilemma of the modern age, given the "mounting evidence of climate change unprecedented in the history of Homo sapiens," as adding to the strong "sense of untethering" in the contemporary world. Part of this untethering is a stressed democracy. The disenchantment of democracy "as a control regime of highly complex societies is already underway because [...] its intelligence is no longer sufficient to tackle the really serious problems in any reasonably adequate way" (Willke, 2014: 162). Helmut Willke's (2014: 7–9) more detailed explanation of his skeptical judgment of the state of democracy stresses that

> although democracy is the best of all available forms of government for modern complex societies — it is not good enough in the face of the new challenges. These new challenges can be summarized in three keywords: globalization, knowledge society and organized complexity. [...] Democracy presupposes that citizens entitled to vote can assess

> and understand what they are voting on. Both constitutive conditions for democracy are increasingly proving to be fiction. [...] The decision-making scope of national parliaments depends on transnational and global conditions. The political management of the huge problems in particular – from the financial crisis to ecology to development – depends on international and transnational 'readiness' and leaves the national democracies only the appearance of decision-making power.

The untethering of democracy, as Dipesh Chakrabarty reminds us, is ultimately due to the long history of humanity's dependence on one form of energy, fossil fuels. Both populists (especially climate deniers) and progressives agree that the traditional political order in democratic countries is being torn apart by multiple secular trends such as globalization, rising social inequality, the depletion of natural resources, and demographic change. Paul Crutzen and Eugene Stoermer (2000: 17), who initially explicated the term "Anthropocene" for the new geological period of human agency, described some of its intimidating consequences in the future present for humanity:

> Mankind will remain a major geological force for many millennia, maybe millions of years, to come. To develop a world-wide accepted strategy leading to sustainability of ecosystems against human-induced stresses will be one of the great future tasks of mankind, requiring intensive research efforts and wise application of knowledge thus acquired. [...] An exciting, but also difficult and daunting task lies ahead of the global research and engineering community to guide mankind towards global, sustainable, environmental management.

Bruno Latour demands an "alternative description" of the relationship between society and nature.[2] An alternative description requires an "inventory, survey, measure, centimeter by centimeter, being by being, person by person, the stuff that makes up the Earth for us" (Latour, [2017] 2018: 94). Everything must be "mapped anew, at new costs. Moreover, this is an urgent task that must be carried out before the sleepwalkers, in their blind headlong rush forward, have crushed what we care about" (Latour, [2017] 2018: 33). The Anthropocene amounts to an admission that the fates of humans and non-humans are increasingly and deeply intertwined.[3] Hence, the Anthropocene also offers at least the backdoor to the readmission of climate determinism. However, only partial re-entry has occurred since climate determinism has denied, for the most part, human agency. The term from Latour's ([2017] 2018: 98) perspective therefore refers to humans raising their voice "on the part of all animate beings."

At present, it is no longer *short-term* conflicts and problems, such as wars, inflation or unemployment, ideological tensions, or generational tensions, that ultimately contributed to the failure of the democracy of the Weimar

Republic which represents the most serious challenge faced by democracies. Instead of loudly talking about the dangers of the future present, there is a growing chorus of voices demanding that prompt action should replace mere chatter. Democratic regimes do not seem to be able to find convincing answers to the urgent questions of our time; answers that are also convincing to the population.

Not all voices issue pessimistic forecasts about the efficacy of democracies in the face of urgent environmental challenges. Ray Kurzweil (2024: 183) is convinced that democracies have a hopeful future given accelerated technological advances:

> History gives us reason for profound optimism, though. As technologies for sharing information have evolved from the telegraph to social media, the idea of democracy and individual rights has gone from barely acknowledged to a worldwide aspiration that is already a reality for nearly half the people on earth. Imagine how the exponential progress of the next two decades will allow us to realize these ideals even more fully.

In general, the ability of the modern state to impose its will despite the good intentions of governments is declining (Strange, 1996; Stehr, 2000). Although democracy is not directly and fundamentally called into question on all sides, at most by politically and nationalistically extreme minorities, approval of the democratic model is as I have noted eroding.

We know that the scientific consensus on human-caused climate change has strengthened, and in several recent studies, far more dramatic and long-lasting consequences of global warming are anticipated compared than such predictions just a few years ago. For example, as a current analysis (Luo et al., 2024) demonstrated, "anthropogenic forcing has increased the risk of longer-traveling and slower-moving large contiguous heatwaves." Heat waves have detrimental impacts on economic ambitions, human health,[4] pollution, the economy generally, agriculture, labor productivity, wildfires, and mortality burden. The impact of climate change on the economy is another apparent but uncertain risk. The risk from climate-related damage to private residences could, for instance, be quite significant.[5] The uncertainty of the expected damage caused by climate change as modeled by economists, although estimations increased in recent studies likely remains quite significant (Berger and Marinacci, 2020).

The Gulf States' economic ambitions and efforts to reduce their dependence on fossil fuels could be jeopardized by more frequent and intense heat waves in the region, which could make it uninhabitable, according to a *New York Times* article:

> The wealthy petrostates of the Persian Gulf have big plans for the future, hoping to increasingly attract tourists and investors, host marquee sporting

> events, build new cities and diversify their economies away from oil. However, they face a looming threat that they cannot easily buy their way out extreme and sometimes deadly heat that roasts their countries every summer, which climate change is expected to exacerbate in the coming decades [...] projections warn of weekslong heat waves with temperatures of up to 132 degrees [55,5 Celsius] during the second half of this century. Temperatures that high can endanger human life.[6]

As reported by the *Copernicus Program* of the European Union in April 2024, the month "March 2024 was warmer globally than any previous March in the data record with an average ERA5 [that is a global atmospheric reanalysis developed by the Copernicus Climate Change Service] surface air temperature of 14.14°C." The ocean surface temperature has also broken records. The *Copernicus Climate Change Service* stated that the average sea surface temperature reached a new monthly high in March 2024.

Despite the growing risks, most societies and the global economy are not prepared. In many countries, employment in carbon-intensive sectors remains relatively high. In the *European Union*, an average of 40% of employment is in carbon-intensive sectors. CO_2 emissions continue to rise worldwide, especially in emerging economies. In *European Union* countries, carbon dioxide emissions are decreasing.

Klaus Töpfer and Günther Bachmann (2010: 56) describe a climate catastrophe, also known as a "total climate meltdown," and the threat it poses to democracy if the bureaucratic 2°C average global warming limit is surpassed:

> Exceeding this limit is often referred to as a climate emergency because of all kinds of impacts that could jeopardize flood control, food security, urban development or climate-induced migration. If they do, this, of course would very profoundly impact every aspect of global interaction, both ecologically and economically. It would potentially destabilize the entire society and contribute to the breakdown of democratic and institutional power known as the failing state syndrome.

Töpfer and Bachmann (2010: 56) further add to their verdict the pessimistic observation that there is "no political idea of how to react to a climate emergency and its impact on the decision-making system, not to speak of democracy."

Although commonly referred to as "global warming" or "climate change," the usual expected multiple impacts are an increase in global average temperatures, a rise in sea levels, and a more frequent occurrence of extreme weather events. However, unanticipated research observations may reverse some of these assumptions, for example, the now-considered normal expectation that sea level rise will inevitably spell the end of the habitable

Pacific islands.[7] Considering these developments, the relationship between climate and society will inevitably change in new and unpredictable ways (see Stehr and Machin, 2019). This is all the truer in the case of a so-called "climate catastrophe" that places the world in a state of emergency. More than 2300 communities around the world, including capital cities such as Amsterdam, Berlin, and Tokyo, have declared a "climate emergency." Environmental organizations are hopeful that a narrative of "climate emergency" will become a driver for more environmental engagement and democratic participation. However, it is also feasible that such a narrative legitimizes the opposite political development and enshrines authoritarianism.[8] In general, as some observers claim, in view

> of the enormous, utterly unprecedented stakes presented by the climate challenge, humanity is failing miserably. Everywhere there are promises, plans and regimes, but nowhere are they remotely adequate to what needs to be done. To put it bluntly, the alligators are winning.
>
> *(Dorman, 2022: 264)*

It is therefore realistic to speak of a fundamental gap between the broadly consensual narrative of climate change in the scientific community, the expressed intentions of many actors in many societies and the actions that would follow from the narrative, and the intentions of actors leading to what might be called "ecological integrity" (Stevenson, 2021: 86), that is, an explicit convergence of narrative, intention, and action.

Can the dangers of climate change be averted? There is no shortage of proposed solutions, which are developed mainly by international teams of engineers, as I have already noted and are aimed largely at reducing greenhouse gas emissions. An entirely separate category of overcoming or averting a warming climate, at least in the northern hemisphere, would be a collapse of the Atlantic Meridional Overturning Circulation (AMOC), which may be due to the injection of fresh water into the North Atlantic (see van Westen et al., 2024).[9]

Talking about climate needs to be replaced by acting[10]

Climate scientists, social scientists, the media, and environmental groups (NGOs) concerned with climate change point to a "future present" with extraordinary circumstances and protest that evolution did not enable us to cope with such problems (Di Paola and Jamieson, 2018). As Dale Jamieson (2014: 61) explains:

> Climate change poses threats that are probabilistic, multiple, indirect, often invisible, and unbounded in space and time. [...] Moreover, climate change can be seen as presenting us with the largest collective action

> problem that humanity has ever faced, one that has both intra- and inter-generational dimensions. Evolution did not design us to deal with such problems, and we have not designed political institutions that are conducive to solving them.

Members of the same groups impatiently complain that no one is listening enough to the diagnosis of the historically unprecedented global risks and dangers of the 21st century.

The constant reinforcement of the threat discourse—many may remember the 1986 DER SPIEGEL cover with the Cologne Cathedral halfway under water—can paradoxically prove to support the opposite virtue, namely as a defense of the present and an encouragement of skepticism toward threat scenarios. I am thinking, for example, of a basic psychological mechanism, a kind of *denial psychology* (Norgaard, 2011), that may be responsible for our blocking out of future unpleasant developments and for our inability to deal proactively with the dangers of the climate crisis today. This psychological mechanism is not unlike the way we deal with extreme weather events daily, which is often interpreted as an affirmation of the normal course of climate (see Stehr, 1997; Stehr and Machin, 2016b; Stehr and Machin, 2019).

Given that climate science has proven anthropogenic climate change, the discourse has necessarily become future-oriented in important aspects. The focus has shifted to the question of how societies can best be governed in the not-too-distant future under the massive effects of global warming. How will it be possible to govern a future present that is likely to be completely different from the societal context in which democratic systems have emerged and flourished in the past? In the cases I discuss below, strong opinions advocating the need to suppress political freedom in the face of profound future environmental changes are no longer uncommon.

In this chapter, I address this disenchantment with democracy, particularly with respect to disappointing expectations and the supposed shortcomings of representative democracy in its currently predominant liberal form. My contribution will focus on the conflict between politics and science. I critically look at the argument that political decision-makers must act in the face of the risks and dangers of climate change, even without a broad public mandate and without legitimacy. The time is very short before the future lies before catastrophic damage occurs. Instead of lamenting the inconveniences of democratic governance, it is important to reflect on how democracy can be strengthened, not despite but in the face of the massive challenges of climate change. As history shows, and as I will argue, major environmental challenges can best be addressed within the confines of democratic rather than authoritarian political systems. In this analysis, climate change is seen as a matter of political governance, not just an environmental or economic issue.

I leave aside the complaint, often heard in recent years from social scientists and some political parties, for instance, that the behavior of states, especially highly centralized states with strong hierarchical structures, is the cause of environmental degradation, manifesting in virtually all countries in their relentless pursuit of economic growth. John Barry (2012: 131) outlines this perspective and agrees with the view that attributes the underlying causes of climate change, in his case, in the first instance to the pursuit of unregulated economic growth under *capitalism*. Similarly, effective and meaningful approaches to combating climate change need to focus on moving away from the economic growth of orthodox policies. Meaningful reductions in emissions are not compatible with "economic prosperity." Hence, "dangerous climate change can only be avoided if economic growth is exchanged, at least temporarily, for a period of planned austerity within [rich] nations and a rapid transition away from fossil-fuelled development within [poorer] nations" (Anderson and Bows, 2011: 41). A period of "*planned economic contraction in the developed or minority world* is inevitable if we are to avoid dangerous climate change" (Barry, 2012: 141).

However, the political system is the primary party to and the prime counterpart of the economic order in modern societies. For example, it is a near-universal phenomenon that calls for state intervention to follow almost immediately upon the onset of a crisis in the capitalist economy. Moreover, these appeals frequently result in action. Theoretically foregrounding the economic order therefore misses a crucial part of the story. Helmut Willke (2014: 42) highlighted the short-sightedness of an exclusive stress on the economic order in general and in the case of environmental problems in particular:

> The usual failings of capitalism, such as inequality, poverty, exploitation or environmental destruction, which are the standard topics in a rather naive critique, appear in this view primarily as a failure of a policy that formally and in principle would have all the means to avoid, correct or at least reduce these failings to a tolerable level.

The Interventions of the state into the economy occur not only based on economic interests. Jürgen Habermas ([1972] 1976: 35) also stresses the intervention of the administrative machinery of the state in market decisions as constitutive of "late" capitalism:

> The state apparatus carries out numerous imperatives of the economic system. These can be ordered from two perspectives: by means of global planning, it regulates the economic cycle as a whole; and creates and improves conditions for utilizing excess accumulated capital.

Widely accepted, it seems to me, therefore, is the accurate view that 20th and 21st century politics is constructed around the economy, as not only the contentious issues of political campaigns in virtually all countries demonstrate but also everyday political conversations.

I focus on the primacy of the economy, particularly the capitalist economic order, political governance, and governments that are committed in principle to constitutional democracy, the scaffolding of the democratic order. Without the rule of law—ensuring, for instance, the peaceful transfer of power—democracy is flawed. As the sociologist Raymond Aron ([2013] 2023) and other scholars before him, such as Montesquieu and de Tocqueville, noted, the judiciary is the most important branch of modern government regarding individual liberty.

I will develop my argument in several steps: First, I address the increasingly common claim that today's democracies are in exceptional situations. Second, I will reflect on the classic and current social science discourse on the erosion of the foundations of democracy. Third, I describe the growing sense of an uncomfortable democracy among climate scientists, other scientists, non-governmental organizations, and the media. Climate scientists propose to overcome the inability of modern democracies to address the catastrophic consequences of climate change by abolishing democracy. The alternative, of course, is to strengthen democracy. In what follows, I discuss the proposed change in the role of climate scientists as policy makers. In the final section, I will examine the serious flaws in the claim that contemporary society is an "inconvenient democracy" in the face of the greatest challenge humanity has faced in history.

The emergence of "exceptional circumstances"

> As never before, the continuity from past to future is broken in our time.
>
> *Niklas Luhmann ([1992] 1998: 67)*

> Democracy, however, desirable, is not an effective device for solving global or transnational problems.
>
> *Eric Hobsbawn (2007: 118)*

Perhaps the most prominent assertion associated with the claim that the climate crisis will create exceptional circumstances is the view that it erases the distinction between democratic and authoritarian political governance. Both democracies and authoritarian regimes face the same pressures to implement drastic environmental policies, potentially blurring the lines between them, as they focus on survival. The emphasis always stresses the benefits of authoritarian governance. The climate crisis erases the distinction between political systems, as both face the same existential threat and must prioritize survival and environmental sustainability. This could lead to convergence in policy approaches and governance styles, especially as the pressure for decisive action increases. As the crisis intensifies, the capacity of a system to deliver results may become more important than how representative or free it is.

Niklas Luhmann's suggestion of a radical break with the past is also the premise of Eric Hobsbawm's contention that the form of democracy as it has emerged in recent centuries is ill-suited to enable humanity to adapt to a radically different future present. The narrative about exceptional circumstances is part of what Ulrich Beck and his colleague Joost van Loon (2011: 114) called "the socially constructed anticipation of natural catastrophes" *in the present*. Joachim Radkau (2017: 372) offers the following account of moving the present of the future into the present:

> If we assume that most people primarily live in the present – and especially in the alternative scenes of the seventies, living fully in the here and now became the new message of happiness! – then it is easy to explain the tendency of many environmental alarmists to stylize future, chronic or partially hypothetical risks – whether radioactive emissions, *Waldsterben*, climate change or overpopulation – into acute catastrophes.

Risk is a matter of anticipating the future (Beck, [1986] 1992). While the dominant climate change discourse points to the future, from a social science perspective, climate change is already present.

The widespread strength but also peculiar assertion about the greater efficacy of authoritarian governance must be noted. The assertion is curious because authoritarian governance may present some appeal in terms of centralized, swift decision-making; it is peculiar to claim that it is inherently more effective in addressing the climate crisis because of the complex and interdependent nature of the problem, which often demands broader participation, innovation, and justice than authoritarian systems typically allow. The assertion is also peculiar in that it relies on the strength of the strong state, on the ability to predict future conditions of action by extrapolating from the present to the future present, and by analogy from past exceptional conditions to future exceptional circumstances.

More specifically, premature doubts about the efficacy of modern democratic nations in responding to pressing environmental limits and related concerns can in part be traced back at least half a century (Paehkle, 2005: 25–43). Concerns about future energy options, the availability of adequate resources, increasing pollution, significant food shortages, a lack of capacity to adapt to multiple emergencies, and/or exponential population growth underscore a growing skepticism about the ability of democracies to address these challenges simultaneously. A future present in which multiple emergencies (Lazar, 2009) must be confronted is compounded by pre-existing stresses such as inequality, poverty, geopolitical conflicts, and resource scarcity. The work of the "eco-authoritarians" of the 1970s is an example of such a pessimistic perspective, which claimed that it would be difficult to restrain economic goals and population growth in democracies.

These environmental concerns could very well spell the end of a moderate, pacified, and tolerant world. The economist Robert Heilbroner (1974: 161) was one of the first notable warners:

> [...] given these mighty pressures and constraints we must think of alternatives the present order in terms of a social systems that will offer a necessary degree of regimentation as well as a different set of motives and objectives. I must confess I can picture only one such system. This is a social order that will blend a 'religious' orientation and a 'military' discipline. Such a monastic organization of society may be repugnant to us, but I suspect it offers the greatest promise for bringing about the profound and painful adaptations that the coming generations must make.

Like many environmentalists in the 1970s, Heilbroner anticipated a coming era of environmental limits and resource depletion, leading, for example, to what philosopher John Passmore (1974: 193–194) saw as the gradual emergence of a "bureaucratic police state." Democratic rules of the game would have to be suspended to achieve a sustainable future present.

Three decades later, Robert Heilbroner ([1974] 2013: 154] revisited his 1974 *Inquiry into the Human Prospect* (written between July 1972 and August 1973) and changed the paragraph with its prognosis from 1974 which gave rise to most of the critical response to his inquiry:

> Given this mighty task, we must think of alternatives to the present order in terms of a system that will offer a necessary degree of social order as well as a different set of motives and objectives. The order that comes to my mind as most likely to satisfy these requirements is one that blends a "religious" orientation with a "military" discipline. Such a monastic organization of society may be repugnant to us, but I suspect it offers the greatest promise of making those enormous transformations needed to reach a new stable socio-economic basis.

However, William Becker (2023: 155) sums up the downside of the doubts about democratic government in times of a massive societal crisis:

> Undemocratic governance may seem more efficient at managing crises than unruly democracies. Perhaps it can be, but if and only if the decision-makers have adequate knowledge of the details of all relevant issues, the

> means to ensure decisions are implemented as intended, and true concern for the public welfare over private interest—standards many autocrats are unlikely to meet [...] Undemocratic systems lack crucial mechanisms of resilience, which will be key for navigating the long emergency.

In retrospect, the early doubts about democracy in the 1970s, fueled by the emergence of the environmentalism of ecological movements, did not debate the impacts of climate change and underestimated the capacity of democratic institutions and civil society to mobilize against environmental pressures. The downward spiral could perhaps still be prevented because the response to the environmental exigencies extended to the democratic mobilization of a strengthening but moderate environmental movement (Nelkin and Pollack, 1977; Schnaiberg, 1980).

Much more recently, addressing the consequences of climate change has involved periods of time and likely societal changes that are clearly beyond the usual human imagination and current political institutions. Apart from individual historical events such as wars, revolutions, economic collapses, or national liberation struggles, there are no human experiences in historical time that climate science could refer to when it begins to think about a "future present" in which massive climate change impacts have begun. This affects all levels of society and its relationships with the outside world, e.g. the way the world produces and uses energy, the virtues of the nation-state, migration patterns, the global economy, and civil societies. In such contexts, crisis conditions favor the creation of emergency powers, the delegitimization of the existing political order, the abolition of freedom and justice, and the introduction of a revolutionary form of government. The past is not an infallible guide to the future. However, it is often the only guide we have.

The justifications invoked are exceptional circumstances, a war-like situation (Lovelock, 2009; McKibben, 2016), a period of major crisis and upheaval or exceptional necessity that requires the suspension of freedoms and the political rise of climate scientists to achieve the necessary transformations. In 2009, the German climate researcher John Schellnhuber in the run up to the Copenhagen climate conference in December 2009, warned that a war-like economy was likely to exist by 2020. That is, if aggressive climate protection (mitigation) is postponed until 2020, emissions would then have decreased by 9% annually, but this could only be achieved "within the framework of a war economy," warns Schellnhuber, i.e. massive state intervention in the economy. Schellnhuber added, this is by no means "an extreme opinion"; it has been "tested a thousand times over."

The case at hand also illustrates that climate science knowledge or climate scientists who move climate science knowledge into the public domain make the clear-cut distinction between scientific knowledge and political knowledge a dubious division. Given the purpose of the discourse, the forms of knowledge are conflated or deliberately mixed. In contrast, Lisa Herzog (2024: 118–119) maintains that

> the topic of climate change, [...] is scientific, not political [... it is] a confusion of different fields of discourse: while 'always showing both sides' is the appropriate strategy for questions of values and interests, for scientific issues that have reached a level of certainty as high as that concerning anthropogenic climate change, it is simply a mistake.

Of course, we are still some distance from such a war-like economy in 2025, and we have learned much more about climate change since 2009.

As the French political scientist Pierre Rosanvallon ([2011] 2013: 184) points out, the centralized nation-state is seen as the only source of security in the face of radical risk. It is the hope that invoking exceptional circumstances, i.e. a threat to the existence of civilization, if not humanity, "might alone be able to restore capacity and [...] energy to a failing or disabled [political] will." Frank Fischer (2017: 54) adds to this by criticizing that "the current political-economic efforts of contemporary democratic systems to deal with problems such as global warming [...] are little more than limited symbolic gestures, especially in the face of time pressures." The problem of global warming and its consequences is not only a question of current democratic governance, but also the lack of commitment of citizens to change their ambitions and behavior. Above all, it requires a future perspective (Lovelock, 2009). The future perspective imposes its own norms on the present (cf. Jonas, [1979] 1984: 143). Hans Jonas believes with William Ophuls (1977: 159) that at times of a major crisis ("the coming severity of a politics"), "democracy must give way to elite rule" and that therefore, democracy "is at least temporarily unsuited, and our present comparative weighing is, reluctantly, between different forms of 'tyranny'" (Jonas, [1979] 1984: 151).

However, how can nation-states be governed well under exceptional circumstances?[11] This question comes up against two opposing forces: that of an *uncomfortable mind* and that of *uncomfortable social institutions*. The former refers to a public that is assumed to be "present-centered" (Skidelsky and Skidelsky, 2012: 130), i.e. satisfied with the status quo, and which justifies imposing its own (superior) ideas on the citizens of future generations (because we truly need to care whether the future public cares). The reference to the inconvenient mind is a play on words that uses the more familiar metaphor of "an inconvenient truth" as an anchor. A simple example of an inconvenient opinion in the case of climate change is the claim that the science of climate change is far too complicated for the average person to understand. A less "neutral" version of the inconvenient opinion would be the claim that the public is intellectually incapable of grasping the idea of global warming and its consequences.

In other words, overcoming exceptional circumstances legitimizes the temporary suspension of freedoms (Hayek, 1944: 108). The reference to unconformable social institutions implies a strong state (that is, harmlessly, in the eyes of some observers, "having an effective government and an efficient government") in the form of a command society. In other words, good governance of society based on citizen participation must be subordinated to overcoming the exceptional situation by almost any means. The invocation of a strong state is often accompanied and justified by a rhetoric of freedom and democracy designed to conceal authoritarian political decision-making. Contemporary regimes in countries such as Hungary, Turkey, India, and earlier Poland are examples of societies in which exceptional circumstances are invoked and in which nationalism blended with anti-immigrant rhetoric as a political strategy is often paired with anti-environmentalism, for example, the dismantling of environmental regulation and climate skepticism (cf. Darian-Smith, 2022). Isolationism, by definition, has a hard time with global issues such as a warming planet.

The temporary suspension of liberties (Hayek, 1944: 189): but is any massive appropriation of power by the state and its representatives reversible in the long run? Are the possible consequences of climate change, in particular *cascading* climatic disasters as the result of a trigger of multiple tipping points[12] that could undermine democratic institutions, perhaps already weakened by multiple, converging crises, such as mass migration, a pandemic, global food shortages, recession, inequality, and political polarization, making authoritarian rule likely?

Such a process is captured by the concept of "polycrisis" introduced by French philosopher Edgar Morin (1993) and co-author Anne-Brigitte Kern in their book *Terre-Patrie*. Morin and Kern describe a polycrisis as a situation in which multiple, interrelated crises and risks converge and reinforce each other, resulting in a complex predicament that is difficult to manage or resolve, creating a whole that is more dangerous than the sum of its parts. This concept has gained renewed attention in the 21st century, particularly in the context of global challenges such as climate change, pandemics, and political instability. In recent years, especially since 2022, the term has gained renewed prominence through the use of scholars such as Adam Tooze,[13] who has used it to describe the interconnected global crises of climate change, pandemics, war, and economic instability (cf. Liu and Renn, 2025).[14] David Held (2022) discusses how global systems have entered a state of "gridlock," in which overlapping crises-such as financial instability, migration, and inequality-undermine democratic cooperation and problem-solving capacity. He warns that this gridlock fosters populism and authoritarianism, drawing parallels to the political climate of the 1930s.[15] Proponents of the polycrisis framework for the current political, economic, and cultural predicaments do not appear to consider the opposite, the possibility of a "polyresolution," which implies that multiple problems can be solved together or synergistically.

More elementary, how can the onset of a state of emergency be determined? The shortcomings of democratic governments are manifold and go far beyond the problem of climate change and its social consequences; however, does this justify such a dismissive conclusion as the diagnosis of an inconvenient democracy? After all, authoritarian and totalitarian regimes have no record of environmental success; nations that have followed the path of "authoritarian modernization/environmentalism," such as China or Russia, have no better record. In post-communist transition nations, climate change policies do not seem to differ from those of the rest of the world (Steves and Teytelboym, 2023). Nonetheless, disillusionment with democracies has persisted and may grow as established climate policies fail to be delivered.

One should therefore not assume that democracies are automatically beneficial for climate policy. Of course, to the extent that domestic politics is primarily a response to and implementation of public concerns and priorities, it is possible that in democratic countries, where public knowledge about climate change is limited[16] and where concern about the impacts of climate change is absent, climate policies will be anything but ambitious. The public may not support strong climate policies because of overriding economic interests associated with a strong fossil fuel industry lobby,[17] cultural preferences (e.g. against limiting mobility), a lack of administrative capacity, or the absence of international treaty obligations.

An inconvenient democracy: the authoritarian turn

> Sadly, it is not entirely clear that democracy is up to the challenge of climate change.
>
> *Dale Jamieson (2014: 100)*

The assertion of exceptional circumstances and the associated demand to overcome an "uncomfortable democracy" is also fed by several new and not new considerations about the lack of resilience of democracies beyond the threats of climate change and leads to different forms of blame with different addressees. The risks facing democracies are the result of external or internal forces or a combination of both. Helmut Willke (2014: 135–145), for example, attributes the risks faced by modern democracies primarily to the forces of globalization and the transition from the industrial to the knowledge society. Dale Jamisson (2014: 101) is convinced that global conditions are changing in ways that threaten the development and sustainability of human life; the existing system of governance is not "up to the task of managing such threats." This calls "for more rigidity and less responsiveness in political institutions, perhaps through such mechanisms as constitutional restrictions."

The erosion of democracy: the classic perspective

In the classical social science literature, the threat to democracy posed by unequal access to and unequal distribution of knowledge in society, e.g. through the emergence of social inequality in society (see Stehr and Machin, 2016a), has, in the eyes of many observers, radically displaced the earlier optimistic enlightenment views on resilience and even the possibility of a democracy on the basis of a general circulation of knowledge in society. Numerous authors, from Max Weber to Robert Michels, have pointed to this and other threats to representative democracy.

Max Weber, for example, feared a kind of pacifism of the bourgeoisie in the face of the unstoppable advance of bureaucracy in modern societies, because how can it be guaranteed in light of the

> growing indispensability and hence increasing power of state officialdom [...] that forces exists which can impose limits on the enormous, crushing power of this constantly growing stratum of society and control it? How is democracy even in this restricted sense *to be possible at all*?
>
> *(Weber, [1918] 1994: 159)*

In his classic study of undemocratic tendencies in the Social Democratic Party, a political organization that strives and fights for democratic goals, Robert Michels ([1915] 1949) points to a quasi "natural" state of incompetence and immaturity of the mass of people in modern democracies: since ordinary people are incapable of "[...] looking after their own interests, it is necessary that they should have experts to look after their affairs" (Michels, [1915] 1949: 93). The grassroots community is rarely willing to break the authority of expert leaders and release them from control. Whether the disillusioned conclusion drawn by Robert Michels ([1915]: 1949: 95) in the face of the tendencies he observed, namely, that "social democracy is not a democracy, but a party struggling to attain democracy. In other words, democracy is the end, but not the means," is inevitable, i.e. universally valid as a kind of iron law, can of course be disputed, although many observers are willing to concede that Michels has discovered one of the few law-like relationships in social science.[18] Many of the classic concerns about the viability of democratic governance are echoed in contemporary reflections on the fragility of democracy.

The erosion of democracy: the modern perspective

I concur with Ralf Dahrendorf's (1997: 139) observation and assessment of recently rising attitudes toward democratic governance, that is, the "international debate about the governability of democracies has brought up, in the last few years, a rather disturbing new demand for authority, even for authoritarian rule." Aside from the last few years, novel threats to democracy have been discovered as a result of the development of artificial intelligence; Yuval Noah Harari (2024) is a prominent observer who anticipates such a threat from AI:

> The rise of unfathomable alien intelligence poses a threat to all humans and poses a particular threat to democracy. If more decisions about people's lives are made in a black box, so voters cannot understand and challenge them, democracy ceases to function. What happens when crucial decisions not just about individual lives but even about collective matters such as the Federal Reserve's interest rate are made by unfathomable algorithms?

The discourse on the erosion of modern democratic governance relates to a number of interlocking factors: for example, deep-rooted pessimism about the psychological nature of human beings; the time-bound nature of human thought; to the inability to mobilize individuals around effective climate policy; the inability of the government to pursue long-term goals in the face of constitutional constraints; to the fragility of the political order[19]; the influence of vested interests on the political agenda; and, in the case of anthropogenic climate change, the dependence on fossil fuels as its material foundation and their continued promotion by the government, business and consumers[20] and, last but not least, the inability of climate science itself to ensure that its message does not fall on deaf ears.

However, climate change communication has been, and perhaps still is in some contexts, based on the belief that once the public knows the facts about climate change and begins to understand the seriousness of the problem, they will raise their voices and demand that our governments and corporations act.[21] As Ulrich Beck (2010a: 175) aptly notes, "climate researchers are great realists, but socially and politically idealists, because they see all people as little climate researchers and therefore cannot understand why their apocalyptic model calculations do not immediately trigger the urgently needed countermeasures."

The reality of climate communication today is increasingly based on the idea that vigorous communication of the "objective" findings of the climate science community is not enough. Among the key conclusions from the perspective of communicating climate science knowledge are (a) simply providing more scientific information does not guarantee public engagement or policy action; (b) messages must be aligned with audience values (e.g. health, safety, justice); (c) emotionally charged social contexts are significant and fear-based messaging can be paralyzing; hope and agency are critical; and (d) co-producing knowledge with communities improves relevance and uptake (cf. Funtowicz and Ravetz, 1993; Hulme, [2009]: 214; Kahan et al., 2011).

Secular trends

> If the 20th century was the story of slow, uneven progress toward the victory of liberal democracy over other ideologies—communism, fascism, virulent nationalism—the 21st century is, so far, a story of the reverse.
>
> *Anne Applebaum, November 15, 2021*

As I have noted, referring to democratic backsliding, the erosion of democratic governance in the first quarter of the 21st century is characterized by often successful efforts to undermine liberal and traditional democratic values. A political process is now broadly recognized that is fueled by ambitious politicians engaging in democratic backsliding to protect their own power, claiming at times that they are busy protecting democracy. This process involves a decline in the quality of democracy, often leading to a more authoritarian form of government. Today's processes

> of democratic backsliding differ in a fundamental way from old assaults on democracy. According to many experts, contemporary democracies are destroyed in gradual and often imperceptible ways, whereas in the past they were victims of coup d'états, wars, and revolutions. Moreover, though many elements of democratic regime are technically left in place, they are manipulated and distorted beyond recognition.
>
> *(Ekiert, 2023: 4)*

Backsliding democracies generally disregard democratic, constitutional norms, often display greater tolerance for violence and hate speech, intolerance toward minorities, hostility toward the media and the press, strong opposition to migration, and, in foreign relations, favoritism toward autocratic political regimes. There is public support for backsliding. To what extent public support is elite induced, employing the power of the new media, for instance, remains a distinctive possibility but is not well researched.

Philipp Lorenz-Spreen and colleagues (2023) published a meta-analysis entitled "Reviewing worldwide causal and correlational evidence on digital media and democracy." This study synthesized 496 studies from 2000 to 2022 to assess the impact of digital media on democracy. The meta-analysis found both democracy-enhancing and democracy-undermining effects of digital media, such as positive associations in emerging democracies. In autocracies and emerging democracies, digital media use was associated with increased political participation and increased information consumption. These effects are considered beneficial for democratic processes, as they can empower citizens and promote civic engagement. Negative associations in established democracies: Conversely, in established democracies, digital media use has been associated with declining political trust, rising populism, and increasing polarization. These trends are seen as detrimental to democratic stability and cohesion. A possible increase in political polarization: The review found that high levels of digital media use were often associated with increased political polarization. This effect was particularly evident in established democracies, where social media platforms contributed to greater divisions between political groups: Increased digital media use correlated with higher levels of populism, as evidenced by increased vote shares for populist parties and the prevalence of populist messages online.

The backsliding has both a supply side, such as the "dashed expectations and defects of representative democracy" (Alonso et al., 2011: 15), and a demand side, for example, curtailing migration. Backsliding is further associated with growing disparities that fuel social unrest and disillusionment with democratic processes, extreme political divisions that hinder compromise and effective governance, government interference with or manipulation of independent media outlets, and attempts to undermine the independence of the judiciary and foreign interventions or disinformation campaigns aimed at destabilizing democratic systems. All the attributes of backsliding listed are impacted by unique contextual societal conditions (Grzegorz, 2023: 8).

With the growing digitalization of the affairs of society as the result of the proliferation of digital technology, a new body of "legal" norms has evolved: the law and the empire of the *code* (Lessig, 2006). The constraining changes of the code are less first and foremost a matter of the changing collective consciousness in modern society, but more the unintended consequence of the massive deployment of new technologies governed by major technology companies. As Daniel Susskind (2024: 105) emphasized,

> what makes this new political power so pernicious is not just its scale—extraordinary and growing though it may be—but the fact that it is *unaccountable* [...] That is not the case with the formal law. In mature political societies, those rules are composed and scrutinized by a legitimate political process; they are interpreted and enforced by an established justice system.

The code lessens liberties, opens new social justice questions, and thereby undermines democracy.

The degree to which environmental issues play a role in the process of democratic or autocratic backsliding is still a largely black box. Environmental problems, present or future present, do not appear to play a notable role in the discussion of the causes and conditions for the erosion of democracy and the dictatorial drift in autocratic regimes. In contrast, it often is in the eyes of many observers the "capitalist economy [that] bends politics to its shape" (Purdy, 2022: 72). The colonization of politics by economic exigencies may well be a matter of the past, soon.

However, as Burnell noted, environmental degradation and climate change can lead to resource scarcity, potentially fueling social unrest and conflict. Scarce resources such as water have the potential to disrupt governance. Conflicts may arise that are difficult to contain, and economic advancement will be hampered. Under these circumstances, "the case for tough action by the authorities to maintain or impose order can look very compelling, increasing the chances that more authoritarian and illiberal measures will be introduced where possible" (Burnell, 2012: 818).

As observed from the *theoretical* vantage point of systems theory, there are a few relevant secular trends associated with the dynamics of modern society that pose risks to democracy and that can be seen as undermining democratic governance, as well as conditions for the possibility of climate policies that transcend the abstract boundaries of social subsystems. Both observations are the result of systemic confusion. Helmut Willke (2014: 140–143) summarized these observations as follows,

1. the risk of an expansion of economic rationality, an economization of all social areas, which thereby lose their specific logic and rationality and instead submit to supposed economic constraints. This represents a confusion in the sense of a merging, an incorporation of different logics. This becomes visible, for example, in the economization of the health system and the economization of the "university system." This risk also appears in more covert forms – for example, in a rather imperial gesture of economics as a science that believes it can be the leading science for all social and human sciences.
2. [The] second risk does not result from the foreignization of society by a single logic, but from the rigorous separation of the various functional areas.[…] The major thematic complexes of social life, such as governing, managing, healing, believing, learning, researching, loving, creating (art) and even traveling (tourism) or sports (sports) develop into specialized topics, specialized semantics and finally specialized languages in long social processes, which differ more and more from other semantics systems.

The outcome of both types of societal development is the restrictive capacity to respond to *societal* issues such as climate change. Everything that falls within the logic of the political system comes to the attention of political

> decision-making and can be addressed by the political system. From a systems theory perspective, democracy in a functionally differentiated society is therefore faced with a problem that is almost impossible to solve: 'Systemic thinking assumes that political control as an external intervention in self-referential social systems, such as the economy, health or education, is in principle impossible and must fail as an attempt at direct control because incompatible and mutually incomprehensible modes of operation and logics come into conflict.'
>
> *(Willke, 2014: 143)*
>
> For Niklas Luhmann (e.g., [1997] 2012: 54–55), *structural coupling* is the answer. Structural coupling, following Humberto Maturana, refers to how different subsystems of society (e.g., the legal system, economic system, and the political system) can interact without becoming indistinguishable from one another. Structural coupling here allows a system to react to changes in another system while preserving its own mode of operation. Helmut Willke (2014: 145) advocates a theoretical extension of the systems perspective beyond the idea of a 'structural, refer to themselves in a self-referential way, seal themselves off through specific meanings and finally condense into operationally closed communication coupling' of subsystems that allows for the *control* of the affairs of other societal systems by the political system: 'the systems theory perspective has to be broadened by taking the functional systems' logics and operational unity seriously, but not excluding political control – through context control.'

The empirical reality shows that the political system does act on environmental concerns and does so by transcending system boundaries. In fact, climate change policies must treat the changes imposed by a warming climate on society as a *total social fact*. The focus on functional differentiation, especially the idea of self-referential subsystems, means that subsystems interact with their environment based on their internal logic rather than direct external pressures; hence, in the case of environmental issues, they are treated as external "irritations" rather than central system-specific concerns. Particularly in his studies on *ecological communication* and *risk*, Niklas Luhmann ([1986] 1989; 1991) emphasized how a functionally differentiated society can have no unitary approach addressing serious ecological risks such as climate change. For example, **scientific warnings** do not automatically trigger political decisions—unless those findings are translated into politically resonant narratives. In other words, Niklas Luhmann would probably view today's climate debate as a communication problem in which different systems have difficulties in adequately processing the complexity of the problem and reacting to each other. He would probably be skeptical of simple solutions and would point to the self-referential processes and structurally complex nature of the problem.[22] In contrast, the differentiation into functional systems indicates that ecological issues can be addressed only within the context of each distinct functional system. Social systems define and defend their own boundaries.

The systems-theoretical perspective not only neglects other forms of social differentiation, such as class, race, gender, and exposure to environmental risks, but also ignores the need for system-indifferent policies. The systems-theoretical perspective also downplays issues of power, inequality, and conflict. However, Helmut Willke's perspective highlights secular trends that impose a system-internal logic on policy, such as the logic of economic reasoning that has a system-indifferent goal, in this case, coping with a warming future present. Willke therefore argues in favor of an extension of the systems-theoretical perspective that allows for the possibility of "political control/management" that transcends subsystems (Willke, 2014: 143). Notably, the "founder" of sociology, August Comte, wrestled with the difference between the dependence and independence of social subsystems more than two centuries ago. Comte criticized the theoretical efforts of his time to isolate the economy from the rest of society and thus advocated the need to integrate the field of economics into sociology:

> Comte's critique of political economy derives from this thought, arguing that political economy isolates its object of study from the rest of social life, aping the methods of the inorganic sciences: by isolating the economic domain from the remainder of social life, the economist is well on the way to forgetting that human action is not only economic, rational and conscious but also contains moral and affective dimensions.
>
> *(Steiner, [2005] 2011: 170)*

On the other hand, the paralysis of politics in the face of climate change is also attributed to the prevailing perspectives of the public.

Blaming citizens

> No matter how spurious a controversy [of the debate about climate science], it remains that for a larger part of the population, there *is* there *is* a controversy, the effect of which we may witness everyday through the total inertia of governments as well as of civil societies that are supposed to exert a pressure on their elected proxies.
>
> *Bruno Latour (2015: 146)*

The blame attributed to the public by social scientists is widespread, as Bruno Latour's observation suggests. The mere idea that there is an argument about the reality of anthropogenic climate change is enough to keep the debate going, at least among some segments of the public. The position articulated by Daniel Kahneman is more substantial and more difficult to address politically. The economist Daniel Kahneman sums up the growing skepticism about the motivation of citizens to engage in behavior that benefits mitigation and/or adaptation to the forces of climate change when he says:

> The bottom line is that I am extremely skeptical that we can tackle climate change. To mobilize people, this has to be an *emotional issue*. It has to be immediate and urgent. A distant, abstract and controversial threat simply does not have the necessary qualities to seriously mobilize public opinion.
> *(quoted in Marshall, 2014; 57; emphasis added)*

It appears that most citizens are reluctant to endorse and adhere to policy options based on scientific evidence, not only because these options may interfere with cherished individual (constitutional) rights but also because an uncritical public has been manipulated and betrayed by vested interests (e.g. Adorno and Horkheimer, [1947] 1977: 120).

A limited number of official occasions would seem to support the thesis that the public is reluctant to support measures designed to combat climate change, even on a small scale. Democratic publics have on multiple occasions voted via referendum against even moderate carbon taxes, at least in the United States and in France (against increased fuel taxes). It is not surprising to see such voting behavior; tax increases are hardly popular anywhere in the world, and given an opportunity to express these sentiments, the voting citizens will rarely miss the chance to act on their opinions. The primacy given to individual autonomy in democratic countries is seen as a burden when it comes to government interventions that restrict individual conduct, for example, in the field of emissions and fuel-efficient standards or diets. Considering these and similar restrictions faced by democratic governments, Ross Mittiga (2022: 2) states that

> liberal-democratic governments (and theorists) must confront the bleak possibility that responding to the existential threat of climate change at this late stage may (soon) require relaxing or suspending adherence to some of the most widely shared CL [contingent legitimacy] standards and embracing authoritarian power.

Mittiga defines contingency legitimation standards as government conduct that involves exercising power in ways acceptable to its citizens.

Kahneman's thesis suggests that the numerous freedoms afforded to citizens in democratic societies, both emotionally and practically, can impede their ability to make responsible decisions. Most citizens are, by nature, inclined to act irrationally (cf. Schumpeter, 1942: 262–263). The climate scientist

Hans-Joachim Schellnhuber[23] gloomily described why communication on climate change is not reaching civil society: "My own experience and everyday knowledge show that convenience and ignorance are the greatest faults of the human character. This is a potentially lethal mix."

More persuasive than Schellnhuber's sober perspective is the view that the public is simply uninformed about the magnitude of the threat to future generations and reluctant to take such a future risk to their everyday life on board, but "while true, this is neither decisive nor inevitable. The lack of correct information," Charles Perrow (2010: 60) argues, "is at least partly the result of the efforts of large polluting organisations [oil and coal corporations]. It is hard to make distant threats salient and we can suffer from 'fear overload' […], but powerful organisations can exploit these cognitive limits." In addition, powerful interests can easily exploit these cognitive limits. As Elke Weber (2006: 104; emphasis added) notes, the "observation that the feelings of members of the general public about such risks as global warming or nuclear power do not coincide with objective outcome information and the risk assessments provided by scientists and engineers" is not unusual. Incessant repetitions of the threat major risks and dangers in a future present may prompt the opposite from the indented, well-meaning reaction: Emotions "like fear or worry motivate us *to remove ourselves* from a dangerous situation or to change the environment in ways that reduce our feeling of being at risk." Hence, the conclusion of what amounts to a psychological reduction in excessive anxiety triggered by threatening communication. Moreover, as Cass Sunstein (2017: 128) observes, based on psychological experiments,

> in the case of information about ourselves (about how attractive others perceive us to be, or how likely we are to get sick or to succeed), people normally alter their beliefs more in response to good news than in response to bad news.
>
> *(also Davidson, 2025)*[24]

Confirming Kahneman's assertion for an emotionally charged context necessary to provide support for the persuasion of long-term risks, "behavioral decision research over the past 30 years strongly suggests that attention-catching and emotionally engaging informational interventions may be required to engender the public concern necessary for individual or collective action in response to global warming" (Weber, 2006: 116).

However, to see the fate of democracy and politics as the outcome of the competence and emotions of individual citizens would be to argue for microsociology without macrosociology. The reference to public perceptions of science and expert knowledge goes beyond the implicit or explicit assumption that the public is fundamentally information and knowledge deficient, perhaps even reactionary, and tends to react to complexity with trepidation (cf. Gauchat and Andrews, 2018). Joseph Schumpeter ([1942] 1950: 145) noted that the "mass of people never develops definite opinions on its own initiative." Such an observation about the lack of independently acquired knowledge of the public is therefore an observation with a long history for modern societies and is asserted frequently by prominent social scientists. I have already mentioned Robert Michels' "Iron Law" of the oligarchy, which was formulated at the beginning of the last century: Friedrich Hayek ([1949] 1960: 372), for example, notes there

> is little that the ordinary man of today learns about events or ideas except through the medium of this [intellectual] class; and outside our special fields of work we are in this respect almost all ordinary men, dependent for our information and instruction on those who make it their job to keep abreast of opinion. It is the intellectuals in this sense who decide what views and opinions are to reach us, which facts are important enough to be told to us and in what form and from what angle they are to be presented: Whether we shall ever learn of the results of the work of the expert and the original thinker depends mainly on their decision.

The seemingly widespread ability to not know what the future might bring can, of course, also be interpreted as a psychological "incentive" to live with the knowledge of limited knowledge about the outcome of events that lie in the future (cf. Gigerenzer and Garcia-Rettamero, 2017). This position also extends to the pragmatic idea that scientific expertise is not about incontrovertible and indisputable facts. From this perspective, science and politics are both mundane activities subject to mistakes, revision, and doubt. To shift "from a science-*versus*-politics to a science-with-politics" is not without its hazards (cf. Latour, 2015: 149). Moreover, political scientists, who are in many ways concerned with the lack of information/knowledge among voters (Innerarity, 2021), emphasize that the democratic political system functions despite the "ignorance" of citizens (Kuklinski, 1990). In his 1922 book *Public Opinion*, Walter Lippmann ([1922] 1997) famously asserted that our conception of democracy is fundamentally flawed. He argued that we are asking voters for something that is impossible: a fully accurate understanding of the world. In an observation still applicable today, Lippmann ([1922] 1997: 16) asserted that

> the real environment is altogether too big, too complex, and too fleeting for direct acquaintance: We are not equipped to deal with so much subtlety, so much variety, so many permutations and combinations. And although we have to act in that environment, we have to reconstruct it on a simpler model before we can manage with it.

Petersen and Aarøe (2013: 289) have more recently documented that, despite the widespread lack of comprehensive political knowledge, citizens "readily form opinions about what is the best and most efficient policy."

An assessment in favor of the political virtue of informed citizens is advocated by Seymour Martin Lipset and colleagues ([1956] 1962): Lack of information, passivity, and lack of interest by ordinary members in the affairs of an organization are in the interest of the powerful and support their ability to maintain power advantages. It seems that it is not the amount of knowledge or information that citizens possess that influences the relationship between democracy and knowledge but rather the importance of democracy-promoting individuals and collectively shared *value orientations*; or, as Robert Dahl (1977: 1) argues, it is "the way we think about ourselves as a people" that supports the existence and stability of democracy. Of course, value orientations and educational success are linked: "Education presumably broadens people's outlook, enables them to understand the need for norms of tolerance, discourages them from adhering to extremist and monistic doctrines, and increases their ability to make rational choices" Lipset (1959: 79).

But education will not inevitably ensure support for the reality of climate change. Research indicates that individuals with higher education levels are not immune to climate change denial. In fact, they may be more adept at engaging in *motivated reasoning*, using their cognitive skills to defend pre-existing beliefs. In developed countries, the positive effects of education on climate change awareness are attenuated by populist ideology. Highly educated individuals who align with right-wing politics exhibited lower climate change awareness, suggesting that political ideology can override educational influences (Marquart-Pyatt et al., 2014; Wang, 2024).

Blaming the (ruling) political class(es)

In the eyes of many climate scientists, not only citizens, as consumers, for example, but also politicians are not prepared to pursue policies that effectively combat climate change: climate activists, climate scientists, some politicians, and many other observers agree that the recent climate summits in Copenhagen, Cancun, Durban, Warsaw, and Dubai have failed. The summits did not result in a new global agreement to limit greenhouse gas emissions. The 2015 Paris Agreement, widely regarded as a historic achievement, seems to mark at least a general scientific and public consensus that human-induced climate change poses a very serious threat to human civilization and its environment.

However, the agreement is not binding. There are no formal sanctions if a country fails to meet its commitments on mitigation, adaptation, or finance, and there is no guarantee of how far-reaching the Paris Agreement will be.

This issue first came to the forefront on June 1, 2017, and again in 2025 when the United States, under the leadership of President Donald Trump, announced its formal withdrawal from the agreement, rejecting the scientific consensus that greenhouse gas emissions are warming the planet.[25] The Trump administration has consistently launched and proposed a strongly anti-environmental agenda. With his announcement to end participation in the Paris Agreement, Trump fulfilled his campaign promise to "end the war on coal" and his alleged goal of reclaiming the sovereignty of the American people and "putting America first." The political situation in the United States continues to illustrate the high degree of politicization of climate and climate change issues in today's world. The interests of the political class in the lack of aggressive, long-term political action on climate issues may be linked to the constitutional framework in the respective nations and their requirements, but the lack of political action on climate issues may of course also be linked to the material interests of the ruling elites in a society.[26]

Indeed, the nature of the relationship between temporality and democracy justifies doubts about the effectiveness of democratic governance in the face of the longer-term future risks and dangers of climate change. Questions of temporality relate to at least two important aspects that are determined by different but interrelated systemic conditions of democratic governance: On the one hand, democratic governance is determined by the immediacy of frequently changing "events" that often come and go quickly, and on the other by constitutional rules of representation that impose relatively short time horizons. The short time horizon of democratic decision-making is seen as a profound shortcoming of democracy:

> If it is indeed rational for politicians to think about the next election and radically discount any other effect decisions may have far beyond that temporal horizon, democracy is necessarily plagued by chronic short, termism, unable to develop and pursue long-term policies.
>
> *(Biebricher, 2018: 92)*

On the other hand, major societal adaptation measures may impose immediate costs, while the public of the future present may enjoy the benefits. This stark dilemma exists, for example, when economic interests, often linked to the reporting cycle of corporations, come into conflict with government intentions, which may be shaped by a somewhat longer cycle of attention.

As David Held and Fane Hervey (2009: 5) explain in more detail,

> the electoral cycle tends to focus policy debate on short-term political gains and satisfying the median voter. The short duration of electoral cycles ensures that politicians are concerned with their own re-election, which may compromise hard policy decisions that require a great deal of

> political capital. It is extremely difficult for governments to impose large-scale changes on an electorate whose votes they depend on, to tackle a problem whose impact will only be felt by future generations.

The short-term election cycles correspond and respond to short-term interests of the voting public, for example, in the case of economic issues, the inflation or unemployment numbers of the day. On the one hand, "it is extremely difficult for governments to impose large-scale changes on an electorate whose votes they depend on, in order to tackle a problem whose impact will only be felt by future generations" (Held and Hervey, 2011: 90). On the other hand, ecological time frames are measured not only in generations but also in centuries.

Similarly, support for democratic politics is often driven not by immediate issues in the lives of individual citizens but by the collective political and economic imperatives of the nation-state. Public perceptions of the urgency of political issues and the agenda-setting of political parties are dynamic and relative. The attention given to climate change depends very much on how important other political issues are perceived to be at any given time, especially the perception of pressing economic issues. The constitutional framework of democracies allows for the organization of interests and the intervention of interest groups in the political decision-making process. Intervention is rewarded or leads to a stalemate in the provision of public goods, such as effective adaptation to climate change.

Are democracies, social institutions, and citizens, constrained by short-term constitutional frameworks and governed by principles of freedom such as the market and immediate gratification, capable of dealing with harms and risks to society that lie in the future present? How can democracies maintain interest in a future present that lies several decades in the future and thus escape the typical political attention cycle of events in the media (Downs, 1972; McDonald, 2009)? The comfortable way in politics and everyday life is to stick to the present and avoid challenges that lie in an uncertain future. Governance processes are often bound by such perspectives. Governance processes are largely designed and responsive to problems that are relatively straightforward, isolated, immediately urgent, and linear. Desynchronized changes in social systems generate problems for governance in general and democratic governance in particular.

For example, the dilemma of short-term political constraints and the awareness of inter-generational problems, especially environmental issues and conflicts in the future present, is also at least indirectly affected by the acceleration of technological developments and the resulting gaps (Ogburn's cultural lag; Simmel's "objective" and "subjective" culture) from a political perspective. The different timescales may lead to increasing demands and actual practices of displacing decisions to nondemocratic procedures and decision-making bodies. The "dilemma of the time lag" (Chester Bernard) refers to a genuine democratic problem; it refers to the discrepancy between immediate adaptive action and the slow process of obtaining democratic consent or approval. In his 1938 work, *The Functions of the Executive*, Chester Barnard examines the "dilemma of the time lag." In this volume, Barnard analyzes the challenges managers and leaders face in making decisions within complex organizations, particularly in situations where there are time delays between actions and their effects. The "dilemma of the time lag" refers to the issue of delayed effects, whereby the immediate impact of decisions is difficult for executives to discern. This presents a significant challenge for leaders, who are required to make decisions without immediate feedback on the outcomes. This can lead to uncertainty in the decision-making process.

Robert Merton's ([1970] 1976: 82) demanding solution for the problem of the time dilemma is that

> democratically organized groups can cope with it only by having their members come to recognize *in advance* that, remote as they are from the firing line of daily decision, there will be occasions in which decisive action must be taken before it can be fully explored and validated by the membership. This comes as a difficulty for democratic organizations whose members often prefer to pay the price of recurrent maladaptations to avoid having their leadership converted into Caesarism or Bonapartism. However, to earn the right for leases of independent decision, democratic leaders must provide for continuing accountability.

The puzzle is how to embed environmental policies, often with long-term aspirations, within the framework of constitutional democracies without creating repercussions that threaten the democratic infrastructure itself. While fossil fuels may have contributed to the creation and maintenance of democracy, as Timothy Mitchell (2010: 231) argues, will the abolition of fossil fuel dependence save democracy in a future present without fossil fuels? Energy determinism aside, what kind of political order will follow the age of fossil fuels? Timothy Mitchell (2010: 252–253) explains,

> forms of democracy that emerged in leading industrialised countries by the middle decades of the twentieth century were enabled and shaped by the extraordinary concentrations of energy obtained from the world's limited stores of hydrocarbons, and by the socio-technical arrangements required for extracting and distributing the energy they contained [...] Democratic politics developed, thanks to oil, with a peculiar orientation towards the future: the future was a limitless horizon of growth.

Observers from Marxist perspectives tend to blame the ruling class (which may be identical with the ruling political class) for defending the capitalist economic order for the broad inaction on the climate front. Andreas Malm (2020: 14), for example, argues that the dominant classes who are responsible for global warming either throw up their hands in resignation or work explicitly against decarbonization: the ruling classes

> are not perturbed by the smell from the blazing trees. They do not worry at the sight of islands sinking; they do not run from the roar of the approaching hurricanes; their fingers never need to touch the stalks from withered harvests; their mouths do not become sticky and dry after a day with nothing to drink [...] The commitment to the endless accumulation of capital wins out every time. After the past three decades, there can be no doubt that the ruling classes are constitutionally incapable of responding to the catastrophe in any other way than by expediting it; of their own accord, under their inner compulsion, they can do nothing but burn their way to the end.

However, what alternative governance model can address climate emergencies? The answer according to Malm is war communism. War communism, as practiced in Russia between 1918 and 1921, "would entail a more thoroughgoing breakdown of private property. It would bury forms of capital for good. It would be something more akin to war communism" (Malm, 2020: 38). Andreas Malm (2020: 178) offers the following more detailed description of the state of war communism:

> The future, then, is ecological war communism, in a figurative sense, this being 'only an analogy – but an analogy very rich in content'. It means learning to live without fossil fuels in no time, breaking the resistance of dominant classes, transforming the economy for the duration, refusing to give up even if all the worst-case scenarios come true, rising out of the ruins with the force and the compromises required, organising the transitional period of restoration, staying with the dilemma. It does not mean cosplay re-enactments of the Russian Civil War.

Malm thus aligns his view from the far left with those who, for other reasons, have serious doubts about democracy's ability to meet the challenge of the risks of global warming.

There is a parallel discourse on democracies in the social sciences, to which I now turn, that expresses strong doubts about the "sustainability" of modern democracies.[27] This is a crisis triggered not only by major environmental problems but also by various structural and secular challenges to contemporary democratic governance. The implication is that democracies are inefficient and incapable of meeting the growing challenges of climate change and its attendant problems; hence, a more coercive form of governance may be needed.

Altogether, the skeptical sentiments about democracy have always, so it seems, accompanied the evolution of democracies. David Runciman (2011: 536) sums up these skeptical sentiments well: "No matter how successful in practice and over time, democracies have always been full of people worried that everything is about to go wrong, that the system is in crisis and its rivals are waiting to pounce." Is it possible that *intellectual* anxieties have merely served as a backdrop for a more significant and real recent transition? Maybe democracies have exhausted their string of luck? Are democracies becoming complacent and literally sleepwalking into disaster?

A position quite sympathetic to such a pessimistic assessment of the fate of democratic governance, gaining followers from time to time, in the face of the climate crisis, can be found in the political theory of Carl Schmitt. But a vigorous response to the alleged deficits of democracy is a critique of the very sympathies Schmitt developed for authoritarian solutions to major political problems.

Carl Schmitt's affinity toward climate authoritarianism

Carl Schmitt, the controversial German political theorist and jurist known for his work on sovereignty, the state of exception, and his critique of liberalism, would likely have a distinct view of climate change that is consistent with his broader political theory, i.e. his affinity for authoritarianism; it sacrifices justice in the name of urgency. From a Schmittian perspective, climate change could be analyzed through several key concepts: Schmitt would likely see climate change as an emergency that reveals the inadequacy of liberal democratic institutions. He might argue that the slow-moving, consensus-based approach of liberal democracy cannot address an existential threat that requires decisive action. He might see climate governance mechanisms—such as climate treaties, cap-and-trade systems, or sustainability metrics—as typical liberal attempts to manage rather than confront conflict. In his view, this avoids the hard political question of power and choice. He would likely question whether climate policy can truly transcend national interests. For Schmitt, the sovereign is "the one who decides on the exception." Climate change may be the ultimate test of sovereignty – who has the authority to declare an ecological emergency, and what extraordinary measures are justified in response? Finally, Schmitt would likely be skeptical of technocratic approaches to climate governance and might instead see climate change as a fundamentally political problem that requires political choices, not just technical solutions or market mechanisms. Carl Schmitt would likely see climate change as a catalyst for conflict and authoritarianism, rejecting cosmopolitan solutions in favor of survivalist sovereignty.[28]

Carl Schmitt's rigid state-centered view of politics is ill-suited to problems that require planetary cooperation and distributed forms of governance. Climate change cannot be solved by isolated nations acting as "heroic sovereigns." Effective mitigation requires binding international agreements (e.g. the Paris Accords), technology sharing, and the pooling of intellectual and material resources-the very liberal multilateralism that Schmitt despised. The consequences of carbon emissions and ecosystem collapse do not respect borders. Schmitt underestimates how scientific reasoning, democratic legitimacy, and participation can coexist. Schmitt's obsession with national sovereignty overlooks how economic globalization is one of the primary drivers of the climate crisis.

Is democracy done?

Perhaps the "buoyant" spin that can be attached to the observation that democracies are disappearing is the idea that democracy has successfully put itself out of business. Massive deregulation, privatization of state property, delegation of state activities to private hands, elimination of state subsidies, promotion of the principle of subsidiarity, and self-regulation of functional systems such as education, finance, and scientific research. Many of the developments in advanced democracies follow ideological commitments, such as the priority given to the free market economy, the gains expected from debureaucratization in international competition, the assignment of state functions to international organizations, or the practical difficulties resulting from a declining capacity to govern a complex political world (cf. Willke, 2014: 132, 135). The sum of these developments results in a loss of authority and power exercised by democratic political systems, but they can be interpreted not so much as an erosion of democracy but as evidence of the success of democracies in putting increasingly more of their traditional activities in the hands of their citizens.

The discussion in climate science about the shortcomings of democratic governance coincides with assessments of the current state and future of democracy in the social sciences. Only a few years ago, political scientists proclaimed the end of history (Fukuyama, 1992) and thus the final victory of democracy. Today, political scientists, including Francis Fukuyama (2018), are thinking much more about the dissolution of democracy. Titles such as "The Future of Freedom" (Zakaria, [2004] 2007), "The Retreat of Western Liberalism" (Luce, 2017), "How Democracy Ends" (Runciman, 2018), "How Democracies Die" (Levitsky and Ziblatt, 2018), "The People vs. Democracy" (Mounk, 2018), and "Can Democracy Survive Global Capitalism?" (Kuttner, 2018) provide strong support for this observation.

Before the now extensive and increasingly vociferous, even desperate, discussions about the likely fate of democracy, doubts about its efficacy in the face of the ecological challenges ahead began to surface explicitly in the 1990s. John Dryzek's (1994: 177) strong views on the incompatibility and failure of capitalism, liberal democracy, and the administrative state are exemplary: "the three institutions are each thoroughly inept to ecology, that any combination of them can only compound error, and that any redeeming features are to be found only in the possibilities they open up for their own transformation." The wholesale negation of each of these social institutions is the only viable solution. Anthony Giddens (2009: 94) agreed with Dryzek's judgment: "Responding to climate change will prompt and require innovation in government itself and in relation between the state, markets and civil society." Whether going beyond existing socio-economic and socio-political institutional arrangements and considering what kind of democracy, if any, does the ecological age permit? (Dryzek, 1994: 177) amounts to an abolition of liberal democracy and capitalism being left open, as far as I can see. Tony Giddens, in any case, favors a strong state, as I will argue later. The struggle over climate change and climate policy is central to the current shift in the debate over the well-being of democracy. However, climate change and climate policy are part of a larger set of immediate or distant political concerns. The general attitude toward global warming is volatile. After all, foregrounding the dangers of climate change, sometimes at an uncertain date in the future present, is difficult, as economic misery or ill health is threatened just around the corner (cf. Radkau 2017: 405).

Major political shocks since the beginning of the 2020s, including a pandemic, the Ukraine war, economic instability, and geopolitical repercussions, have influenced what people consider important political issues. In 2023, according to a survey carried out by the OECD, climate change ranks way behind other urgent political issues. The uncertainties during this period are reflected in significant concerns about the economy (Figure 4.1).

The general conclusion drawn by political scientists is often that (existing) democracy is losing legitimacy in the eyes of its citizens in the face of a multitude of social and economic changes—without democracy as an ideal value being rejected. But the ideals of democratic governance, as I have argued, are also being questioned as a value to be prioritized above all others. These transformations include foremost high inflation, large migration flows, transformative health issues such as the pandemic (infectious diseases), generational divides, the rise of populism, mistrust of the power of large institutions, and robust patterns of social inequality. The distrust obtained in an OECD survey for the year 2023 varies significantly from country to country. The greatest lack of trust occurs in the Czech Republic, and the highest or moderately high level of trust is observed for Switzerland (Figure 4.2).

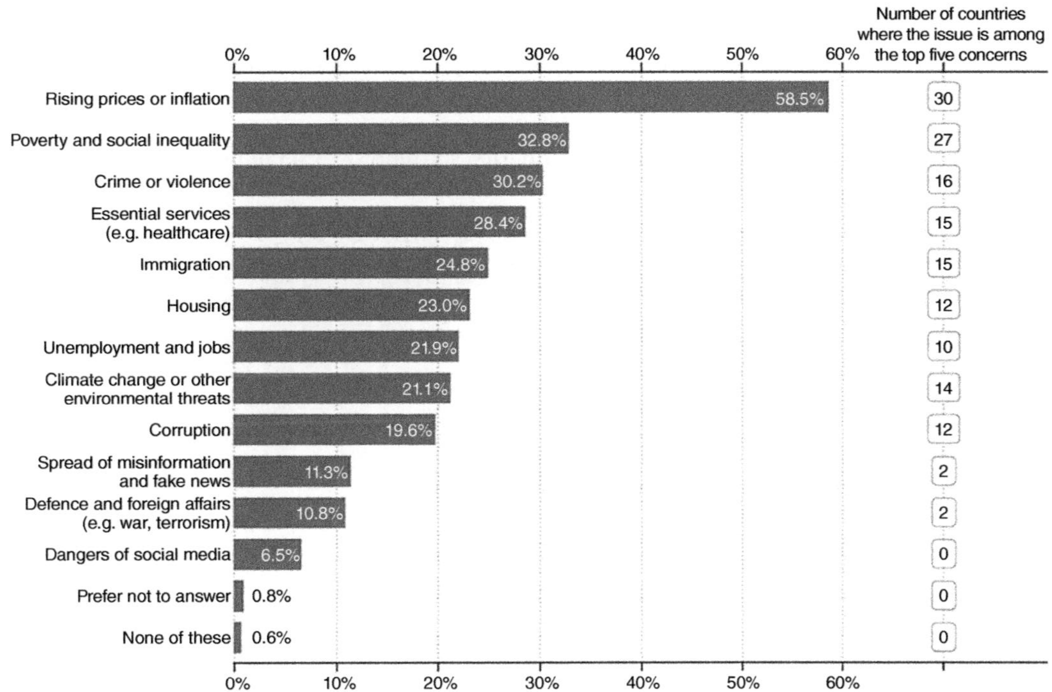

FIGURE 4.1 Share of population that views policy issues as among the three most important ones facing their country (OECD, 2024).

Notes: The figure presents the unweighted OECD average of responses to the question "What do you think are the three most important issues facing [COUNTRY]?". Immigration was not a response option in Mexico or Norway. The listed number of countries where the issue is among the top five concerns relates to the number of countries where the issue has among the five highest proportions of mentions among the respondents.

Source: OECD Trust Survey (2024).

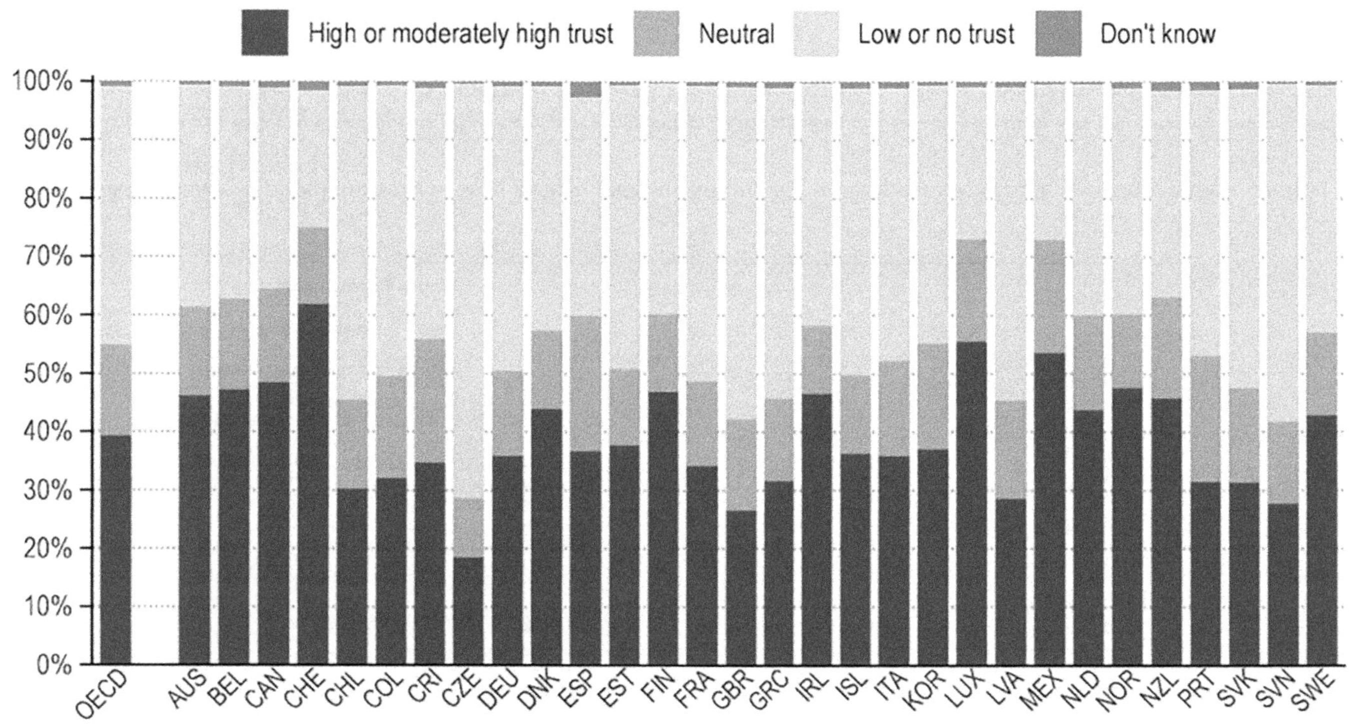

FIGURE 4.2 Share of the population in OECD countries with different levels of trust in their national government (on a 0–10 scale), 2023.

Notes: The figure presents the within-country distributions of responses to the question "On a scale of 0 to 10, where 0 is not at all and 10 is completely, how much do you trust the national government?". A 0–4 response corresponds to "low or no trust," a 5 to "neutral," and a 6–10 to "high or moderately high trust." "OECD" presents the unweighted average across countries.

Source: OECD Trust Survey (2024).

The fear of social science observers therefore is that today's democracy—in many respects, whether deliberately or because of structural economic, political, and moral changes—fails to deliver and might well be on the way to autocratic forms of government. The erosion of democracy manifests itself, for example, in processes of depoliticization, the substitution of politics by techniques of management or the restriction of public space (cf. Rosanvallon, 2006: 228; also, Swyngedouw, 2011): "in an erosion of citizenship, the marketization of the public sector, the soul-destroying targets and audits that accompany it, the denigration of professionalism and professional ethos and the erosion of public trust" (Marquand, 2004: 172). Democratic governance is increasingly muted by the rapid abolition of democratic principles of political equality and even replaced by autocratic forms of governance reminiscent of Robert Michels' ([1915] 1949) one hundred-year-old iron law of oligarchy.

What distinguishes the discussion about the poor state of democracy between social scientists and climate scientists is the remedy that both sides advocate: on the one hand, social scientists discuss efforts that could restore democracy, such as rebuilding a "society of similar individuals" (Rosanvallon, [2011] 2013), through the active participation of large numbers of citizens who help shape the agenda of public life. On the other hand, climate scientists and other observers of global climate change doubt the ability of democratic governance to address major environmental problems effectively and therefore call for a more authoritarian state and/or a state in which decision-making by technical experts is paramount. However, then democracy is supposedly dismantling itself.

For example, Colin Crouch (2004: 4) described the transition from democracy to post-democracy as follows: "In the conditions of a post-democracy that increasingly cedes power to business lobbies, there is little hope for an agenda of strong egalitarian measures to redistribute power and wealth or to contain powerful interests." Post-democracy is also accompanied by a rapid erosion and denial of democratic rights and values, as Richard Rorty (2004: 10) argues:

> At the end of this process of erosion, democracy would have been replaced by something quite different. This would probably be neither a military dictatorship nor an Orwellian totalitarianism, but rather a relatively benevolent despotism enforced by a gradually hereditary nomenklatura.

In some images of "post-democracy," the state of the state—a return to an aristocratic society—has already been achieved: self-appointed elites claim to carry out the wishes of the masses.[29] In short, as Pierre Rosanvallon (2006: 228) noted, politics has been replaced "to make way for a single actor on the stage: international society, which unites the champions of the market and the prophets of the law under the same banner." This is a political development that is very much welcomed by the representatives of climate science.

What the public thinks about and how they act toward climate and climate change should have important implications for how governments in democratic societies respond to climate change. The attitudes toward climate change and climate policies differ from country to country and from time to time. In Europe, for example, there are country differences in elements of individual climate citizenship,[30] including climate concern, perceived responsibility, and willingness to support and take climate action; Pohjolainen et al. (2021: 14) in their analysis of attitude data in 23 European countries in 2016 and 2017 find that the link between climate concern and perceived climate responsibility is stronger in more democratic countries.

In the following excursus, I explore one aspect of this relationship, namely the likely impact of weather extremes on the public's perceptions of climate.[31] Reference will be to the *collective consciousness* of societal beliefs about climate and typical social action toward nature in general and the local climate in particular. The main point will be that existing cultural frames find confirmation in extreme events. As Maurice Natanson (1962: xxxvii) stressed in his introduction to the *Collected Papers* of Alfred Schutz, "men living in the paramount reality of everyday life are enmeshed in situations as *they* define them in the context of their lives. It is idle for the neutral observer to point out to committed actors the 'objective' situation." The reference to *weather extremes*[32] for collective climate consciousness is, in addition, important because the speed of an increase in the average global temperature translates into a rapid *rise and swings* in weather extremes.[33]

The role of weather extremes and the attribution of extremes to climate change in scientific and public discussions has changed rapidly. Bert Bolin who was among the founders of the *Intergovernmental Panel on Climate Change* (IPPC) describes the origins and development of the IPCC and the 1995 Berlin conference in his memoirs and criticizes the German Minister of the Environment Angela Merkel for unnecessarily stressing the frequency of extreme events: "On this occasion," Bert Bolin (2007: 108) remarks, "the German minister responsible for environmental issues, Angela Merkel, also spoke. She was forward looking but referred to the probable increasing frequency of extreme events which had already become more common, although evidence for this was very meagre." Today, such an inference has become almost common knowledge, and attribution research of single extreme weather events is a flourishing subfield of climate science (e.g. Otto, 2023a).

As the structural theory of culture suggests, most social contexts and most social actions are characterized by nothing more than routine attributes and habitual conduct. The French sociologist and anthropologist Gabriel Tarde ([1890] 2003) introduced the notion that most human action is based on imitation. *Habitual action* has, of course, the constructive function of stabilizing human conduct, enhancing the predictability of social action, and opening avenues free from the pressures and constraints to act. Whether habitual conduct or imitation is always a carbon copy of previous social conduct (in that limited sense, all conduct is a modification of previous conduct) is not at issue; what is at issue is the overwhelming constraint in social life to repeat and therefore get on with life.[34] The likelihood is that the master constraint of everyday life, which is to move along well-trodden paths, could amount to what Anthony Giddens (2009: 2) calls a straightforward paradox (as a matter of fact, the "Giddens paradox"). The paradox states that

> since the dangers posed by global warming aren't tangible, immediate or visible during day-to-day life, however awesome they appear, many will sit on their hands and do nothing of a concrete nature about them. Yet, waiting until they become visible and acute before being stirred to serious action will, by definition, be too late.

In the following excursus, the paradox is investigated through the lens of the public's response to weather extremes. The excursus delves into the phenomenon of nature's role in fortifying the collective consciousness. The focus of this analysis is on the factors that primarily contribute to the normality of normal weather patterns. The focus is on *weather* extremes, as opposed to terms such as "climate extremes" or "catastrophes" and their long-term and massive impact on societies.[35] Weather extremes appear to be real, but at the same time they are socially constructed (see Grundmann und Stehr, 2000).

My excursus relies on historical quotations, literary references, and theoretical frameworks, but, as might be expected, lacks large-scale, longitudinal (generational), or cross-cultural empirical data to substantiate claims about contemporary public perceptions of extreme weather events. Some opinion polls (e.g. Eurobarometer, Pew Research) are cited, but they do not make up for the lack of more systematic and comparative empirical information.

Notes

1 The opposite view can now be heard more loudly and perhaps with greater political influence: "Having once positioned himself as singularly capable of fixing global warming, Musk has more recently called climate warnings 'overblown,' and told Donald Trump—his prospective new boss at the so-called Department of Government Efficiency—that 'we don't need to rush' to reduce emissions" (Kate Aronoff, "Elon Musk is why we shouldn't cast CEOs as climate saviors,"

The New Republic, December 12, 2024; https://newrepublic.com/article/189275/elon-musk-climate-change-industrial-policy?utmsource=newsletter&utmmedium=email&utmcampaign=tnrdaily&vgoee=TkRzhWdp%2BCilv7eer%2BpO7hju35fzsJNIUzQ4Nh8GoDe%2Bp%2FDIKRs%3D%3AjhtdkNaKJh5QRTvcjlgg%2BG2EpcNdwIHl.

2 Thomas Luckmann (1974) questions the widespread case made for the natural classification of reality in nature and culture and points out that the separatist world view with its exclusion of nature is typical of more complex societies.

3 In spite of its drawbacks, the concept of the Anthropocene offers, as Latour (2015: 145; also Chakrabarty, 2015) notes, "a powerful way, if used wisely, to avoid the danger of naturalisation while ensuring that the former domain of the social, or that of the 'human', is reconfigured as being the land of the Earthlings or of the Earthbound."

4 Cf. M.B. Mahon et al., "A meta-analysis on global change drivers and the risk of infectious disease," *Nature* (2024); https://doi.org/10.1038/s41586-024-07380-6.

5 Cf. The leader: "Global warming is coming for your home: Who will pay for the damage?," *The Economist*, April 11, 2024; www-economist-com.login.ezproxy.library.ualberta.ca/leaders/2024/04/11/global-warming-is-coming-for-your-home.

6 Ben Hubbard, "Rich Gulf states have huge ambitions. Will extreme heat hold them back?," *New York Times*, June 27, 2024; www.nytimes.com/2024/06/27/world/middleeast/gulf-states-heat-climate-change.html.

7 An analysis by Dura et al. (2025) warns of "a process rarely considered in future sea-level rise scenarios is sudden (over minutes) land subsidence associated with great (>M8) earthquakes, which can exceed 1 m. Along the Washington, Oregon, and northern California coasts, the next great Cascadia subduction zone earthquake could cause up to 2 m of sudden coastal subsidence, dramatically raising sea level, expanding floodplains, and increasing the flood risk to local communities."

8 A paper by Lena Partzsch's (2025) explored the dual potential—empowering and authoritarian—of climate emergency declarations, examining how such narratives shape political responses to climate change. Partzsch concludes that the effect of climate emergency declarations is context-dependent. They can serve as catalysts for democratic transformation or tools for authoritarian consolidation. The challenge is to harness their disruptive potential without falling into the trap of technocratic or top-down control. Instead of seeking "one inconvenient truth," societies must embrace a plurality of narratives and solutions to address the climate crisis equitably and democratically.

9 Not a new prediction in principle, but better documented is the following prediction: "The AMOC collapse dramatically changes the redistribution of heat (and salt) and results in a cooling of the Northern Hemisphere, while the Southern Hemisphere slightly warms. Atmospheric and sea-ice feedbacks, which were not considered in idealized climate models studies), further amplify the AMOC-induced changes, resulting in a very strong and rapid cooling of the European climate with temperature trends of more than 3°C per decade. In comparison with the present-day global mean surface temperature trend (due to climate change) of about 0.2°C per decade, no realistic adaptation measures can deal with such rapid temperature changes under an AMOC collapse" (van Westen et al., 2024: 7).

10 See Martin Wolf, "Action must replace talk on climate change," *Financial Times*, May 4, 2021; www.ft.com/content/3fa154f3-84e7-4964-9a21-d3dbd41e1470.

11 I emphasize the limitation of my consideration of the consequences of extraordinary necessities to national policy systems. The emergency policies of international organizations beyond nation-state societies would be another aspect (cf. Kreuder-Sonnen, 2019).

12 A more nuanced picture of the feared uncontrollable cascading effects in the climate system is painted by an analysis by Deutloff, Held, and Lenton (2025). The research group confirms that global warming could bring key tipping elements of the global climate system dangerously close to their thresholds in certain scenarios. They investigated the likelihood of a total of 16 important Earth systems tipping in various future scenarios, depending on how global greenhouse gas emissions develop by the year 2500. The focus was on the thawing of permafrost, which can release large amounts of methane and carbon dioxide, and the Amazon rainforest, whose drying up and death would also release large amounts of carbon into the atmosphere. The results show, on the one hand, that the changes already taking place are not irreversible. On the other hand, they put into perspective the likelihood of an uncontrollable domino effect, at least for those tipping elements that directly affect the global carbon cycle, such as permafrost or the Amazon rainforest.

13 Adam Tooze, "Welcome to the world of the polycrises," *Financial Times*, 2022. Available at www.ft.com/content/498398e7-11b1-494b-9cd3-6d669dc3de33; also T. Homer-Dixon, M. Lawrence, and S. Janzwood, "Dismissing the term 'polycrisis' has one inevitable consequence: reality always bites," *The Globe and Mail*, 2023. Available at www.theglobeandmail.com/business/article-lessons-from-the-pandemic-adam-tooze-on-what-covid-teaches-us-and-what/the-term-polycrisis-has-one-inevitable-consequence-reality/.

14 Huan Liu and Ortwin Renn (2025) highlight the inadequacy of traditional risk management methods in addressing the complexities of polycrises and propose advanced approaches like agent-based modeling, network theory, and resilience-based strategies. They stress the need for trans-sectoral cooperation, inclusive governance, and effective communication to build adaptive and resilient systems. Liu and Renn advocate interdisciplinary perspectives and practical tools to better understand and mitigate the impacts of polycrisis and systemic risks in an increasingly interconnected world.

15 Jürgen Scheffran (2025) is looking into the complex interplay between human development, planetary boundaries, and geopolitical conflicts in the Anthropocene. He examines how the "Great Acceleration" of human activities is straining the Earth's ecological limits and has led to interconnected crises (polycrises) such as climate change, biodiversity loss, and social instability (see also Heinberg, 2024). Marxist critics argue that proponents of the polycrisis model have not paid enough attention to the role of capitalism (see Albert, 2025).

16 Steves and Teytelboym (2013) report, based on their empirical study that "popular knowledge of climate change is a powerful driver of climate change policy adoption [...] This means that even controlling for democratic institutions, the public's concerns about climate change are reflected in climate policy."

17 Steves and Teytelboym (2013) also find that "the relative size of the carbon-intensive industry is significantly and negatively associated with climate change policy adoption."

18 For recent studies by economists, sociologists, and political scientists that adopt Michels' thesis of the inevitability of oligarchic tendencies in organizations, see Williamson (1975, 1985, 1995), Granovetter (1985), Foucault ([1981–1985] 2005), and Stehr and Adolf (2018: 321–324).

19 An important source for this observation is Scott (1998); also, Stehr (2001).

20 In early December 2023, the Secretary General of OPEC, Haitham Al Ghais, urged the oil-exporting countries participating in the Dubai climate conference to block a demanded phase-out of fossil fuels: "The OPEC Secretary General reportedly referred to an initial version of a [final] document being negotiated at the Dubai climate conference. In it, various options for the future handling of fossil fuels were mentioned, including an 'orderly and just withdrawal' as a variant in the document": "OPEC urges members to veto fossil fuel phase-out," *Frankfurter Allgemeine Zeitung*, December 9, 2023; www.faz.net/aktuell/politik/ausland/cop28-opec-draengt-mitglieder-zu-veto-auf-klimakonferenz-19373198.html. Banks around the world continue to promote climate-damaging projects with trillions in loans (see "Dirty money," *Der Spiegel*, November 4, 2023; "Banks pumped more than $145bn into companies running 'carbon bomb' projects in 2022," *The Guardian*, October 31, 2023). The Saudi state, "the world's top oil exporter is the fiercest opponent of a new deal at the UN climate summit, according to people close to the talks," *New York Times*, December 10, 2023, "Saudi Arabia Is Trying to Block a Global Deal to End Fossil Fuels, Negotiators Say."

21 Siehe A. Revkin, http://dotearth.blogs.nytimes.com/2014/04/16/a-risk-analyst-explains-why-climate-change-risk-misperception-doesnt-necessarily-matter/?php=true&type=...

22 In an essay on Niklas Luhmann's epistemic sociology, Gert Verschraegen (2023: 88) notes Luhmann's paradoxical perspective, viz. "while functional differentiation has allowed society to scientifically register and visualise this problem, as well as to highlight scientifically backed solutions for tackling it, it has also dismantled other systems' ability to tackle the problem by simply 'following the science'."

23 Climate researcher Hans Joachim Schellnhuber in an interview with DER SPIEGEL (issue 12, March 21, 2010, p. 29) on the question of why the messages of science are not reaching society.

24 Bebra Davidson (2025: 10) explores, for example, "the unique attributes associated with pro-climate action, by integrating the extensive but disparate empirical literature on conditions supporting agency in response to the climate emergency, including norms of responsibility; efficacy; and social belonging, which entails a return to important emotions like shame, pride and empathy."

25 The initial justification for the withdrawal can be found here: Donald J. Trump, "Remarks announcing United States withdrawal from the United Nations framework convention climate change Paris Agreement," June 1, 2017, the *American Presidency Project*, presidency.ucsb.edu/node/32.

26 In *The Smoke and the Spoils: Anti-Environmentalism and Class Struggle in the United States*, John Hultgren (2025) explores the enduring alliance between elite interests and segments of the working class that sustains anti-environmental politics in the United States. He argues that while the benefits of fossil fuel production accrue to a small elite, these interests have historically forged cross-class coalitions to resist environmental regulations. By tracing this dynamic from the 19th-century frontier to contemporary "Trump Country," Hultgren situates

environmental politics within broader class struggles, asserting that meaningful environmental progress hinges on a revitalized, justice-oriented *working-class* movement.

27 The concept "sustainability" itself is an essentially contested concept (cf. Connolly, 1993); cf. Lally et al. (2019) discussion of the sustainability in the Anthropocene. Most theories of sustainability development have been framed in the socio-economic context of developed countries. Albertone (2025) recalls earlier work from the developing world. In fact, "more than a decade before the publication of the Brundtland Report (1987), classical development theorists—principally Latin American—had already engaged in the theoretical attempt to reflect on the environmental dimension of development and, more generally, on the meaning of the ecological crisis for peripheral development" (Albertone, 2025: 2). Ignacy Sachs's (1977a and 1977b) work on eco-development represents such an effort.

28 Some scholars (e.g. Mann and Wainwright, 2013) apply Schmitt to theorize a coming ***climate authoritarianism***—a planetary sovereign willing to suspend rights to "save" civilization. Schmitt might see this as inevitable.

29 Hans Jonas' ([1979] 1984: 147) sober response to such an assertion is quite appropriate and worth quoting in this context: "If [...] only an elite can ethically and intellectually assume the kind of responsibility for the future that we have postulated - how is such an elite created and recruited, and how is it endowed with the power to exercise it?"

30 Also defined as the individuals' commitment to rights, entitlements, and obligations in relation to the environment and particularly to climate change.

31 My approach therefore differs from and does not compete with the science of [individual] extreme weather event attribution to anthropogenic climate change as a subfield of research of climate scientists (e.g. Otto, 2017, 2023b).

32 A small amount of research has been devoted to understanding how "climate disasters" (should be weather disasters) shape attitudes toward climate change and climate policy (e.g. Arias and Blair, 2024). Although this research tends to lack a definition of what constitutes a "climate disaster," one can infer a definition from the instances typically given: For example, hurricanes or forced migration. Examples of *climate* disasters therefore refer to more serious, longer-lasting, and more damaging events than "extreme weather events." Moreover, attributing "disaster to climate can divert attention from these place based vulnerabilities and their socio-political causes" (Lahsen and Ribot, 2022: 1).

33 As a report in the *New York Times* (April 22, 2025) about a *Nature Communications* article recounts (www.nature.com/articles/s41467-025-58544-5): "A September heat wave switching into a snowstorm over one day in the Rocky Mountains. Winter snowfall suddenly melting and saturating fields of dormant crops, before refreezing and encasing them in damaging ice. Early spring warmth prompting plants to blossom followed by a cold snap that freezes and drops their petals"; www.nytimes.com/2025/04/22/climate/temperature-swings-climate-change.html.

34 For as Emile Durkheim ([1912] 1965: 479) perceptively observed: "Life cannot wait" (cf. also Gehlen, [1940] 1988, 296–297). In their discussion of expertise and how expertise can be justified, Harry Collins and Robert Evans (2002: 241) make an analogous observation about the essential dilemma encountered in the public domain and its typical time frame when one has to wait for expert

advice: "Decisions of public concern have to be made according to a timetable established within the political sphere, not the scientific or technical sphere; the decisions have to be made before the scientific dust has settled, because the pace of politics is faster than the pace of scientific consensus formation."

35 For discussion of "catastrophes," in the context of climate change, see Kitcher (2016).

EXCURSUS

Climate change and weather extremes – anchor in an uncertain world?[1]

The relentless November weather is predictably unforgiving.

Charles Dickens, Bleak House, 1852–1853

The awareness of the constancy of the climate is deeply rooted in the people and is expressed in the certain confidence that the unusual weather of one season or one year must be compensated for by that of the following year.

Eduard Brückner (1890: 2)

Everyone knows that the climate is important, especially the extremes.

William F. Ogburn (1943: 785)

The snow has fallen on the cedars. In fact, it was only the seventh white Christmas in six decades, and Vancouver residents, who are used to spending their holidays rather damp, were a little puzzled by the white lashing they've been getting lately [...] Rest assured, the rain will come here eventually.

Globe and Mail, December 26, 1996

My excursus is about climate and weather. In a "sort of introduction" (his own words) to his successful novel *The Man Without Qualities*, Robert Musil ([1930] 1995: 3) described the weather on an August day as a remarkably unremarkable day, in keeping with the title of the novel:

> A barometric low hung over the Atlantic. It moved eastward toward a high-pressure area over Russia without as yet showing any inclination to

DOI: 10.4324/9781003682066-6

> bypass this high in a northerly direction. The isotherms and isotheres were functioning as they should. The air temperature was appropriate relative to the annual mean temperature and to the aperiodic monthly fluctuations of the temperature. The rising and setting of the sun, the moon, the phases of the moon, of Venus, of the rings of Saturn, and many other significant phenomena were all in accordance with the forecasts in the astronomical yearbooks. The water vapor in the air was at its maximal state of tension, while the humidity was minimal. In a word that characterizes·the facts fairly accurately, even if it is a bit old-fashioned: It was a fine day in August 1913.

Robert Musil's wonderful description of a beautiful August day in 1913 is not necessarily a praise of the constancy of weather phenomena, but a recognition that the weather can be predicted with some certainty for the future present. This passage links meteorological terminology (barometric lows, isotherms, humidity) with a simple, subjective judgment: "a beautiful day." Musil's juxtaposition ironically comments on the fact that, despite all scientific accuracy, experience can always say more than mere data—or vice versa. This motive constitutes a fundamental observation in the subsequent excursus.

With even greater certainty, based on the Earth's rotation around the sun, it is possible to predict the coming and going of the seasons. The understanding of weather events and seasons is deeply embedded in the culture of a region or society at large (Horn, 2024: 340–364). My basic observation will be that the collective consciousness about nature recognizes the volatility of nature, but that nature always recovers from its periodic volatility. And that this basic observation also applies to weather phenomena.

The excursus is therefore about the stability of world views, because as Thomas Luckmann (1970: 96) summarized, the

> socialization processes lead to a high degree of congruence between the socially objectivated conception of the world and subjective reality. In other words, the plausibility of the institutionally stabilized world view is pervasive as well as compelling, and traditions change imperceptibly.

However, the excursus is not about the nature of climate and weather. It is about how people and science have imagined the climate in relation to weather and society over the last few centuries, and how extreme weather events may confirm preferences and beliefs. The practical importance of this question can be measured by the fact that the outcome of the climate crisis in the coming decades is not least a function of our understanding of the climate and short-term weather events.[2]

Does the climate ever speak to us? Climate does not speak to us directly. It speaks to us through our consciousness, our measuring instruments, including our bodies. However, in that way, it clearly demonstrates that we live in a world of nature and society (Latour, [1999] 2004; Mitchell, 2010: 239–254). Weather extremes such as heavy precipitation, extreme heat, floods, extreme cold, forest fires, and violent storms and floods provide people with information about the climate. The climate is not voiceless. For example, extreme drought in northern Italy has doubled in the last two decades, creating a climate that increasingly resembles that of Ethiopia and the Horn of Africa. As reported by the British newspaper *The Guardian* (October 14, 2023), climate researchers from English universities have observed another phenomenon of climate extremes: how

> global warming is causing a 'whiplash effect' that leads to unpredictable extremes. Analysis of climate data [...] shows that under extreme climate pressures, areas that used to be frequently hit by droughts are now more frequently affected by floods, whereas other regions that were affected by floods in the past are now more frequently hit by droughts.

Global warming, as can be generalized from these observations, has very different consequences in different regions of the world. Should we listen to the climate?

Weather extremes can give people the impression of both a constant and a changing climate. Weather extremes signal that we can trust the "local" constancy of the highly variable global climate but also that we may have to fear progressive climate change. Climate scientists therefore also consider weather extremes to be "instructive moments."[3] Instructive experiences that convey and underline the reality of anthropogenic climate change. The fact that my diagnosis of climate understanding in historical times, between fear and trust, is by no means a contradiction in terms will be explained in more detail in this excursus.

More than a century ago, the Viennese glaciologist and geographer Eduard Brückner was one of the first climate scientists to analyze anthropogenic climate change, climate variability, and its effects on human society. He was increasingly concerned about the social impact of climate change.

Eduard Brückner ([1889] 2000] in a lecture held at the German Geographers' Conference in 1889 entitled "To what extent is today's climate constant?" points out the following:

> The question of climate change in historical times has always been of great interest to the widest circles, as it is undeniably of great practical importance. For once it has been proven that the climatic conditions of our globe have undergone a change before the eyes of mankind, we must also reckon with the possibility that changes will take place in the near future. But such changes cannot remain without the most profound influence on

> the whole life and activity of the human race. Thus, to a certain extent, the general answer to that question contains a prognosis for the future destiny of man and his works.

Nevertheless, Eduard Brückner's further thesis, which will be analyzed in more detail here, that the general understanding of climate in modern societies is based on a deeply rooted belief in the normality or a fundamental order of climate, is just as remarkable.

Based on considerations comparable to those of Eduard Brückner, one of the first scientific treatises in English, George P. Marsh's ([1864] 1965: 29) *Man and Nature, draws* attention to the fact that "nature, if left unaltered, so shapes its territory as to preserve an almost invariable constancy of form, outline, and proportion." Marsh and Brückner's assumptions reflect a widespread view in Western thought and in the scientific community in the 19th century and well into the following century that natural conditions are constant and permanent. The relationship between society and climate is static. Deviations from constancy are exceptional phenomena.[4]

In his work *Le Contrat Naturel*, the French philosopher Michel ([1992] 1995: 3) emphasized: "The climate system varies greatly and yet relatively little, because it is relatively unchanging in its fluctuations: fast and slow, catastrophic and mild, regular and chaotic. Rare phenomena are therefore conspicuous, but they should not surprise us." In the eyes of the inhabitants of a habitat, the constancy of the climate guarantees the maintenance of reliable living conditions.

Climate as a reliable background

> Since man is not an independent substance, but is connected with all the elements of nature, he lives from the breath of the air, as from the various children of the earth, food and drink; he processes fire, as he absorbs light and pollutes the air; waking and sleeping, at rest and in motion, he contributes to the change of the universe, and should he not be changed by it?
>
> *Johann Gottfried Herder (1794: 87)*

> SPIEGEL: "You'll never be safe from night frost until Sophie is over." Can a farmer today still do something with this country saying? Brömse: "He does not need it. He has his weather app." SPIEGEL: "Does the rule apply valid?" Brömse: "In general, yes. Over decades, centuries, it has often gotten colder again, and that is still the case." [5]

The observations of natural scientists about the widespread social notion of the constancy of the climate are thus confirmed by the classical sociology of the time. The climate problem hardly plays a role. In the classical social theories of Max Weber, Georg Simmel, or Èmile Durkheim, one searches in vain for the topic of a *changing* climate. Georg Simmel ([1908] 1992: 518), for example, emphasized that "it is not space, but the structuring and summarizing that takes place from the soul that has social significance." Climate issues are analytically negligible.[6] Climate is an influential but overall constant natural environment. A helpful definition of the "natural" from a social science perspective comes from Luc Boltanski ([2009] 2011: 162):

> The objects that are called "natural" are characterized by the absence of reflexivity, in particular by their indifference to the representations that are given to them and to the descriptions that their modes of being offer, by ordinary people or by professionals who equip themselves with science.

Human "natural capital" is a constant, steady, and reliable background for human action. This also applies to so-called climate determinism, for example in the geography of this period (see Stehr and von Storch, 2023), "this exaggerated anthropogenic approach" (Le Roy Ladurie, [1967] 1971: 8).[7] A return to the crude theories of climate determinism of the 18th, 19th, and 20th centuries is inconceivable today.

At the same time, the 20th century until the recent past of the present century was a century of profound ecological change. It was a time

> ranging from the effects of massive urbanization to the large-scale extraction of resources needed to sustain populations in cities of unprecedented size [... and] in which the consequences of our lifestyles were barely understood or considered – with the result that the environmental and climate changes of the present and the future will be shaped by what has already happened in the past, not by the choices being made today.
>
> *(Frankopan, 2023: 512)*

This excursus examines the ideas that people, whether scientists or laypeople, have developed about the climate in the course of their many years of encountering and engaging with nature. The idea of the cosmos as part of our "environment" places us in a special relationship with the climate. It is often the image of the climate as a faithful servant and asset, not as a source of fear. A harmonious order that was also divine.

In reference to early religions, using the Mediterranean as an example, Ellen Churchill Semple (1931: 510), for example, stated that "the chief gods of the time were weather gods, who had the power to regulate climatic conditions and to give or withhold water." Whether ancient weather gods were merely calm and predictable or whether their temperament fluctuated in inexplicable ways is unclear. If we follow the observations of Semple (1931: 51), then "the meteorological functions of the chief gods were primitive, basic and constant." In her study of the *Influences of Geographic Environment*, Ellen Semple ([1911] 2023: 51) emphasized her thesis of the permanence of geographic factors explicitly and in detail:

> The geographic element has had a strong and persistent effect in the long history of human development. Therein lies its importance. It is a stable force. It never sleeps. This natural environment, this physical basis of history, is basically unchanging compared to the other factor in the changing, plastic, progressive, retrogressive humanity.

Against this background, this excursus focuses specifically on the systems of representation that have emerged in relation to the climate. Analytically, this is not an easy task. The complexity of the issue, as well as the sources of credibility of the contingent social construct of climate and weather in different societies, are obscured by the routine, ease, and frequency with which we use these terms in a variety of contexts. The ease with which we use these terms, for example, suggests that we have an almost "natural," intuitive understanding of these phenomena. However, the centrality of these terms also masks ambiguity, fragility, change, and a lack of real understanding of human natural capital. From a philosophy of science or social theory perspective, the *constructivist* approach used here is in the tradition of the French sociologist Emilé Durkheim (e.g. Durkheim, [1912] 1965). Durkheim's approach assumes that the world as we know it does not codify and categorize itself but is actively ordered, organized, and classified by observers. Relevant knowledge about the environment in general and the climate in particular arises in a specific social context, which decisively shapes knowledge, or more precisely, the social understanding of "climate."[8]

The social context determines whether our climate semantics are generally determined more by fears or by trust in the consequences of climate change. In addition, the cause of our trust or our fears was not directly attributed to nature. As we have seen, it can be not only religious beliefs but also political or ideological views.

From the perspective and experience of everyday life, the term "climate" is understood as a largely unchanging phenomenon that has remained reliably constant over decades and centuries.[9] The weather is constantly changing. The weather is directly experienced by the inhabitants of the Canadian city of Vancouver in the winter months of 1996 at Christmas time. At least secondary, if not primary, element of any conversation about extreme weather events is the routine reference to a different future present for the weather and a past present. The unpleasant consequences of weather extremes are made more bearable, if not even nullified, by reference to a reliable climate. Not least, the literary, musical, and poetic praise of the seasons supports this everyday climate awareness, in which the reliability of the fluctuations of joy (e.g. spring) and sorrow (e.g. winter) plays a prominent role. Weather extremes are, in many cases, emotionally charged events (see Brosch, 2021). An event that leads to intense identification with the traditional semantics of the climate.[10]

The way in which society addresses and explains extremes thus provides an important clue to the semantics of environmental representations in general. They make it possible to understand both ordinary and unusual natural events. In contrast to other contemporary social and cultural constructions, such as photography (Bourdieu, [1991] 1994) or modern landscape images (Greider and Garkovich, 1994), our understanding of climate is less predetermined by decisions made by the state or its representatives.

As Eduard Brückner emphasized more than a hundred years ago, the practical and political significance of societal climate semantics is obvious. The societal climate picture not only influences society's trust in the climate and its willingness to behave toward the climate in the future but also influences the population's interpretation of the scientific constructs of climate and climate change. The robustness, certainty, and durability of common sense beliefs about climate change are becoming crucial elements in the precarious equation of knowledge, politics, and everyday life. I believe that the trust of people in modern societies in the regularity and constancy of a self-regulating climate is greater than their trust in science.

How is the climate perceived in society?

> The power to become habituated to his surroundings is a marked characteristic of mankind [...] We assume some of the most peculiar and temporary of our late advantages as natural, permanent, and to be depended on, and we lay our plans accordingly.
>
> *John Maynard Kenyes ([1919] 1920: 18)*

Habitual conduct is unsurpassed as John Maynard Keynes ([1919] 1920) notes in the introductory chapter of his *The Economic Consequences of the Peace* as a bargain for the persistence of social conduct and as a barrier against social change. On the other hand, anthropologists, ethnologists, and sociologists widely assume that man's experience of nature has changed fundamentally over the course of history. This change is described by scientists as a process of cultural evolution, social expansion, or civilization.

However, it is difficult to judge whether common sense's perception of the environment in general and the climate in particular has become more emotionally neutral over time or less relevant and therefore emancipated. Norbert Elias ([1939] 1982) refers to this "emancipation" as a "civilizing process." According to Elias, the human perception of things and people has become more affectively neutral, less directly controlled by human wishes and desires, and more strongly determined by experiential orientations. Individuals are forced by changed social conditions to curb their affective impulses and focus more on control, restraint, postponement of gratification, etc., in their behavior.

On the other hand, the experiences of earlier civilizations with nature are not only more direct but also more comprehensive and more developed. Social behavior is less emancipated from the demands of nature. Nature is more of a source of life and is probably often seen as an area of risk and danger. In any case, for earlier civilizations, nature is an area with which every human being is confronted directly and often unexpectedly daily.

As societies emancipate themselves from their natural environment, including the climatic conditions that influence them, in the course of social development—leaving aside for the moment the question of the extent to which the process of civilization has changed the climate—this "liberation" from the climate took place in different ways and with varying degrees of success. Until the recent past (and to some extent until today), some occupational groups, e.g. the rapidly declining professions of seafarers, mountain farmers, etc., continued to be exposed to the dangers and risks of climate and climate extremes. In contrast, most people today spend their lives in closed, air-conditioned rooms. However, the frequent preoccupation with and fascination with weather, climate, and their effects, especially for immediate events in the future, is not diminishing but is perhaps even becoming more characteristic.

Whether preoccupation with catastrophic or highly improbable future events, such as epidemics, technological accidents, or natural disasters, is a particular phenomenon of prehistory or even modernity is an empirical question. If this is the case, however, the question arises as to how such interests can be explained.

Niklas Luhmann (1991: 3) suspects that such fascination with the possibility of catastrophic events cannot be explained solely by the fact that modern technology provides such eventualities. The explanation for an interest in extremes, especially in the media, is probably related to the fact "that in today's case it is people or organizations, i.e., decisions [made in the present or irreversible], that can be identified as the triggering causes." The social system of representing nature can thus become more object-oriented or "scientific," as Norbert Elias expected. Nature now becomes visible in a new sense: for example, nature becomes anesthetized; it develops into a place of relaxation and mental and physical renewal. In addition, such a transformation of the perception of nature maintains the trust in nature gained earlier.

The people "out there in the country": opinion polls

The public perception of climate and climate change is highly important for several reasons. For instance, the potential support for climate policies may be triggered or delayed because of such perceptions. Additionally, the willingness to alter or refrain from environmentally relevant behaviors is likely to be influenced by climate attitudes. However, there is a complex issue regarding the relationship between social action and societal attitudes. Importantly, attitudes and behaviors may not necessarily converge. Hence, it would be premature to suggest that "climate change itself may change people's attitudes and behaviour" (Rüttenauer, 2023: 1246).

An empirical study carried out in Munich by Marc Keuschnigg and Fabian Kratz (2018), using a mail survey format with responses from a total of 1,057 respondents (32% response rate), for example, demonstrated that environmental attitudes only matter as drivers of environmentally friendly conduct when costs are low and when there is no strong social norm prescribing the behavior in question and no strong social norm dictates behavior or greater willingness to participate in recycling exists under conditions of increased normative support.

> for appropriate household waste disposal not only increase the likelihood of recycling but also promote additional effects on pro-environmental behavior if social visibility permits tracing of individual disposal strategies [...] In cases of weak informal support, however, normative appeals to voluntary contribution will be most effective in low-cost situations.
>
> *(Keuschnigg and Kratz, 2018: 1084)*

Otherwise, even people who are less concerned about the environment will take up the behavior.

European opinion polls show that citizens recognize climate change as a reality and want drastic reforms if they are not affected themselves.[11] People are worried about the climate. However, they are more concerned about themselves. This observation is particularly true for selected government institutions that are responsible for climate policy. A brief look at the latest results of selected opinion polls in various countries on the subject of climate change and global warming is revealing. According to a Eurobarometer survey[12] from spring 2023, 93% of Europeans believe that climate change is a serious problem worldwide. In Germany, 89% of the population shares this opinion. Sixty percent of Germans and more than half of Europeans (58%) believe that the transition to a green economy should be accelerated. This is particularly necessary due to rising energy prices and concerns about gas supplies following the Russian invasion of Ukraine. From an economic perspective, 73% of Europeans believe that the cost of the damage caused by climate change is significantly greater than the investment required for the green transition. Three quarters (75%) of Europeans also agree that climate protection measures will lead to innovation. A study from the USA, which summarized numerous American surveys from 1999 to 2017, concluded that

> the continued rise in temperatures will likely lead to increased public concern about climate change in the future. However, a warming climate alone is unlikely to lead to a consensus among the public about the threat posed by climate change.
>
> *(Bergquist and Warshaw, 2019: 686)*[13]

However, the responses of respondents in European Union member states to the question of the perceived experience of weather extremes are interesting given the broad agreement that climate change is a serious problem. The voice of the climate is generally reserved, at least thus far—if one measures this reservation by the frequency of experiences with extreme weather among respondents to the Eurobarometer survey. Only 18% of respondents in Germany state that they are personally exposed to weather risk. The extent of this perceived risk can strengthen confidence in the climate-related stability of living conditions in one's own region. It is not insignificant that modern societies have thus far been better able to address the consequences of extreme weather events and thus keep the damage and costs of extremes comparatively low.

A *Pew Research Center* 2024 survey asks "Americans what policies they support to address weather-related damage. We also ask people who recently lived through *extreme weather events* if they think climate change contributed to them and how their lives were affected." Reports by respondents about weather extremes are common. The survey revealed that

> seven-in-ten Americans say that in the past year, they've experienced at least one of five types of extreme weather we asked about: severe weather, such as floods or intense storms; unusually hot weather; droughts; wildfires; or rising sea levels. Democrats are more likely to report these experiences than Republicans.

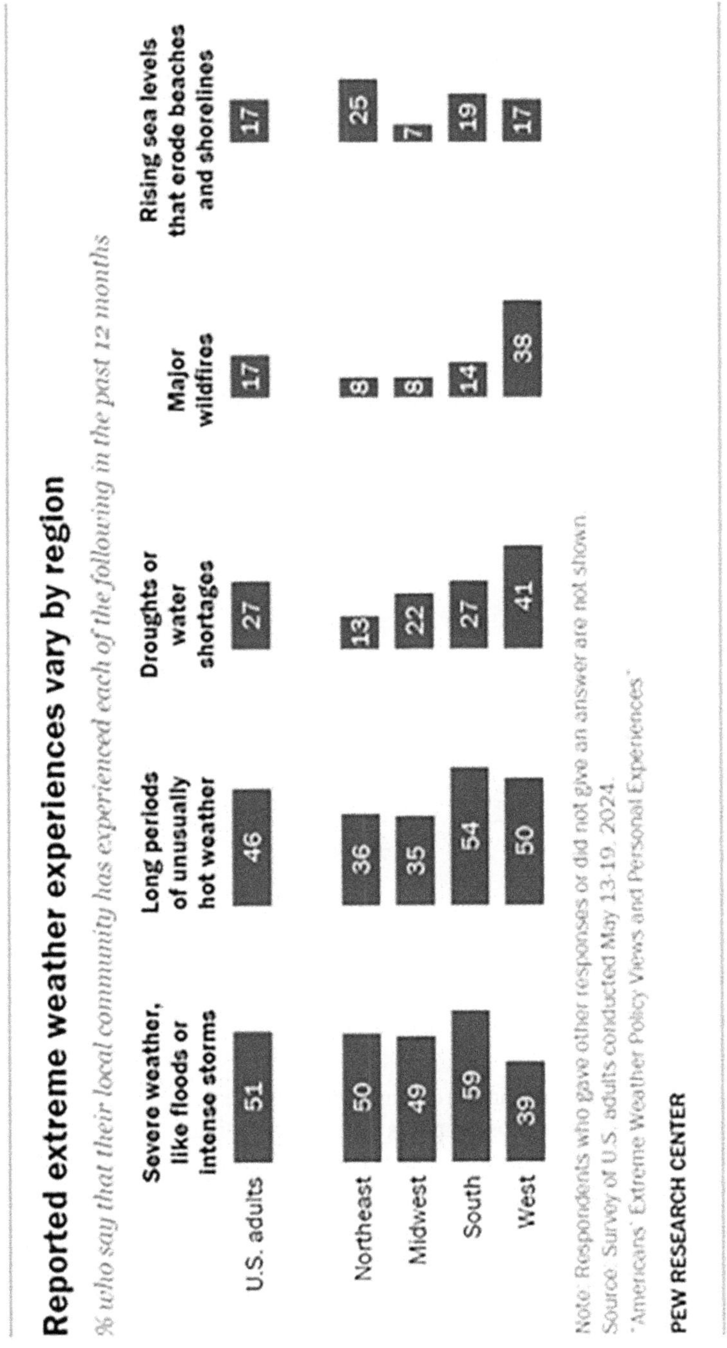
Reported extreme weather experiences vary by region
% who say that their local community has experienced each of the following in the past 12 months
Severe weather, like floods or intense storms
Long periods of unusually hot weather
Droughts or water shortages
Major wildfires
Rising sea levels that erode beaches and shorelines
U.S. adults
51
46
27
17
17
Northeast
50
36
13
8
25
Midwest
49
35
22
8
7
South
59
54
27
14
19
West
39
50
41
38
17
Note: Respondents who gave other responses or did not give an answer are not shown.
Source: Survey of U.S. adults conducted May 13-19, 2024.
"Americans' Extreme Weather Policy Views and Personal Experiences"
PEW RESEARCH CENTER

Modest shares of Americans say climate scientists understand key aspects of climate change very well

% of U.S. adults who say that climate scientists understand each of the following ...

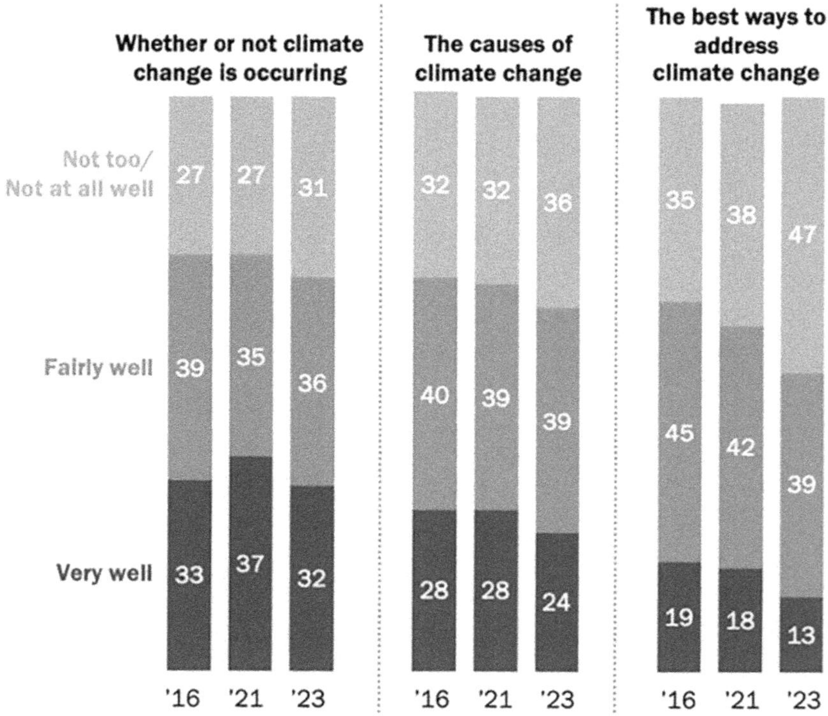

Note: Respondents who did not give an answer are not shown.
Source: Survey of U.S. adults conducted Sept. 25-Oct. 1, 2023.

PEW RESEARCH CENTER

EXCURSUS 2 Only a small proportion of Americans believe that climate researchers understand the most important aspects of climate change very well.

Source*:* PEW Research Center, October 25, 2023; www.pewresearch.org/short-reads/2023/10/25/americans-continue-to-have-doubts-about-climate-scientists-understanding-of-climate-change/.

Among those who say they have experienced any of these extreme weather events, "a large majority say climate change contributed at least a little. Most Republicans – and nearly all Democrats – say climate change played a role." However, these survey results should be interpreted as showing that sensitivity to weather extremes as signals of climate change is on the rise; whether the response to the survey also indicates that basic confidence in the reliability of the climate, for example, basic climate features such as seasonal variations, is also on the decline is difficult to infer.

The public's perception of climate change and climate policy issues differs from the public's trust in the science of the scientific community. The Pew Center in the United States asked the following question in a 2023 survey: the share of Americans who say that climate scientists understand the causes of climate change very well decreased slightly from 28% in 2021 to 24% today. Only 13% of Americans now say that climate scientists understand very well the best ways to address climate change, down from 18% in 2021.

Social behavior and trust

One of the central problems facing society is the uncertainty of the future. How does a society deal with a future about which hardly anything can be said with certainty? One possible starting point for a better understanding of the case of everyday climate semantics is the thesis that an essential feature of everyday concepts of nature is basic trust and benevolence, the idea of a good nature, which can always refer affirmatively to the rhythm and reliability of natural processes, such as solar radiation, as a constant source of heat or to the basic temperature balance of the globe.

It can be assumed that such trust is neither uniform nor independent of social and cultural conditions and is less directly determined by natural environmental conditions during the transformation of society (Stehr, 1978: 1–17). This basic trust in a reliable environment is socially and historically more widespread and can therefore be assumed to be more universal than, for example, people's trust in major social institutions such as religion, science, or the state. Basic trust in the climate as a reliable background for human action is not necessarily constant but, like the value of money in an economy, can be subject to inflationary or deflationary fluctuations (cf. Parsons, 1963). According to the English sociologist Anthony Giddens (1990: 92), such basic trust is an element of the individual "need" for ontological security. In phenomenological terms, ontological security has to do with "being-in-the-world." Indeed, such a form of trust "in the reliability of non-human objects" may be based on a "more primitive belief in the reliability and care of human individuals" (Giddens 1990: 97). Sociologists who identify with the so-called functionalist perspective tend to emphasize the crucial role of trust in social relations in general and in *social* order in particular.

Trust serves to maintain social solidarity and strengthens social control. According to this view, trust establishes mutual expectations and ensures the predictability and permeability of social situations. Encounters with a climate should have different qualities and characteristics than encounters with technology, especially with modern, rapidly changing technology, which can often be experienced as a source of uncertainty and even destruction in everyday life.

Trust is indeed a basic building block of social behavior. It is the glue that holds society together. It signals reliability, security, and even predictability. Without mutual trust, social behavior is not possible. Trust in trust is not limitless. It is not to assume

> that risk and the grounds for trust will be weighed up rationally before doing anything. Trust can also be shown to be thoughtless, careless and routinized, and this acquire no necessary expenditure of consciousness, especailly if expecation approaches cretainty. [...] Trust meges grdually into expectations of continuity, which are fomed as the firm guidelines by which to conduct out everyday lives.
>
> *(Luhmann, [1973] 1979: 24–25)*

This means that trust can also be a risk (Luhmann, [1973] 1979: 27). It can be assumed that trust extends not only to social institutions and individuals but also to environmental conditions, even if this means that the environment is not seen as dangerous and hazardous from time to time.

At the same time, trust in climatic and other ecological processes offers important relief (*Entlastung*) from immediate weather concerns and provides scope for action and reflection on other pressing issues of everyday life. Paradoxically, relief leads to an increase in freedom, as it makes it possible to focus on other, more complex tasks instead of being confronted with existential challenges such as weather extremes. The practical function of trust in dealing with climatic conditions is similar to that of trust in social life in general. Trust serves to reduce and cope with, but not eliminate, the complexity of a phenomenon that is always present in the background of everyday life (cf. Luhmann, 1973).[14] The reduction in complexity through trust becomes more important as everyday life becomes more structured and complicated. However, the centrality of concepts such as trust also harbors ambiguity, fragility, and perhaps a lack of real understanding. Peter Berger and Thomas Luckmann (1967: 57) use the category of *relief* in reference to Helmut Plessner and Arnold Gehlen's philosophical anthropology as one of the central categories of their sociology of knowledge:

> Habituation [...] frees the individual from the 'burden of decision' and provides psychological relief, the anthropological prerequisite of which

is the undirected instinctual apparatus of man. Habitualization provides precisely the direction and specialization of action that the biological make-up of man lacks and in this way reduces tensions that come from undirected instincts.

As a result, habitualization or routinization forms a "secure background against which human activity can take place"; psychologically, routine activities may even result in pleasure. Barry Barnes (1988: 167) also stresses the importance of the ubiquity of routine or matter-of-course conduct in everyday-life: "routines may be thought of as co-ordination equilibria in societies of calculative agents, patterns of activity from which no significant number of individuals would wish to deviate so long as most individuals continue to conform."

Trust, as Georg Simmel (1906: 450) emphasized, is often a mixture of knowledge and ignorance (see also Stehr and Adolf, 2018).[15] Nevertheless, trust is involved when the confident expectation of, for example, an even progression of the seasons over the course of a year affects the decision-making of individuals and collectives. Trust in future states excludes certain possibilities or alternatives for social action. Trust becomes a rational anticipation of future states. Trust bridges time, eliminates uncertainty for the time being, and generates trust in the future present. Only if the future conditions of social action are fully anticipated and known is trust not necessary as a basis for practical action. Such anticipation is, of course, practically impossible, or inconceivable.

In the course of social action, trust becomes institutionalized or routinized and stable but can become "dysfunctional" under transitional conditions, such as those caused by climate change. To the extent that trust in the stability of climatic conditions itself becomes reflexive, to the extent that our understanding of the foundations and functions of trust becomes conscious and transparent, it could cast doubt on its suitability as a guideline for future social action. The trust that societies place in climate and the trust that climate engenders through the repeated occurrence of expected effects do not mean that the same weather-related phenomena cannot be interpreted completely differently by different communities. For example, Mary Douglas ([1970] 1975: 234–235) reports observations from her field research in Africa of contrasting interpretations of similar climatic features by neighboring communities. One community perceives the hot, dry season as unbearably hot and longs for rain. The neighboring community on the other side of the river perceives the summer weather as pleasantly cool and dreads the onset of the rainy season. In both cases, the perception of the weather and its effects is based on trust in the balancing forces of nature.

More generally, what are the key relationships that best describe our typical encounters with climate phenomena? The relationship to natural events, in which we are primarily spectators, and the experience of nature as an impartial reality stand out. If one roughly distinguishes between the possibilities available to us to assume or be forced to assume the role of an active agent or that of a spectator in a social action context, the typical role that humans are willing to assume in relation to climate is that of spectator. Unlike other parts and features of the environment, which have always been actively managed, climate has not been seen as influenceable by human behavior. It may have been controlled verbally or cognitively, but for centuries, the typical human relationship with climate was passive. At the same time, many communities throughout history were convinced that they could be more than mere spectators with respect to the weather. Techniques for influencing the weather are widespread. However, their success and belief in their effectiveness are highly dependent on beliefs about the normality of climate patterns, i.e. attempts to induce rain are likely to be linked to beliefs and "objective" signals (increasing cloud cover) that indicate the proximity of rain.

In general, social behavior is highly stratified. Nevertheless, there are social institutions that at least try to be neutral and impartial in their behavior toward members and clients and communicate this emphatically. However, the physical conditions and context in which social interaction takes place can achieve a degree of impartiality that is otherwise unmatched in social behavior. Georg Simmel ([1908] 2009: 606) stresses the importance of space for social behavior: "space as a basis of organization possesses that impartiality and regularity of behavior that makes it a correlate of governmental power with its characteristic behavior just as suitable to all its subjects." *The modern state and* the judiciary in democratic societies strive for impartiality in dealing with their subjects. However, whatever social differentiation and inequality develop in relation to space, e.g. place of residence, it appears to be more "rational" than a differentiation linked to ascriptive factors such as kinship or unearned income. However, local or regional emancipation from the climate does not preserve the necessary impartiality toward the climate. In contrast, efforts to standardize the climate, e.g. through air conditioning, lead to a stratification of climate experiences. The effect of the climate now depends on the individual's ability to pay for the partial deprivation of the natural climate and the creation of an anthropogenic climate. The emancipation from certain characteristics of the climate (temperature, flooding, humidity) also has profound demographic, social, and economic consequences. One example of this is the population growth in the South and Southwest of the USA in the 1970s and 1980s, which can be attributed to the widespread use of air conditioning.

The transformation of nature into a socially constituted environment in modern society has led to the climate rightly being experienced as extremely dependent and serving as a reliable backdrop for many of society's actions. The extent to which weather and climate are directly associated with "outdoors" has decreased significantly. As a result, the climate filtered or shielded by the creation of "total environments," e.g. at home, at work, and on the way in between, has become a largely self-evident, constant framework for us.

Extreme events as coincidences and triggering events

> The world is on track for a "hellish" 3°C of global warming, the UN has warned ahead of the crucial Cop28 climate summit starting next week [November 2023] in the United Arab Emirates. The report concludes that today's carbon reduction measures are so inadequate that 3°C warming would still be achieved this century.
>
> *The Guardian, November 20, 2023*

The contingent social construct of climate is part of the collective consciousness (or *conscience collective*) of society. According to the classical definition of Emilé Durkheim (|1893] 1933: 79), the collective or common conscience is the "totality of beliefs and sentiments common to average citizens of the same society forms a dterminatre systems which has its own life; one may call it the *collective* or *common conscience.*" The *collective conscience is* therefore independent of the conditions in which the individual finds himself; it is more than the sum of its parts, and it compels the individual to act in this sense. The sense in which Emilé Durkheim introduced this term was that any apparent deviation from accepted social, shared normative standards is a kind of offense and is interpreted by society as a moral breach that must be healed. Durkheim ([1893] 1933: 81) refers to the example of crime to illustrate that an act is not criminal because it is criminal but because it offends the collective conscience: "We must not say that an action shocks the common conscience because it is criminal, but rather that it is criminal because it shocks the common conscience." Durkheim noted that the effect of punishment in terms of deterrence is quite limited. The actual function of punishment is to strengthen and maintain social cohesion.

Not necessarily by strict analogy, but in line with Durkheim's thinking, it can be argued that climatic extremes confirm the normality of climatic conditions. Weather extremes are not disappointments but rather confirm typical weather and climate conditions.[16] Nature is therefore not a silent world. Extremes are events that reinforce the conviction that the climate is constant and that a return to normal can be reliably expected. Normality balances out the extremes; it cements the common-sense awareness or frame in which we can rely on the constancy of the climate. In addition, there is something like a basic psychological mechanism, a kind of psychology of denial, which may be responsible for our blocking out future unpleasant developments and which is responsible for our inability today to deal proactively with the dangers of the climate crisis.

A characteristic feature of the reinforcement of belief in the normality of climatic conditions in the face of major weather extremes is the reactions of political actors who are not convinced that the climate crisis is real, dangerous, and risky. Extreme events do not falsify their beliefs. Two major hurricanes that hit Florida in the United States just weeks apart in September and October 2024 are examples. The hurricanes did not change the response of most Republicans in the U.S. Congress when it comes to supporting tougher action on climate change. Most Republican politicians do not see climate change as a crisis, and the storms did not change that: "As Governor [Ron] DeSantis noted [...] major hurricanes are a natural phenomenon that have been a part of Florida life dating back to the 1800s," said Representative Daniel Webster (R-Fla.), whose district was affected by hurricanes Helene and Milton.[17]

Christopher de Bellaigue (2023) recalls an observation relevant in this context by Leo Tolstoy in his novel *War and Peace* in the face of the advance of Napoleon's armies on Moscow:

> There are always two voices speaking with equal force in the human soul: One tells man, quite reasonably, that he should think about the nature of the danger and the means of escaping it; the other, still more reasonably, says that it is too distressing and painful to think of the danger, since it is not in man's power to foresee everything and avert the general course of events, and it is therefore better to ignore the painful until it comes and to think of the pleasant. In solitude, a person usually listens to the first voice, but in company to the second [...] It has been a long time since people in Moscow have been as cheerful as they have been this year.

Tobias Rüttenauer (2023: 1) carried out a complex empirical study in the United Kingdom investigating how extreme weather events, in this case, floods and heatwaves influence "people's concern about climate change and their pro-environmental behaviour."[18] One of the core assumptions of this study is that personally experiencing climate change can influence attitudes and behavior related to climate issues. Moreover, extreme weather events should "have a stronger impact on climate change concern and pro-environmental behaviour, the closer these events are in time and space" (Rüttenauer, 2023: 4). The reported findings indicate that

> floods and heatwaves indeed increase the belief in climate change. The magnitude of this association increases with spatial and temporal proximity. [...] However, updating attitudes does not change peoples' pro-environmental behaviour. Extreme weather events have a negligible

impact on peoples' behaviour, and this finding is consistent across floods and heatwaves, across different behavioural measures, as well as across several subgroups.

(Rüttenauer, 2023: 3)

These results are consistent with the findings and theories of mainstream social-psychological literature about action and intention (e.g. Ajzen, 1985, 2011). It is possible that "soft denial" of climate change is still active despite experiences with weather extremes in one's geographical vicinity.

To place my reflections on weather extremes in a broader societal context, a context that is not necessarily directly related to the public's response to weather extremes, I will point to some core characteristics of modern society that should be seen as supporting a sober response to weather extremes. Essential , however, is that to adequately study climate change, sociocentrism must be transcended.

It is a matter of continuity

The history of social science discourse alternates, and many shades in-between, a strong interest in the future of society, the potential of social transformation underway or visible in the not-too-distant future, as progress toward a better society, or dismissing concerns about the future as a hopeless enterprise since the future is unknown. As a result, the potential of the social sciences to contribute wavers equally between hope and dismissal. Many social scientists became weary of future narratives. Instead, Norbert Elias (1987), for example, urged social scientists to engage more actively with the past. In contrast, Huber, Boldt, Bell, and Mau (1971: 23) had earlier made an equally strong appeal for social scientists to consider alternative scenarios and possibilities for the future present.

The shifting interest among social scientists in the future present also reflects a corresponding change in societal sentiment about the future. In the years immediately following World War II, there was a pervasive sense that the future was irretrievably lost. The devastation wrought by the Holocaust and the specter of nuclear annihilation undermined the conventional wisdom that humanity was progressing toward a desirable future. Social scientists' empirical findings corroborated this widespread sentiment, confirming a sense of future opportunity being foreclosed. In contemporary societies, this apprehension about a threatening future persists, particularly among younger generations.

Apprehension about the state of the future present comes into conflict with another fundamental expectation about the future present in modern societies, which can be characterized almost neutrally in that the nature and motor of the ongoing transitions of modern societies can generally be described as the *extension and enlargement of social relations and social rewards*.

The extension or evolution of social relations and social rewards points both forward into the future and backward into the past—affirming the law of the simultaneity of the non-simultaneity of social phenomena.[19] For example, with each successive stage of capitalism over the past few centuries, the accumulation of wealth has reached unprecedented levels, whereas poverty never disappeared. Enlarging the economic pie could be seen,[20] as is often argued, not only in political campaigns as a key to sustaining actors who struggle with the social order. For example, there is evidence, presented by at least some observers, that political democracy requires not only "civil society" but also economic growth as a (pre)condition (cf. Lipset, [1960] 1980: 31; Burkhart and Lewis-Beck, 1994). As Adolph Lowe (1971: 570) observes, there can be little doubt that economic growth has been "an instrument for easing social frictions, and that any constriction of our standard of living would exacerbate the struggle for shares that is masked or muted when consumption steadily rises." Political theorists of all stripes, except for "de-growth" advocates (e.g. Soener, 2021; Saito, 2023), endorse the idea that economic growth is politically needed. Therefore, the widely supported conclusion, as expressed, for example, by Michael Mann (2013: 366), is as follows: "All modern states have scarificed the environment to GDP, regardless of regime type." Michael Mann's observation, however, rightly raises the question posed by Michael Th. Graven ([2009] 2020: 166): "Can Western democracies survive a prolonged period of relative or even absolute decline in individual and collective welfare as democracies?" The widespread promise (in the West) to live "a life of our own" (Beck, 2000: 164) includes the expectation of enhanced life chances and hunger for an enlargement in social experiences and rewards. Correspondingly, a part of the expectation to live one's own life is the *rejection* of obeying commands, getting into line, and *making sacrifices*.

Ralf Dahrendorf (1979: 122) sums up these expectations and demands of the citizens of modern society:

> modern societies are based on the principle of giving as many people as much as possible. There are several ways of formulating this principle – the utilitarian way of 'the greatest happiness of the greatest number', the Marxian way 'to each according to his need, from each according to his ability', or more generally as 'the full development of the capacities of all' – but in one way or another, the twin assumption, that society should enhance the life chances of its members.

Ralf Dahrendorf (1979: 122) adds, "The purpose of equality is inequality; the purpose of general rights is different individual lives."

Weather extremes and climate policies

Recent climate research has shown that human-induced climate change is increasing the frequency and intensity of extreme weather events such as wildfires, severe storms, floods, and droughts. But as I have emphasized, extreme weather events are more episodic in terms of their impact on promoting changes in the collective consciousness about climate narratives. By the same token, climate policy changes tend to be a slow process. But it is of course possible that weather extremes, and perhaps depending on the magnitude of the damage that extremes cause, may have implications for climate policy—representing windows of opportunity for significant societal change that new policies may require, for example, incentives for new regulations regarding mitigation efforts or adaptation policies. Weather extremes may at least be a necessary but not a sufficient condition for political movement on the climate change policies.

Empirical evidence is scarce. As far as I can see, among the only empirical studies of a relation between weather extremes and climate policies is a study by Lauri Peterson (2021). Her study examines whether extreme weather events influence national climate change mitigation policies, focusing on the role of democracy as a moderating factor. Using data from 58 high- and middle-income countries (2008–2017), Peterson examines the relationship between climate-related disasters (e.g. floods, storms, wildfires) and policy ambition (that is, policy outputs regarding greenhouse gas emissions), as measured by the *Climate Change Performance Index*. Among the key findings are that extreme weather events alone do not significantly increase climate policy ambition across all countries. The hypothesis that such events universally spur policy change is rejected, as the effect is weak and short-lived.

Peterason finds that highly democratic countries (using the *V-Dem Electoral Democracy Index*) are more likely to enhance climate mitigation efforts after extreme weather events. Democracies, with their accountability mechanisms and public engagement, respond more robustly to disasters by prioritizing climate action. Autocracies, lacking public pressure, show no significant policy changes. India and the UK exemplify democracies where extreme weather (e.g. floods, heatwaves) correlated with increased climate policy measures. In contrast, less democratic nations like Thailand exhibited minimal policy shifts despite severe disasters.

The study by Lauri Peterson includes several control variables to account for other factors that might influence climate change mitigation policies. These controls are used to isolate the effects of extreme weather events and democracy. Here are the key control variables: Quality of Governance, GDP per capita, economic growth, carbon-intensive industry, federalism, and Government Ideology. This approach strengthens the study's conclusion that democracy amplifies the impact of extreme weather on climate policy, independent of economic or structural factors.

Extreme weather events alone do not significantly increase climate policy ambition across countries. The effect is weak and short-lived, suggesting that such events are insufficient to drive substantial policy change without other factors. Influential voices that question what most climate scientists know about changes in our natural environment may be strengthened by widespread conviction that the climate is essentially reliable. However, the observations about the political impact of weather extremes highlight the critical role of democracy in translating climate crises into policy action, underscoring the importance of institutional frameworks in addressing global warming. While disasters alone are insufficient drivers, democratic institutions amplify public demand for mitigation. The analysis focuses on policy ambition (outputs) rather than emissions reductions (outcomes) and excludes low-income countries. Results may not capture long-term effects beyond the studied decade.

A world unhinged

> In the depths of winter, I finally realized that I had an invincible summer inside me.
>
> *Albert Camus ([1953] 1957: 157)*

> Extreme weather is hitting people's wallets more than ever before. Whether it's the price of air-conditioning or soaring flood insurance rates, the financial cost of enduring today's climate seems to grow each year.
>
> *New York Times, December 28, 2024*[21]

Consistency seems to be a seal of quality in many areas of life. This is undoubtedly true of politics, the family, or the world of work, not to mention religious life. Continuity, one might say, is one of the guarantees of psychological well-being. Continuity provides relief. I am using the term "relief" in the broad sense of Arnold Gehlen's ([1940] 1988: 54) concept of *Entlastung*. *Entlastung* is usually translated into English as "relief" or "unburdening." In Gehlen's philosophical anthropology, relief is the process by which humans create structures, institutions, tools, and routines to "unburden" or "relieve" themselves of constant decision-making or direct engagement in survival tasks. In other words, relief represents the way in which humans reduce the burden of everyday challenges through these external supports, allowing for a more efficient and directed use of energy. By reducing the mental, physical, and emotional "load" of constantly responding to environmental pressures, humans achieve a degree of stability and predictability that allows for cultural, intellectual, and social development.[22] This concept is essential in Gehlen's view because he sees humans as *Mängelwesen*, or "deficient beings," who lack strong instincts for survival and therefore rely on external aid to compensate. Through relief, humans construct systems—such as technology, habits, and social norms—that "relieve" or "unburden" them of some of the challenges that would otherwise require constant, conscious attention. In the context of climate change, the seasonal continuity of climate is the driving force behind the perception of climate as unchanging. Paradoxically, seasonal change guarantees that the climate is not changing. Arnold Gehlen's basic philosophical conception can be traced to the forgotten work of Paul Alsberg (1922)—which was unacknowledged by Gehlen. Consistency protects against loss of trust. Consistency cures the weariness of change.[23] In addition, as it appears to many observers of the climate, nature follows some sort of *self-regulation.* The emphasis on the continuity of seasonal and, ultimately, climatic conditions perhaps inadvertently plays up human exceptionalism and the causal primacy of social processes (sociocentrism). Proponents of sociocentrism then engage in essentially contentious debates about which social phenomenon or process is the fundamental level. Human exceptionalism allows nature to be overlooked or viewed as a benign background condition (Connolly and Macdonald, 2015: 264). However, a climatic world in which seasonal variability is significantly intensified by climate change argues against such a comfortable constellation (see Moore, 2024).

How will socio-environmental concepts, now widely part of the collective consciousness of the public and developed long ago and sustained over centuries by increasing liberation from risky climatic conditions, respond to an uncertain future? In the future present, when certainty and confidence in climate normality may no longer be the norm. Global warming increases the likelihood of extreme weather events such as droughts, heat waves, and heavy rains. As NASA data (in Excursus 3) demonstrate, the monthly intensity of droughts and floods by affected area (in square kilometers) increased significantly between January 2003 and December 2024:[24]

Monthly intensity of droughts and floods by area affected, Jan 2003 to Dec 2024

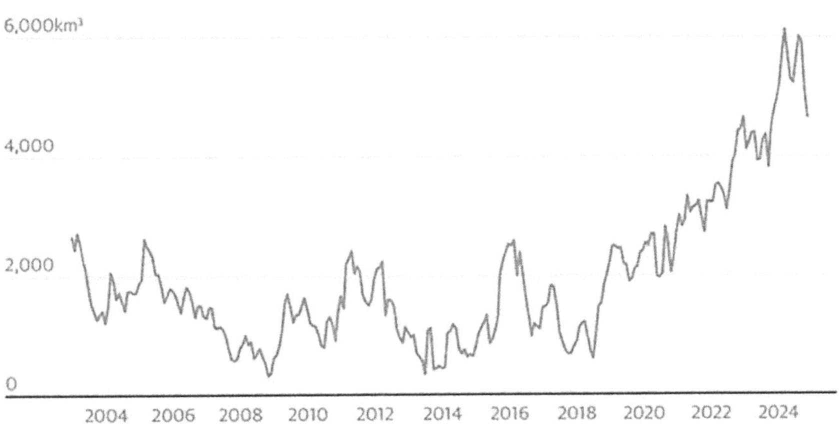

Guardian graphic. Source: Nasa Grace/GWI, Rodell and Li

EXCURSUS 3

In 2023, Canadian wildfires were so large that they released more carbon dioxide into the atmosphere than all but three other countries: the United States, China, and India. The public consciousness (conscience collective) in the face of widely accepted scientific evidence also signals that scientific claims and reasoning are not the only "combatants" involved in complex political issues (see Hulme, 2009; Brown, 2014).[25]

We may have "forgotten" nature, but nature increasingly reminds us of its existence (Serres, [1992] 1995: 29). Will it be possible to develop a new common understanding of the climate, let alone a policy, under conditions that may have destroyed centuries-old environmental ideas? How can we increase our sensitivity to our interdependence with other agents? How can we overcome the division into strict tribal areas within the sciences? In any case, a long-term perspective is needed to answer these questions.

My assessment suggests that (1) modern environmental awareness is based on trust in the normality of the climate and thus on the idea of extremes as random events, as well as on the observation that (2) people experience natural events primarily from a spectator's perspective and that (3) they are convinced that natural events are indispensable, while the shaping of future society is becoming increasingly decision-dependent and decision-intensive (Luhmann, 1991: 6; Stehr, 1994). The social trust extended to the climate means, as it does in the case of the trust extended to one's fellow citizen or a social institution such as the stock market, that the climate will "behave" as it ought to behave most of the time. In modern society, for these reasons, the natural climate is perceived by large sections of the population as not relevant to every-day decision-making. Climate is not an important determinant of well-being, as *permanent* living conditions have little influence on well-being (see Kahneman, 2011: 393–397).[26] It will be correspondingly difficult to make and implement appropriate political decisions that extend to the natural climate of societies.

My emphasis on the role of social trust is based on the conviction that climate has significant, reliable attributes that stand in a peculiar contrast to or even contradict the often-heard decline in public sentiment toward representative democracy (Norris, 2011). In particular, the assertion that Western democracies suffer from a declining level of trust among their citizens is accompanied by demands for political reforms. As Simon Tormey (2015) maintains, the level of trust in politicians has reached an all-time low. There has been a notable decline in voter turnout and membership in political parties, accompanied by a rapid reduction in public interest in parliamentary politics. Even opposition and radical parties, which should benefit from the public's growing disenchantment with politics, are experiencing difficulties. However, alternative forms of political activity, including instant politics, direct action, and insurgent politics, are emerging to fill the void left by representative politics. Many societal attributes of the trust crisis have been identified. More recently, climate change has been introduced as a contributing factor to the discussion.

One of the fundamental questions that environmental research and climate policy can therefore confront is the possibility that the foundations of trust, which have thus far played a key role in determining our everyday relationship with nature, may be called into question not only by the current research findings of climate science but also by the reality of climate change. One can only speculate about what might lead to a sudden or rapid loss of trust, and what would happen if the basis on which trust was built were suddenly removed (cf. Coleman, 1984). Research must address the possible fear-inducing effects of climate change. How exactly does climate anxiety arise? How does mental health change because of emotional reactions such as increased anxiety to climate change? What role do extreme weather events play?[27]

To summarize, the double threat to trust in nature and the partially shaken trust in the credibility of science (see the backlash to the political handling of the COVID-19 pandemic) is a potentially aggravated problem of the climate crisis.[28] To reduce mistrust in science, the public needs to develop a better understanding of the scientific community and how it works. In summary, social awareness of a stable climate is a form of relief for the fundamental fragility and indeterminacy of human existence. We are often forced to act without knowing it. Preestablished trust in a stable future provides relief and security. However, for how much longer? Trust or confidence can, depending on the circumstances, move beyond the rational and rapidly become overconfidence and self-defeating (cf. Akerlof and Shiller, 2010).

A reference to the nature of frequent and extensive media coverage of extreme weather events should count as evidence for my argument and as complementary theoretical as well as empirical perspectives. Paradoxically, it has been argued that normality only manifests itself in contrast and is thus simultaneously asserted: as coverage of extreme weather events all too often demonstrates, "I've lived here for 30 years and I've never seen weather like this," is the typical statement of a local resident in the face of an extreme event. The extreme is ordinary. As so often in everyday life, consistency is a virtue. As a *Nature Climate Change* editorial noted, "many more factors than just meteorological extremes are necessary to make a climate disaster,"[29] and many more factors than just meteorological extremes are necessary to support the public's view that climate change is a real threat. As Luisa Neubauer, the climate activist of *Fridays for Future* Germany, accurately observed, the rise in extreme weather is not generating political support for climate action.[30]

A reference to journalist, biographer, and novelist Gore Vidal's phrase "United States of Amnesia" can also serve as an interpretation of the rapid return to normality following weather disasters. Gore Vidal coined the term

"United States of Amnesia" to criticize what he perceived as a pervasive cultural and historical forgetfulness in American society. Vidal commented on what he saw as Americans' preference for an optimistic, forward-looking narrative, often at the expense of historical awareness and critical reflection. He believed that this collective forgetfulness served the interests of the political and economic elite by preventing the populace from learning from past mistakes and holding leaders accountable. Finally, Harold Innis's (1951: 76) seminal *Bias of Communication* offers another perspective on the consequences of communication patterns and substance in the modern age as a way of understanding the pervasive "forgetfulness" in modern societies. With information flooding the public sphere, people are falling victim to what Innis called the "present-mindedness." Overwhelmed by volume and the immediacy of new information, the public lacks the incentive to view information in a broader and critical context and engage in the reflexive work of interpretation: "Enormous improvements in communication have made understanding more difficult" (Innis, 1951: 31).

The next chapter examines the argument that scientific expertise should inform climate policy, potentially replacing democratic governance with technocratic rule. It explores how trust in 'abstract systems' and expert knowledge has become central to modern societies, with some advocating authoritarian governance to enforce scientific consensus on climate change. Supporters such as climate researchers and philosophers claim that the slow processes of democracy are inadequate for addressing urgent environmental crises, and propose "enlightened leadership" by scientists with specialized knowledge. This concept is similar to Plato's idea of an epistocracy, where governance is reserved for the experts. However, I highlight the critical limitations of this science-centered approach. Technical knowledge alone cannot eliminate political uncertainties or address fundamental issues such as resource constraints, competing interests and differing worldviews. A comparative study of 22 countries reveals that, while public trust in scientific experts remains high, people often overlook the importance of social and ethical expertise in climate governance. The chapter concludes by arguing that, while climate science can inform policy positions, knowledge alone cannot ensure effective governance or replace democratic deliberation.

Notes

1 The idea of "*climate disasters*" is based on the notion that they are more severe, long-lasting, and damaging than "*weather extremes*" (cf. Seymour, 2924: 58–78). My excursus builds on a paper published under the title "Trust and Climate" (Stehr, 1997). I also acknowledge the monograph *Society and Climate* (Stehr and Machin, 2019: 136–140), in which we examined the social phenomenon of climate awareness in everyday life.

2 Economists have examined large aggregate "shocks" such as the financial crisis or the pandemic because "shocks may help "explain new trends (such as the rise of populism and anti-globalization) and because shocks help test alternative theories or conjectures concerning the formation of preferences and beliefs" (Giuliano and Spilimbergo, 2024). Weather extremes therefore differ from shock events due to their shorter duration.

3 See, for example, Zanoco et al. (2019). As Joshua Ettinger and co-authors (2023: 829) point out, "further research is needed to explore how environmental communication practitioners can use the occurrence of extreme weather events as an opportunity to promote individual and policy action to reduce future risks." The developing field of extreme event attribution research in climate science attempts to disentangle "different drivers of extreme weather from human-induced climate change and hence provides valuable information to adapt to climate change and to assess loss and damage" (Clarke et al., 2022).

4 The scientific conviction of a fundamental basic order of the world can also be understood with the logician Kurt Gödel as a reference to the rational order of the world. As Goedel emphasizes in a letter to his mother dated 23 July 1961: "Is there any reason to assume that the world is organized rationally? I think so. For it is absolutely not chaotic and arbitrary, but - as natural science shows - there is the greatest regularity and order in everything. Order is indeed a form of rationality" (in Englert, 2023).

5 Agricultural meteorologist Andreas Brömse from the *German Weather Service* on May 4, 2024 in *Spiegel* Interview, No. 19, p. 40.

6 In my opinion, there are two further reasons for the disinterest of classical social theories in climate issues or human natural capital: (1) the necessary intellectual separation of developing social theory from the natural sciences as a guarantee of the autonomy of the social sciences, and (2) the previously widespread idea that humans can adapt to any climatic zone. The global patterns of settlement prove and guarantee that humans are capable of living and forming societies in all climatic zones of the earth.

7 Alexis de Tocqueville (2010: 453) described his observations of the French-speaking inhabitants of North America in a letter from 1832 (during his trip to America) as follows: "If you meet people who tell you that climate has nothing to do with the constitution of nations, assure them that they are mistaken. We have seen the French in Canada: They are a quiet, moral, and religious people; in Louisiana, we left other Frenchmen who were restless, dissolute, and negligent in everything. There were 15 degrees of latitude between them; that, in truth, is the best reason I can give for the difference" (letter to Ernest de Chabrol, January 16, 1832).

8 "An important part of the objection to the sociology of knowledge, as David Bloor ([1976] 1991: 8) points out, "stems from the conviction that some beliefs need no explanation or require no causal explanation. This feeling is particularly strong when the beliefs in question are regarded as true, rational, scientific or objective." In other words, the constructivist approach doubts a kind of "autonomy" of knowledge based, for example, on the thesis of the possibility of an "objective" representation of nature.

9 This can be seen, for example, in the routine response of a member of the public, prompted by a local media representative reporting on the extreme event, e.g. a

hurricane that has just passed: "I have lived in this region for 40 years and have never experienced such an event," implicitly hoping that the extreme event will not occur again for another 40 years.

10 In this context, compare Gernot Böhme's (1992: 56–76) very useful and illustrative treatise on such intuitive properties ascribed to nature at different points in time.

11 According to the Eurobarometer survey conducted between May and June 2023 and cited below, only a third of respondents feel *personally* responsible for climate change. Worldwide, economic issues and concerns are generally at the top of national priorities.

12 Source:https://germany.representation.ec.europa.eu/index_de?prefLang=en.

13 Bergquist and Warshaw's conclusion is based on a meta-study of over 400,000 respondents from 170 surveys conducted in the US between 1999 and 2017.

14 Niklas Luhmann ([1973] 1979: 24–26) elaborates a form of trust that is relevant in our context by suggesting that "trust ... can turn out to be thoughtless, careless and routinized, requiring no unnecessary effort of consciousness, especially as expectation approaches certainty [...] *Trust gradually transitions into the expectation of continuity, which emerges as a firm guide for our everyday lives*[... But] trust remains a risky endeavor (emphasis added)."

15 Georg Simmel (1906: 450) points out: "Trust as a hypothesis of future behavior, which is certain enough to become the basis of practical action, is, as a hypothesis, a mediating condition between knowing and not knowing another person. The possession of full knowledge makes the necessity of trust superfluous, while the complete absence of knowledge obviously makes trust impossible."

16 We can be confident that such a response to weather extremes will be aided by the shrinking half-life of public and media attention to rapidly changing agendas.

17 "Hurricanes do little to move Republicans on climate," *Politico*, October 12, 2024; www.politico.com/.

18 The study uses, first, "individual-level panel data of nearly 36,000 individuals and link those to recorded flood outlines in England and weather data across the UK based on participants' place of residence. Second, we apply panel-data methods and within-person estimators to rule out person specific confounding. Third, the study investigates two types of extreme weather events – floods and heatwaves – increasing the robustness and generalisability of results. Fourth, it tests the impact of extreme weather on climate change belief and various forms of behavior, which allows us to test if attitudinal changes translate into more sustainable behaviour. Fifth, it differentiates the impact of extreme weather events by spatial and temporal distance to the respective events. Finally, we explore heterogeneous effects according to partisanship and initial climate scepticism, which provides additional insights into the psychological pathways of the relation" (Rüttenauer, 2023: 2–3).

19 The methodological and theoretical significance of the law of the simultaneity of the non-simultaneity of social phenomena is discussed in more detail in Stehr (2022).

20 In a more technical sense, the thesis is linked to the welfare economics theorem of Vilfredo Pareto optimality, that is, one is aiming for an economic condition where some groups of people would be better off without anyone being worse off.

21 Shannon Sims, "Weather's impact on your wallet: See how much you know," *New York Times*, December 28, 2024; www.nytimes.com/interactive/2024/12/20/weather/extreme-weather-costs-quiz.html.

22 A concrete part of a relief process in the event of extreme weather is the straightforward conviction among the public that one is not residing in a danger zone for extreme weather events, for example, in a wildfire zone: "Millions of Americans believe they're safe from wildfires in their cities. New research shows they're not," *Guardian*, April 5, 2025; www.theguardian.com/us-news/2025/apr/05/us-wildfires-cities-dangers.

23 John Keane (2022: 4–6) summarized the importance of the conceptual certainty/uncertainty twin very well: "We usually deal with uncertainty by pushing it aside and hiding from it when we can. [...] Certainty about ourselves and the world comes from our efforts to make things certain, to certify them." Uncertainty, the twin of certainty, cannot be banished from human affairs.

24 Source: www.theguardian.com/world/2025/jun/17/nasa-data-reveals-dramatic-rise-in-intensity-of-weather-events.

25 For example, as Göran Sundqvist and Linda Soneryd (2024: 510) note in their encyclopaedic summary of the relationship between scientific knowledge and governance, "sociologists have [...] recurringly shown that what is unique with environmental problems, compared to many other policy areas, is their relation to science. This means that any constructive revitalization of a democratic governance of environmental issues also needs to take the relation between science and democracy seriously," hence, foregrounding the significance of scientific knowledge in the political process (that is, "listen to science") while minimizing if not neglecting the politics of scientific knowledge (also Yearley, 1992).

26 In this context, the economist Daniel Kahneman (2011: 393) speaks of the "focusing illusion," i.e. "Nothing in life is as important as you think it is when you think about it."

27 See, for example, Clayton (2020) and Tam, Chan and Clayton (2023). An online survey conducted in Germany in mid-March (i.e. during the COVID crisis) (Hajek and König, 2020) indicates a rather low level of climate anxiety in Germany. In addition, "some correlates of higher climate anxiety, such as a higher fear of the coronavirus or a greater fear of war, were found." On the other hand, both authors (Hajek and König, 2023) report a "correlation between climate anxiety and mental health" in another study.

28 Compare America's divergent attitudes in "The strange new politics of science" by M. Anthony Mills and Price St. Claire (2025) of the system of expert knowledge. Survey data demonstrate that particular domains of science linked explicitly to political power, climate change, and public health, are among the most divisive fields of expertise among the public.

29 "Beyond extremes," *Nature Climate Change*, December 14, 2024: 1211; https://doi.org/10.1038/s41558-024-02214-9.

30 Ajit Niranjan, "Extreme weather failing to encourage political climate action, says activist Luisa Neubauer," *Guardian*, January 24, 2025; www.theguardian.com/environment/2025/jan/24/extreme-weather-failing-encourage-political-climate-action-luisa-neubauer.

5

SCIENCE AS THE SOLUTION—POLICY FOLLOWS SCIENCE

In the context of a looming climate catastrophe, a prominent political position is to argue that science is the solution or that policy should follow the scientific expertise. Moreover, scientific and technical solutions have the potential to remove politics entirely from environmental policy. Scientific consensus exerts normative authority. In the area of a "post-political" regime (Swyngedouw, 2009), expert scientists assume the role of the masters of society. These observations about the role of scientific expertise in modern societies follow Anthony Giddens' (1990: 21; emphasis added) observation about the disembedding logic of "abstract systems" [bracketing of time and space by social systems] in modern societies: "Abstract systems disembed social relations by lifting them out of local contexts of interaction and restructuring them across indefinite spans of time-space. Two main types of abstract system are symbolic tokens [like money] and *expert systems*." Trust is a form of faith in abstract systems; it is trust in the correctness of principles—technical knowledge or procedures—rather than in the individuals who operate them; however, it is more than merely placing confidence in a cognitive understanding. It is therefore not uncommon for people in modern societies to place their trust in systems and experts rather than in personal relationships. Individuals and groups still act reflexively—constantly adapting, but on the basis of advice from abstract systems. Following the science, in short, is a reasonable response, perhaps even an unavoidable requirement, in the face of complex risks and hazards in modern societies.

DOI: 10.4324/9781003682066-7

Scientists to the political rescue

> Do we really think that our judgement – or that of anyone else would be as good as that of a scientist whose lengthy immersion in these issues leads to the admittedly imprecise assessment offered?
>
> *Philip Kitcher (2011: 34)*

The philosopher Philip Kitcher (2011: 40) is persuaded that value judgments are a constant part of scientific discourse. Given this premise, they cannot be eliminated, and one should not wait to offer advice for contexts that allow values to be factored out. Moreover, the advice offered by scientists, as Helmut Willke maintains, should be authoritative in our public deliberations, even though value judgments figure in their decisions (Willke, 2023).[1]

The radical conclusion drawn by some observers, especially those who advocate and promote the role of experts and expertise as a form of enlightened leadership, is that democracy itself is inadequate and that the slow procedures for implementing and managing specific, politically relevant scientific knowledge led to massive, unknown risks and dangers. Civilization as we know it could come to an end. Assuming that it is not already too late, an appropriate environmental policy must look very different. To create a globally sustainable way of life, we need, in the words of German climate researcher Hans Joachim Schellnhuber (see WGBU, 2012), an immediate "great transformation." Part, if not the core, of the required great transformation appears to be a new political regime and new forms of governance: for example, the Australian (natural) scientists David Shearman and Joseph Wayne Smith (2007: 12) do not mince their words in their study *The Climate Change Challenge and the Failure of Democracy*: "We need an authoritarian form of government in order to implement the scientific consensus on greenhouse gas emissions."

This perspective is not new but has ancient roots. The philosopher Plato strongly articulates such a view in the *Republic*. Political governance should be left to those with specialized knowledge. The ability to acquire specialized knowledge through special training is limited to a small number of individuals. Modern philosophers, political scientists, and economists are not averse to agreeing with Plato's "epistocracy." Economist Bryan Caplan (2007), for example, argues that "irrational voters" should have limited political influence in contrast to the knowledgeable elite. Institutional opportunities in democratic governance must provide for a division of political power: "A test of voter competence is no more objectionable than a driving test [...] a palatable way to raise the economic literacy of the median voter is by giving *extra* votes to individuals and groups with greater economic literacy" (Caplan, 2007: 197). How can we justify such a conclusion?

The political scientist Mark Beeson (2010: 289) argues along the same lines when he brings the concept of *good authoritarianism* into play:

> [...] given the unprecedented and irreconcilable nature of the challenges we collectively face [...] forms of 'good' authoritarianism, in which environmentally unsustainable behaviors are simply prohibited, may become not only justifiable but essential to the survival of humanity in anything approaching a civilized form.

Another suggestion relates to an openly political role for climate scientists. At the heart of the grand narrative of the great transformation are, in all instances, scientists who are "represented as the ecological vanguard of the world. Not only do they appear as the spokespeople for the Earth but also as shepherds of a public opinion that is ignorant and helpless [...] a geo-government by scientists" (Bonneuil and Fressoz, 2017: 80).

Similarly, in the broad field of climatology and climate policy, there is growing frustration with the virtues of democracy and an increasing invocation of exceptional circumstances and promotion of the role of scientists and experts in policy making. The impatience with democracy and the changing understanding of the role of scientists can be seen in a change in the function of the *International Panel on Climate Change* (IPCC). Although the IPCC is cautious as an advocate of particulate policies in its summary reports, the IPCC nevertheless no longer sees itself exclusively as a scientific organization with the task of offering alternative policy options for political discussion and decision-making, but rather as an expert body that calls for the rapid implementation of the policy options it has identified. Not only are climate scientists affiliated with the IPCC, but scientists in general are now increasingly involved in environmental governance (see Baber and Bartlett, 2021a).

Robert Stavins, director of Harvard's Environmental Economics Program and co-author of the IPCC Working Group 3 report, noted that a

> bottom-up demand, which we usually always want and rely on in a representative democracy, is unlikely to work in the case of climate policy as it does for other environmental issues [...]. It will need enlightened leadership, leaders who take the lead.[2]

Given the severity of the problem of global warming, social scientist Evelyn Fox Keller (2017: 107) argues, in an essay written in 2007, emphatically for a directly effective, practical political role for climate science. Following Sheila Jasanoff's (1990: 230–231) emphatic assertion to relinquish the idea of a strict separation between science and politics and echoing Hans Jonas'

([1979] 1984) ethic of responsibility, scientists should cross the unfortunate science/politics divide:

> There is no escape from dependence on experts; we have no choice but to turn to those (in this case our climate scientists) who have the necessary expertise [...] Moreover, for the particular task of breaking the current impasse, I argue that climate scientists may be the only ones capable of taking the lead. [... and] given the tacit contract between scientists and the state that supports them on the other side, I [...] also argue that climate scientists are not only capable of taking the lead, but that they are obliged to do so.[3]

Evelyn Fox Keller's recommendation has often been advocated but not tested by overconfident technocrats. Fox Keller's Position im Jahr 2007 scheint das Versprechen von John F. Kennedys "New Frontier"-Ära widerzuspiegeln, die von problemlösenden Experten bevölkert war, deren vorausschauende Ideen das Land zu nie dagewesenen Höhen führen würden.

Warnings against recommending a political takeover by experts/scientists were also clearly heard before the environmental crisis. It is worth noting, for example, that in his famous farewell speech in 1961, President Dwight D. Eisenhower warned about the "military-industrial complex," saying that "in holding scientific research and discovery in respect, as we should, we must also be alert to the equal and opposite danger that public policy could itself become the captive of a scientific-technological elite." A well-known scientist has taken up these concerns. In an essay in the *New York Review of Books* (November 18, 2004: 38), molecular biologist Richard Lewontin asserts that

> 'the knowledge required for political rationality, once available to the masses, is now in the possession of a specially educated elite, a situation that creates a series of tensions and contradictions in the operation of representative democracy'. A renowned social scientist, on the other hand, points out that most people 'capacity for individual rational judgment either about the quality of the evidence proffered or about the tightness of the theoretical reasoning applied to the analysis of the data. The 'harder' the science, the truer this is.'
>
> *(Wallerstein, 2004: 8)*

Does democratic governance have to be cast aside? What is the alternative? One alternative, not too far removed from the grand narrative of the great transformation, is a replacement of leadership with the rule of the *knowledgeable class* and with it the imposition of *technocratic* solutions to climate change. The idea of changing political leadership is not only to put science and scientists at the center of governance but also to depoliticize the issue of climate change (see Swyngedouw, 2010; Aitken, 2012). Another alternative, which I do not analyze further here but which has already become apparent during the pandemic crisis (Maschewski and Nosthoff, 2022), is the management of the climate crisis by large technology companies (Amazon, Apple, Google, and Meta).

Science, knowledge, and democracy

In many countries, climate science succeeds in providing governments with the authority to take a factual position on climate change. The wide diffusion of climate science knowledge among the public gives democratic societies a distinct advantage over authoritarian societies, with their typical restrictions on diffusion and access to specialized knowledge. Access to socialized knowledge improves the capacity to govern. Public access to technical knowledge potentially increases the legitimacy of political decisions based on such knowledge. Open access to knowledge strengthens the democratic order.

However, climate science and individual access to climate science knowledge alone cannot ensure that governments act in accordance with science. The strong desire to achieve certain policy outcomes dictated by climate science leads many to assume that scientific knowledge is somehow directly performative or an immediately persuasive form of knowledge and has the virtue of removing uncertainties from the political agenda. This view of knowledge privileges knowledge as a political tool and ignores the essential limits of knowledge's power (Stehr, 1991, 2025; Prewitt, 2010; Sarewitz, 2010). It is not the lack of existing or greater knowledge among relevant actors and organizations that is the problem. The problems are multiple issues that transcend knowledge and its implementation. The fundamental problem with a technocratic or science-centered approach to political decision-making is that technical or scientific knowledge does not eliminate the essential attribute of political action, namely, uncertainty with respect to multiple aspects of political decisions. In fact, relying on technical knowledge creates new socio-political problems. Uncertainties tend to proliferate (Callon et al., 2009). The problems are political issues such as a lack of resources, the influence of entrenched interests, the priority presented linked perspectives, prevailing world views, disputes about how we prefer to live, or the simple absence of political will.

Livia Fritz et al. (2025) conducted a comparative study based on survey results across 22 countries (weighted toward Global North perspectives) that explored public perceptions of experts' roles in governing climate change. The study particularly focused on intervention technologies, like carbon dioxide removal (CDR) and solar radiation modification (SRM). Using surveys (N = 22,222) and focus groups (N = 323), the authors examined the public's perceptions of experts, the roles they expect these experts to play, and the ways in which trust and distrust in scientists are expressed. The results suggest that the public generally perceives natural scientists and engineers, particularly geologists, chemists, and atmospheric physicists, as relevant experts. It seems that this reflects public views, but perhaps the study could have explored more deeply why the publics seem to overlook the importance of social, political, and ethical expertise, which is critical for governing climate interventions.

Trust in scientific institutions is generally reported as high, especially compared to politicians and industry actors, though it varies across demographics and countries: "Transparency, independent funding sources, objectivity in assessing both risks and benefits, integrity and declaration of no competing interests are established by focus group participants as key features of trustworthy experts" (Fritz et al., 2025). While the study mentions the need for careful communication about uncertainty, it does not offer strategies or best practices for how experts should manage public expectations realistically. With regard to the potential democratic implications of technocratic practices through a stronger emphasis on expert-led governance in complex, valuable policy areas, the authors point out that the implications of "distributing decision-making power from elected representatives to appointed experts, are little problematized by publics in the focus groups. This includes those with participants in countries with different systems of government – whether other types of democracies or less democratic regimes" (Fritz et al., 2025).

Given the debatable assumption that climate change knowledge on its own is the key solution to coping with climate change, it is not surprising that climate scientists are at least sympathetic to the suspension of the democratic process. However, the position of inconvenient democracy has a few weaknesses that I would like to explore in more detail below.

The following chapter examines the intersection of knowledge, politics, and climate change policy through the lens of "knowledge politics"—the idea that scientific knowledge is contested in social and legal contexts. It advances the argument that both nature and knowledge about natural phenomena are socially constructed and subject to litigation and control. The text illustrates how geoengineering technologies are undergoing a "patent land grab," with numerous applications filed for solar radiation management and carbon dioxide removal technologies. This creates tensions between innovation incentives and accessibility, which particularly affects developing countries' access to green technologies. A key argument challenges the assumption that scientific knowledge directly translates into political action or control. Instead, knowledge is described as a "generalized ability to act" that only becomes effective under specific social conditions. The chapter critiques how intellectual property rights, particularly patents, create barriers to global climate action, limiting developing countries' access to essential green technologies and potentially undermining international climate cooperation and equity.

Notes

1 Helmut Willke (2023: 78) explains, "A more sensible strategy would be to try to strengthen participation and deliberation not in general, but selectively, and to bring it to bear optimally where the relevant expertise for specific problem areas is gathered, in order to use the distributed intelligence of many knowledge centers and professional communities in a more complex, but democratically justifiable and legitimate decision-making process" (also Willke, 2016).

2 Quoted in Andrew Revkin, "A risk analyst explains why climate change risk misperception doesn't necessarily matter," *New York Times*, April 16, 2014.

3 Evelyn Fox Keller (2017: 120) demands that scientists to become political "activists" and is "urging climate scientists to take on responsibility for bridging the ever-widening gap between expert and lay reader partly because they are the only ones in a position to take it on – there is no one else– but also, and more importantly, because their responsibility qua scientists obliges them to do so."

6

KNOWLEDGE POLITICS

> Nature is a social category. That is to say, what is considered to be nature at a particular stage of social development, how this nature relates to humans and in what form humans engage with it, in other words, what nature means in terms of form and content, scope and objectivity, is always socially determined.
>
> *Georg Lukács (1923: 201)*

Law poses a problem for climate change policies. Georg Lukacs' observation of the embeddedness of nature in society implies, of course, that knowledge about nature and natural phenomena can be litigated. That climate change is a wicked problem can also be documented in the context of what I have called the emerging political field of knowledge politics (Stehr, 2005).

Politicians, scientists, critics, social movements, and individual citizens who are alarmed by and urgently demand control over research activities, especially the use of novel scientific knowledge, are mobilized to issue calls for legal or political action based on effusive promises about the potential consequences of research efforts, most recently, of course, in the fields of biotechnology and artificial intelligence (AI).

The field of geoengineering will likely become an intense area of litigation and control efforts as promises about its efficacy in curbing global warming grow and as fears about unintended consequences mount. A decade ago, the *United States Patent Office* (USPTO) had already received hundreds of applications for patents of geoengineering technologies, mainly pertaining to SRM (solar radiation management) techniques and CDR (carbon dioxide removal) technology. The number of patents granted at the time rose dramatically; climate engineering appears to be undergoing

DOI: 10.4324/9781003682066-8

a "patent-land-grab" typical for a field early in its development. Prior to 2010, hardly any patents were granted. (Chavez, 2015: 9). The regulation of climate knowledge extends beyond the patenting of knowledge and techniques. International regulations for climate engineering technologies have been implemented.

Existing national and international legal norms may incidentally be relevant to geoengineering, such as CDR efforts (carbon dioxide removal), which may be impacted by obligations to protect and preserve the marine environment under the United *Nations Convention on the Law of the Sea* and rules that prevent transborder harm under customary international law (Brent et al., 2018). These developments in the field of geoengineering as a form of knowledge politics raise the usual concerns and promises of legally protecting and restricting knowledge (Stehr, 2023).

Anthony Chavez (2015: 35) concludes his review of the USPTO patents records of geoengineering technology patent applications and patents—in light of the geoengineering-patent land-grab that has already begun, and that these technologies [...] may become critical to society's response to climate change—that

> the United States needs to ensure that these patents do not deter innovation or prevent these technologies from being available for implementation. Specifically, it should develop unique procedures to approve these applications and form a geoengineering patent pool that will facilitate both innovation and accessibility.

Needless to emphasize, perhaps, but such efforts to control the control of novel technical and scientific knowledge will be vigorously contested by vested interests.

Calls for restrictions on new scientific knowledge extend to planned radical interventions in and reconstructions of nature, including (defective) human nature, or to devastating threats to human society as we know it by an autonomous agency of AI In the field of genetics and biotechnology, and in the case of rapidly developing AI, one encounters radically opposed forms of discourse. One perspective champions genetic engineering as a promising contribution to agriculture and human health, especially in the context of deadly diseases. Conversely, there is a narrative that sees genetics as a tool of totalitarian conspiracy to rid the world of disabled people (Shakespeare, 2015). A similar contradictory discourse can be observed in the field of artificial intelligence. However, knowledge politics, as a way of controlling the implementation of new scientific and technical knowledge, in the most consequential case by outlawing knowledge, also extends, as noted, to legal efforts to monopolize knowledge, especially aided by patent law. Intellectual property law poses a dilemma for climate change policies.

Conflating climate science and climate policies

Analyses of climate knowledge and its potential role in political contexts often reveal a lack of understanding of the practical function of scientific findings. Scientific knowledge is neither directly performative (knowledge is synonymous with control and represents practical reason) nor directly persuasive (i.e. knowledge persuades unencumbered). Knowledge alone does not generate profit or score goals (cf. Van Dijk, 2014). One of the fundamental flaws in the portrait of an uncomfortable democracy is the failure to recognize the social character of knowledge in general and the contested and often ambivalent character of political knowledge. The recognition of the actual function of knowledge ensures premature political closure, i.e. the depoliticization of the issue of climate change and climate policy.

It is more appropriate to describe knowledge not as *something that is* but as a generalized *ability to act* on the world, as a model *for* reality, or as the ability to set something in motion (Stehr, 1994; Grundmann and Stehr, 2012; Stehr and Adolf, 2018). The German term that best describes knowledge as a generalized ability to act would be *Handlungsvermögen*. The verb *vermögen* signals "to be able to do," whereas the noun *Vermögen* in this context is best translated as "ability" (and not "fortune" or "wealth").[1] The ability to act—the ability to set something in motion—also extends to the ability to generate "symbolic action." Symbolic action can include, for example, the ability to formulate a hypothesis, perform a ritual, find a new metaphor for an established term,[2] evaluate "facts," organize the literature on a topic or defend a thesis against "new facts." The ability to act thus refers not only to the ability to accomplish something in the sense of a material and physical achievement, but also to the ability to make a fire or drive a car. The capacity to act also refers to *intellectual* abilities as well as the production of *meaning*, as found in the detailed description of the bundle of abilities that I refer to as *knowledgeability* (cf. Stehr, 2016a). This is probably also the reason why Norbert Elias (1984: 252) defines knowledge as "the social meaning of man-made symbols, such as words or numbers, in their *capacity as a means of orientation*" (my emphasis). Knowledge as a generalized capacity for action only becomes "*active*" (i.e. used) in social action under certain circumstances, namely, where social action does not follow purely stereotypical (effortless) patterns (Max Weber) or is otherwise strictly regulated. Under the conditions of ritualized social action, there is no break in the continuity between the past and future. Past and future events are secured by self-evident sequences of events. The conception of knowledge in a knowledge society advanced here does not imply therefore that the expansion of knowledge encroaches somehow on "ideology" and produces a society less affected by ideological thinking.

Niklas Luhmann's ([1992] 1998: 67) elementary reflections on the conditions for the possibility of making decisions at all perhaps allow an even more comprehensive understanding of the use of knowledge but also confirm my description of the probable usefulness of knowledge only under conditions of degrees of openness of the circumstances of action: The need to decide "is only possible if and insofar as what will happen is uncertain."

The circumstances of action I am thinking of can also be described as the ability of actors to change or stabilize a particular reality. However, the ability to make *something happen*, to change and influence reality, as well as the ability to intervene in a context that would otherwise change, is not symmetrical with the ability to act (knowledge). Knowledge and control are not symmetrical: Foresight and control are highly fragile. The progress in knowledge does not necessarily lead either to an improvement in foresight or to an improvement in control (Tenbruck, 1977). The ability to do something depends on control over the conditions of action. The lack of control over the conditions of political action is an apt description of the social role that applies to the position of climate scientists today and will continue to be the case if they have not usurped political power.

Knowledge plays a multifaceted role in understanding, communicating, or responding to climate change. The intellectual and economic weights of knowledge multiply in modern society as its economy evolves into a knowledge economy. A further step in the evolution of the economy in modern society is reached with the emergence of "knowledge capitalism," i.e. a point at which the control and exploitation of additional knowledge, supported by legal norms, becomes the core framework within which capitalism develops (Stehr, 2022).

The term "knowledge capitalism" is justified in this context because of the *persistent* division of ownership between actors and social collectives in knowledge societies, in this case, the ownership of patented or protected knowledge, and social relations with those who own only their labour. The designation for the emergent economic order as a *form of capitalism* continues to apply because "the central function that knowledge plays in the production process *does not in itself* bring about a fundamental change in the economic and social system" (Gorz, 2002: 4; emphasis added).

In this context, intellectual property rights play an important role in the development of the economy and society. It is therefore necessary to consider the relationship between climate change, particularly climate change policies, and the political economy of nations (Adler, 2009). The intersection of intellectual property rights (IPRs) and climate change policy has received limited attention. While IPR can drive innovation in climate technologies, it also poses challenges for equitable access. Tackling climate change effectively requires balancing property rights with global cooperation and equity.

Intellectual property rights and climate change policies

> The right of property consists in essence in the right to withdraw a thing from common usage. The owner may use it or not use it; that is a minor consideration [...] The right of property can be far better defined negatively than in terms of positive content, by the exclusion it involves rather than the prerogatives it confers.
>
> *Èmile Durkheim ([1950] 1992: 142)*

Two prominent sociologists, Werner Sombart and Max Weber, stand out among social scientists who provide a comprehensive analysis of the historical development and fundamental characteristics of capitalism. Sombart and Weber advanced their framework of capitalist order, as it prevailed throughout western and northern Europe by 1850. In the decades that followed, capitalism spread to much of the rest of the world. Without wanting to overstate the possible stages in the development of capitalism, it is useful to refer, as Peter Drucker (1993: 53) does, to the period after World War II, as a new step in the evolution of capitalism in which "knowledge is [...] fast becoming the *one* factor of production, sidelining both capital and labor."

The broad socio-economic framework within which intellectual property rights and future climate change issues intersect can be briefly described as follows: First, I agree with Peter Drahos (2020: 14.13) who comments that the preoccupation with the discussion of the tragedy of the commons "has obscured the more serious tragedies of commodification in capitalism," namely, of knowledge, as the most important "path to a wealth held in common." Knowledge has become "capitalist's most precious source of commodification" denying one of the central norms of science as specified by Robert Merton ([1942] 1957: 556), the norm of communism.

Robert Merton ([1942] 1957: 556–558) defines *communism* as a salient attribute of the ethos of science:

> 'Communism,' in the non-technical and extended sense of common ownership of goods [...] The substantive findings of science are a product of social collaboration and are assigned to the community. They constitute a common heritage in which the equity of the individual producer is severely limited. [...] Property rights in science are whittled down to a bare minimum by the rationale of the scientific ethic. The scientist's claim to 'his' intellectual 'property' is limited to that of recognition and esteem which, if the institution functions with a modicum of efficiency, is roughly commensurate with the significance of the increments brought to the common fund of knowledge [..] The communism of the scientific ethos is

incompatible with the definition of technology as 'private property' in a capitalist economy.

Second, corporations that are based on the principles of knowledge capitalism seek to generate *monopoly returns* by controlling legal rights to the resources that knowledge represents as a means of production. Third, another significant feature typically attributed to the economic order of capitalism also persists under knowledge capitalism, namely, the ongoing tendency toward a *concentration* of market power reinforced by corporate practices of secrecy. The great majority of patents granted are applied for and awarded compared with the past epochs by (large) corporations rather than individual inventors. Fourth, patent owners *prevent* competitors and interested parties from using these novel technologies. Filippo Belloc and Ugo Pagano (2012: 445) draw attention to the economic and social consequences of *internationally* enforceable patent law:

> at the macro level [...] countries rich of IPRs [Intellectual Property Rights] can prevent countries poor of IPRs from using their knowledge for production, with the possible consequence of generating a 'snowball dynamics' according to which IP-rich countries exploit larger investment opportunities and acquire new (proprietary) knowledge, and IP-poor countries tend to stagnate in a low- investments/low-patents equilibrium.

There is a very low level of patenting in developing countries. One of the most important intersections between IPRs and climate change is the transfer of innovative climate technologies to developing countries. Moreover, it is a realistic assumption that the future present of responses to climate change will depend on the challenge of innovation and the way in which these intellectual achievements are shared globally or fail to become common means of combating the conditions of climate change and its consequences (cf. Athreye et al., 2023).

The transfer of technologies has been a key pillar of the *United Nations Framework Convention on Climate Change* (UNFCCC) since its inception. Article 4.5 of the UNFCCC requires developed countries to "take all practicable steps to promote, facilitate and finance, as appropriate, the transfer of, or access to, environmentally sound technologies and know-how to other Parties, particularly developing country Parties." In recent years, however, developing countries have argued that these efforts may be hampered by poorly designed IPR regimes, particularly in relation to patents.

Many existing green technologies, such as solar panels, wind turbines, and energy-efficient systems, are patented by companies or institutions in developed countries. Strict intellectual property rights (IPR) protections around these technologies drive up costs, making them unaffordable for developing countries. This limits the ability of developing countries to transition to cleaner energy systems and achieve sustainable development goals. International property rights frameworks, such as the TRIPS Agreement, provide some flexibility for countries to circumvent patents in emergency situations (e.g. compulsory licensing). However, these mechanisms are complex and often designed for pharmaceuticals, not climate-related technologies. Adapting to international IPR standards requires significant investment in administrative and legal systems, diverting resources that could otherwise be used to address climate-related problems such as rising sea levels, droughts, or extreme weather events. This lack of tailored rules prevents developing countries from producing or modifying green technologies locally.

Beyond immediate and obvious consequences, there is an indirect effect: the developing countries are less likely to agree to tough obligations in emission reductions,

> since meeting those obligations may entail their making large transfers to the developed countries that own the IPR necessary to meet those obligations. Of course, without an agreement, the entire world is put at risk.
>
> *(Baker et al., no date: 50)*

To address these disparities, global treaties should incorporate provisions for climate emergencies, such as expanded compulsory licensing or patent pools, allowing free or low-cost access to green technologies for mitigation and adaptation in developing countries. Developed countries should commit to transparent and unconditional technology transfers under the *United Nations Framework Convention on Climate Change*. International collaborations should prioritize funding and infrastructure development in the Global South, fostering innovation that meets local climate adaptation needs.

Chapter 7 conceptualizes climate change as a "wicked problem"—a persistent, complex, and interlinked crisis that defies definitive solutions. Unlike typical environmental issues, climate change involves political, economic, and societal dimensions and requires continuous management, not resolution. The chapter critiques the overreliance on technical fixes and large-scale planning, highlighting their failure to consider local knowledge and social complexities. Democratic governance, despite its shortcomings, is presented as more adaptable than authoritarian regimes. The chapter emphasizes the need for participatory, inclusive, and pragmatic policy approaches guided by the "magic triangle" of effectiveness, participation, and compliance. It warns against the depoliticization of climate debates and the misplaced faith in top-down scientific authority. Ultimately, I argue that meaningful climate action must be grounded in democratic processes that build trust, empower citizens, and navigate uncertainty—not bypass them.

The chapter concludes by stressing the urgency of responsive, multi-level, and democratically anchored climate governance.

Notes

1 In his discussion of money as a generalized code, Georg Simmel ([1907] 1989: 276) uses the concept of wealth to describe the fact that money is more than just a medium of exchange; his definition of money thus goes beyond a purely functional understanding of its social capabilities.

2 In this context, I refer, for example, to the considerations of Donald Schon ([1963] 1967) in *Displacement of Concepts* (see also Haldane, [2009] 2013).

7

CLIMATE CHANGE AS A WICKED PROBLEM

Anything we can actually do, we can afford.

John Maynard Keynes ([1942] 2013: 270)

A riddle is solved. A mystery is unsolved. It reveals itself to you, and then it hides again. You are finished with a riddle when you have solved it. You are never finished with a mystery. The big questions of life belong to the category of mystery: What am I living for? Why do I suffer? How can I withstand death? What is love? Where do I find forgiveness? What can I hope for? These are not questions that can be solved and answered once and for all. However, they accompany you until you die. Perhaps that is why they are the most important questions. They must be asked even though they cannot be solved. They want to be lived. Climate change is just such a problem. The real hope and the driver are, as Keynes expresses, that what can be done can be afforded.

One of the key assumptions of critics of climate science is a misunderstanding of the climate problem and a misleading framing of the policy process.[1] This misunderstanding of the climate problem and the climate policy process is a fundamental error in framing *climate change as a conventional environmental "problem" that can be "solved."* It is neither one nor the other. Climate change and climate policies are not *single* problems to be solved; they must be understood as a permanent condition—not unlike the chronic disorder diabetes or A.D.H.D.—that need to be managed and can only be managed well to some extent. The climate problem is part of a larger complex of such conditions, which include population, technology, wealth disparities, public values, and resource use. Therefore, it is not just an "environmental" problem. It is just as much an energy problem, an economic

DOI: 10.4324/9781003682066-9

development problem, or a land use problem and can be better addressed through these *different avenues* than through the problem of controlling the behavior of the Earth's climate by changing the way humans use energy.

As I have stressed, climate change is a "*wicked*" problem. "Wicked problems" are problems that are often formulated as if they were amenable to a simple, unilinear solution, when they are not. A wicked problem is the impossibility of definitively formulating the political problem: the information needed to understand the problem depends on one's idea of how to solve it. Moreover, with wicked problems, there is no stopping rule: we cannot know whether we have sufficient understanding to stop searching for more understanding. There is no end to causal chains in interacting open systems, of which climate is the best example. As a political problem, climate change is not solvable. *Action on climate change is best embedded in comprehensive policy perspectives that* address climate change *indirectly*, for example, by accepting that decarbonization will only succeed if other goals that are politically attractive and pragmatic are achieved. The great need is time, even though time is becoming increasingly shorter.[2]

Life in fragile times

Social relations in modern society are generally characterized by a significant degree of fragility and indeterminacy. The fragility of modern society has increased over the course of the development of modernity. In what sense is the fragility of modern society and its social institutions relevant to climate change and climate policies? However, first, what are some of the relevant examples of the fragility of social relations? Fragile are political coalitions in democratic societies; precarious are human partnerships, "now burdened with expectations greater than ever before" (Bauman, 2000: 50); unhinged are traditional loyalties, ethical obligations have become molten; extensive deregulation of economic affairs; and fragmented orientations. The upside of the indeterminacy and fragility of modern society is its openness to the future present. A future present that is largely unknown.

Based on a similar diagnosis of liquid social relations in modern society, the passage from a "solid" phase to "liquid" phase, Zygmunt Bauman (2000: 7–8) acknowledges the emergence of a "fluid modernity." The human condition of fluid modernity signals, on the one hand, a highly "individualized, privatized version of modernity" amounting to a fragility of human bonds and fostering individual uncertainty, fear, and insecurity; on the other hand, it generates "unreachability of systemic structure," the inability of large social institutions to enforce their will. Institutions, finally, because of their short life expectancy "cannot serve as frames of refence for human actions" (Bauman, 2007: 3). This amounts, in the end, to a disintegration within the national boundaries of "effective agencies of collective action" (Bauman, 2000: 14) as much of the local power moves to a global space. Bauman observes a separation of (national) power from politics. At the individual level, among the new emerging contradictions is, for instance, "the deepening imbalance between individual freedom and security. Supplies of security provisions shrink fast, whereas the volume of individual responsibilities (assigned if not exercised in practice) grows on a scale unprecedented for the post-war generations" (Bauman, 2000: 170). Bauman's perspective on "liquid modernity" ranks among the most challenging sociological commentaries of the contemporary social order. His open narrative, because it is a commentary, is itself fluid and therefore open to basic criticism.

Claus Offe offers a related perspective on the intransigence of modern "complex" societies, which have become rigid precisely because of the enhancement of individual freedoms. The "more options we open up for ourselves, the less available as an option is the institutional framework itself with the help of which we disclose them," resulting, last, in a "self-liquidation of freedom of choices" (Offe, 1987: 9). The more

> democratic rights the individual citizen enjoys, and the more citizens are drawn into the enjoyment of these rights, the more is the growth of organizations and prerogatives indirectly simulated, in the face of which these rights (at least when they are used in conjunction with average material, informational, and temporal resources) decline in potential influence.
>
> *(Offe, 1987: 10)*

Nonetheless, the steering capacity or efficiency of major social institutions has declined with obvious implications for climate policies. Even the political steering of modest proportions appears

> hopelessly difficult and time-consuming undertakings that the demand for coordination involved with any serious attempt to manage such problems narrows almost automatically to short periods, small circles of interested parties [...] Coordination of extremely large steering problems which, corresponding to the nature of their object, cannot be managed within the framework of the nation-state – i.e., the military-strategic questions of peacekeeping, the management of problems of poverty in the third world, and the maintenance of minimal ecological balances – thus seems completely futile.
>
> *(Offe, 1987: 11)*

Moreover, policies in modern democratic societies are increasingly "made in venues that are more complex and more insulated from direct electoral oversight" (Hacker et al., 2022: 21). In many instances, well-organized and resourceful interests (lobbies) have a strong hand in shaping or preventing regulatory matters by governments. Voters have a fragile function in shaping the political agenda. Finally, the role of legal expertise and the judiciary often has a decisive effect on policy-making or the revision of enacted policies.

The complexity of climate policies

The prevailing policy approach focuses almost exclusively on a single effect to be achieved through governance, namely, the reduction of greenhouse gas emissions and the necessary measures to adapt to climate change. It excludes other, more complex forms and conditions of action. By focusing on the goals of political action rather than its conditions, the controversial issue of climate change is often reduced to scientific or technical questions. Socio-political questions are neglected. The increase in the complexity of social relations in general in modern societies is accompanied by an "ingenuity gap," that is, the "ingenuity gap [...] is a way of thinking about the very real chasm that sometimes looms between our ever more difficult problems and our lagging ability to solve them" (Homer-Dixon, 2002), in other words, the volume of practical knowledge (Stehr, 1992).

However, the dilemma is epistemological; that is, it is

> a characteristic of open systems of high complexity and with many ill-understood feedback effects, such as the global climate classically is, that there are no self-declaring indicators which tell the policy maker when enough knowledge has been accumulated to make it sensible to move into prescriptive action. Nor, it might be argued, can a policy-maker ever possess the type of knowledge – distributed, fragmented, private; and certainly not in sufficient coherence or quantity – to make accurate 'top down' directives. Hence, the frequency of failure and of unintended consequences.
>
> *(Hayek, [1949] 1960: 27; Scott, 1998; Prins et al., 2010)*

The politicization of climate science has led to the depoliticization of climate change. The issues relevant to the public are permanently removed from politics (see also Jasanoff, 2012). However, the relevant issues are not merely scientific matters answered by experts or technical issues answered by engineers. In the final analysis, they are matters "about power, morality, authority, and trust, which is to say they are questions for democratic citizens" (Sandel, 2020: 107–108).

Equally inadequate in this context is the focus on a *single approach* to combating climate change (cf. Boykoff, 2019),[3] namely, the reduction of greenhouse gases, especially CO_2. Focusing climate policy exclusively on reducing emissions ignores what Roger Pielke Jr. (2010) calls the "iron *law*" of climate policy. The iron law simply states that while people are often willing to pay a certain price for environmental goals, this willingness has its limits. The exact limit varies from place to place and from household to household. The massive resistance of the "yellow vest" protests in France against a regular increase in fuel tax, which the French government wanted to use to combat global warming, in the early winter of 2018 is a perfect example of Roger Pielke's general law: "when environmental and economic objectives are placed into opposition with one another in public or political forums, the economic goals win out" (Pielke, 2010: 59). The protests of the yellow vest movement forced the government to withdraw the tax increase. Public support for climate policy measures is decreasing depending on the impact of these measures on budgetary costs. Convergence of environmental and economic policies is not impossible. However, such convergence is likely to lean toward the economic part of the equation when measures to reduce emissions collide with economic growth or labor market policies. Much more pessimistically, systems theory, as exemplified for example by Niklas Luhmann's perspective, maintains that the functional differentiation of modern society makes it impossible for different social systems (politics and the economy) to stand in for each other. Niklas Luhmann explains the difficulty in *coordinating* effective climate policy across social systems—like politics and the economy—due to ***functional differentiation***. Luhmann maintains that social systems are **operationally closed**—they communicate internally using their own logic and cannot be directed by other systems from the outside. The climate issue is not only a never-ending political problem, but also a boundless societal problem without end.

The reference to hazards and risks in a future present helps to further understand the complexity of climate policy. What is common to all areas of anthropogenic hazards is that no one can predict with certainty how great the danger really is. The modern question of risk involves an "irresolvable ambivalence."[4] Not only is uncertainty being produced on an unprecedented scale, but all the attempts to solve the possible problems make us even more aware of how fragile modern societies happen to be. The language of risk and risk research must therefore reflect these new uncertainties in society, which, in the form of a conscious perception of the future as contingent on the present. Seen in this way, risk means that possible damage is already attributed or incorporated into decision-making attitudes, even though it is impossible to know the extent of the damage or whether it will occur at all. This ignorance, the unpredictability of the consequences of a decision, becomes part of the decision-making process. The only certainty is that a decision must be made, leaving us with the dilemma of with the dilemma of making decisions under conditions of great uncertainty. In short, the positive or negative consequences of decisions related to ecological change are associated with great uncertainty, so that ultimately there can only be plausible opinions, scenarios, and scenarios—less plausible opinions, scenarios, etc., about what to expect in the future. Because we sense uncertainty but would prefer to be certain, we turn to science. We turn to science—sometimes denying the need to embed or defer such decisions in democratic governance structures—with a persistent conviction of its superior rationality, and with a once-unimpaired confidence in its ability to manage, plan, and design the modern world, and in the feasibility of doing so. However, these beliefs are being seriously and increasingly challenged and undermined by the problem of risk, technically, socially, and temporally.

The myth and the urgency of large-scale planning

> **It is difficult, if not impossible, to completely abandon the concept of steering and simply let the future unfold as it will.**
>
> *Niklas Luhmann (1988: 324)*

Fourth, the generally pessimistic assessment of the ability of democratic governance to respond to, manage, and control exceptional circumstances is linked, if only implicitly, to the optimistic assessment of the potential of large-scale planning in terms of massive social engineering efforts as part of socio-political planning. Planning on any scale is difficult to accomplish. Not only the ability of governments but also the general possibility of planning the future present of societies is rather limited, perhaps even non-existent (see Tenbruck, [1967] 1971, 1977: 138). Economic and social planning concepts that were widely discussed positively decades ago, for example, during the period of World War II and in Soviet style societies, have fallen into disrepute, as Anthony Giddens (2009: 94–100) noted. In the decades after the war, planning in democratic countries amounted to the nationalization of core industries. In many instances, central planning as well as partial state planning proved to be disastrous in many cases and rarely lived up to its promise to outperform capitalism. Nonetheless, in reflecting on the future present, the idea of planning, for instance, in the sense of "long-term thinking," often reappears.

Plans to improve human conditions have often failed, as James Scott (1998) emphatically demonstrates, case by case, in his book *Seeing like a State*. James Scott argues that large-scale government interventions often fail because they are overly simplistic and do not consider the complexity of local conditions. He uses several case studies to illustrate how top-down plans can lead to unintended and often disastrous results when they ignore the practical knowledge and practices of local populations. For example, the Prussian state sought to maximize timber production in its forests via scientific and efficient methods. Foresters simplified the complex natural ecosystems of forests by clearing them and planting a single species of tree (often pine or fir) in neat rows, with the idea that this would optimize production. This approach ignores the biodiversity and intricate interdependencies that exist in natural forests. At first, this method seemed successful as timber yields increased. Over time, however, it has led to soil degradation, increased susceptibility to pests and disease, and a decline in forest health. Monoculture forests lack the resilience of diverse ecosystems, and the long-term sustainability of timber production is compromised. Local knowledge, which understands the importance of biodiversity and ecosystem balance, has been overlooked in favor of a simplified, top-down approach.

Additionally, for a cautious assessment of the ability to plan, the once active academic program and enthusiastic support for futurology about desirable futures have disappeared (Seefried, 2015). Modern de-centered, functionally differentiated societies rule out de-differentiated, whole-of-society social planning in principle (Luhmann, 1976, [1992] 1998). However, as noted, effective climate change policies oriented toward a future present likely must transcend the self-referential logic of societal subsystems and incorporate more than *individual* knowledge.

The consensual perspective, ultimately linked to Friedrich Hayek's (1945: 529) "problem of knowledge" and detailing the impossibility of central planning, is beginning to be revisited in the context of unique, contemporary ecological challenges. There is a renewed interest in economic planning with a focus on planetary ecological issues (Kapczynski and Michaels, 2024; Sorg and Groos, 2025). Hayek's minimal state approach is challenged by those who believe in a more active government role in regulating markets and providing public goods. Jonathan Benson (2024), in his paper "Epistemic Problems in Hayek's Defence of Free Markets," questions Friedrich Hayek's classical liberalism argument that free markets allow individuals the greatest opportunity to achieve their ends. His paper develops an internal critique of this claim. Benson contends that when externalities—costs or benefits affecting third parties—are considered, the same epistemic challenges that Hayek identified as obstacles to government intervention also undermine the ability to assess the welfare outcomes of different market regulations.[5] This suggests that Hayek's framework may not adequately justify a preference for less regulated markets. Given the prevalence of externalities in modern economies, Benson asserts that Hayek must either refrain from making comparative welfare claims about market regulation levels or relax his epistemic assumptions to allow for more government intervention than his classical liberalism typically permits.

Despite all the cautious assessments of planning today, Anthony Giddens (2009: 92–94) provides a detailed blueprint for a strong state and encourages businesses, third sector groups, and individual citizens to engage in *long-term planning* with future climate change contingencies in mind. Whatever success can be achieved in tackling climate change, Giddens (2009: 91–94) argues that directly addressing policy makers will depend heavily on the government and the state and, in turn, on widespread political support from citizens. citizens. The state must be the prime mover and facilitator on the following fronts, ensuring in abstract terms that

- help us to think ahead
- manage climate change and energy risks in the context of other risks faced by contemporary societies
- promote political and economic convergence
- make interventions into markets to institutionalize ‘the polluter pays’ principle
- act to counter business interests which seek to block climate change initiatives
- keep climate change at the top of the political agenda
- develop an appropriate economic and fiscal framework for moving toward a low-carbon economy
- prepare to adapt to the consequences of climate change
- integrate local, regional, national, and international aspects of climate change policy.

Whether governments are proactive on all these fronts, although decarbonization, climate as a global issue, or the issue of climate justice appear to be missing from the list provided by Anthony Giddens, clearly represents a daunting set of tasks for governments confronted, in all likelihood with a vast range of competing tasks and demands. Additionally, the list does not include the demand for an economic and societal de-growth perspective.

The at times paradoxical position of the economic de-growth perspective with respect to massive social planning is highlighted by Daniel Susskind (2024). On the one hand, the de-growth position demands that radical socio-economic changes be instituted in societies to relieve them from a commitment to incessant economic growth.[6] Susskind (2024: 147) describes the result of the required economic changes, under the heading of “The fantasy of perfect policies,” as implying

> that political leaders, to build support for the idea, would need to either persuade most of those in poverty to remain destitute, or alternatively convince the vast majority of the developed world to permanently reduce their incomes by an enormous amount. In other words, they must ask the poor to stay poor forever or the rich to become much poorer forever.

At the same time, the de-growth position rejects alternative policy avenues as impractical, for example, large taxes on fossil fuels; however, they retain faith in large-scale planning, initiating a massive recession about realizing the ways that favor a de-growth approach.

The inappropriate conflation of society and nature

Fifth, in the argumentation of the impatient critics of democracy, one notes an inappropriate conflation of nature and society. The uncertainties (about climate) that the natural sciences have supposedly eliminated and the authoritative consensus that the sciences have thereby achieved are simply transferred to the realm of social processes. A consensus on the evidence, it is argued, should motivate a consensus on political action. What should be sought is a rational shaping of the social order "consistent with the scientific understanding of the laws of nature" (Scott, 1998: 4), e.g. a comprehensive planning of human settlement and production. The top-down design of society is schematic and ignores the essential realities of any existing social order: the constitutive uncertainties, the fragility and complexity of social, political, and economic events, and the difficulty of anticipating the future present are treated as minor obstacles that can be circumvented as quickly as possible through a top-down approach, of course, by implementing the policies that faith in scientific knowledge prescribes. This undermines the dignity, plurality, and conflict inherent in today's knowledge societies.

Finally, advanced capitalist democracies, which faced major 'shocks' from their beginnings in the early 20th century to one of the most turbulent modern centuries, are remarkably resilient. Democracy is a more effective adaptive organism than other forms of governance (Kaase, 2007: 793; Luce, 2017: 87). Although the past is not necessarily a sound basis for predicting future states, the probability of rich democracies relapsing into authoritarianism is close to zero (Iversen and Soskice 2019; see also Przeworski and Limongi, 1997). Obviously, there are exceptions. Only the state of the future present will tell whether the United States will be a counterexample, as some already fear. We do not yet know whether the exceptional circumstances of climate change in the future present will be of such magnitude that the past is not indicative of the future health of democracies. Climate policies are a measure of future policies (Darian-Smith, 2024).

The magic triangles of climate policies

To summarize my observations about the wicked problem of climate change and the complexity of climate policy, I would like to point to a *dual magic triangle* of climate policy at all levels of socio-political action: individual, local, regional, national, and global. The mobilization of civil society[7] would be a powerful condition for the possibility of effective climate policies; hence, the *institutional* element of the magic triangle consists of and aims at "participation, effectiveness and compliance" (Held and Hervey, 2009: 11) as the key to successful climate policies. The climate discourse should be multi-institutional and multilateral. Climate policy should also be linked to another set of magic triangle elements, namely, *policy objectives*. These goals are set out in our *Hartwell Paper* (Prins et al., 2010). These goals

> should be *politically attractive*, meaning an approach which allows us to take a few small steps which offer rapid and demonstrable pay-back, thus helping to sustain the effort. It should be *politically inclusive*, meaning an approach which is pluralist in instinct. And it should be *relentlessly pragmatic*, meaning an approach which prizes progress that can be measured in the short as well as long terms.

The institutional and political elements of each magic triangle are interlinked. In Weberian terms, there is an "elective affinity" (***Wahlverwandtschaft***)[8] **among the elements of magic triangles.** Most importantly, the institutional and policy elements of the dual magic triangles are linked, unlike markets, which focus on outcomes and center on inputs. Democratic controls "aim largely at process, at inputs [...] democratic controls usually specify not result but process" (Lindblom, 1995: 686). Moreover, the political goal of inclusiveness is, of course, linked to the institutional process of the extensive political participation of the population. A process that is most difficult to achieve, as the historical record clearly demonstrates (e.g. Page and Gilens, 2017). However, applying Evgeny Morozov's (2024) perspective on the desirable conditions for the development of AI to climate change politics, the primary task would be "to endow individuals, irrespective of class, race or gender, with access to institutions, infrastructures and technologies that augment their creative autonomy and allow them to develop their capacities to the fullest."

Each of the goals or institutional processes of the magic triangle of effective climate policies is clearly afflicted by fundamental political dilemmas, including the need for a fundamental shift in *temporal* political *orientation* caused by climate change (Ejsing et al., 2024). Most of the voting public in democratic societies, based on an even curtailed assumption about their knowledgeability are not well positioned for effective participation if participation is contingent on broad political knowledgeability. Is this truly a handicap for political participation, and exactly what capabilities are required as the public confronts highly complex and contingent political issues such as climate change?

As previously stated, there is a prevalent concern regarding the dearth of political awareness and political knowledge among the voting public in contemporary democratic societies. Concurrently, a prevailing assumption in political philosophy and political science is that the capacity of the public to engage in critical examination of political processes is an indispensable prerequisite of meaningful political participation. Giovanni Sartori's (1987: 13) normative democracy criterion is a relevant expression in this sentiment: "While democracy is more intricate than any other political form, paradoxically enough it may not survive if its principles and mechanisms are not within the intellectual reach of the average citizen." However, what degree of political knowledge can be reasonably anticipated, and what level of knowledge among citizens is truly necessary for democratic governance? How can the Sartori criterion still be guaranteed in view of the increasing reflexive "complexity" of political contexts? Is a kind of fundamental pessimism about the countertendencies of organizational systems justified, as already described by Robert Michels in 1911 as the "iron law of oligarchy"? Political unintelligibility as a modern condition can be squarely blamed on

> our inability to process information, the complexity of topics, and the contingency of decisions [...] Public opinion that does not understand politics and is unable to judge it can be easily weaponized, or it may send equivocal signals to the political system. This confusion explains many of the most regressive political behaviors: populist simplification, the inclination toward authoritarian decisionism, or the passive consumption of a politics that is staged by the media.
>
> *(Innerarity, 2021: 266)*

The diagnosed unintelligibility, "is *politically relevant* and that erodes democracy is the lack of comprehension of those aspects of political life that citizens need to know to exercise their functions of monitoring and control" (Innerarity, 2021: 268; emphasis added).[9]

A partial answer to the lack of the "necessary" competencies of the public in democracies to participate in political deliberation and decision-making in the face of complex political issues can be glimpsed in analogy to the scepticism that Pierre Bourdieu ([1972] 1979) expresses about "public opinion," especially as it is generated by public opinion polls: not everyone has an opinion, not all opinions are of equal value, and it is unlikely that there is agreement on what the problem is in the first place. Some evidence suggests that the "wisdom of the crowds" phenomenon may be capable of delivering helpful decisions in certain circumstances. With respect to the wisdom of crowds, there is no mention of superior knowledge as a prerequisite. Moreover, the thesis of a lack of political competence tends to neglect the importance of elementary *relational emotions* and *values* in political behavior: "Facts alone cannot resolve hard questions about food safety, occupational health, pandemics, or climate change" (Sunstein, 2021: 121).[10] Fear is a lever of coercive authoritarian leadership and trust in responsive democratic leadership. Democratic leadership "provides for the care and feeding of the self-esteem of members of the system [...] for grounded trust in others [...] and for enabling people [...] to actualize their capacities for effective and responsible action" (Merton, [1970] 1976: 81). However, Innerarity (2021: 287) is not entirely pessimistic about the participatory potential of democratic governance; the complexity "of modern societies does not necessarily condemn us to a loss of substance in democracy." Citizens must have the means to generate opinions: "it is not so much that democracy requires political competence as that political competence requires democracy" (Innerarity, 2021: 287).

The "magic" aspect of the triangle commonly highlights at least a couple of ideas (1) achieving the right balance among the three components can lead to optimal or exceptional outcomes, often greater than the sum of their parts, or, (2) reaching the three goals of the triangle proves to be most difficult if not impossible to reach given multiple interactions among the corners of the triangle.

The idea of a magic triangle is a prominent concept in economics exemplifying the potential tension of its policy components: In economic policy, the magic triangle refers to three main objectives, (1) economic growth, in particular, a steady increase in a country's output or *gross domestic product* (GDP) without large fluctuations; (2) price stability (inflation control), specifically, maintaining price stability by keeping inflation at a moderate level, avoiding rapid price increases or hyperinflation; and (3) **low unemployment**, that is, keeping the unemployment rate low, meaning that a high percentage of the labor force is employed. These objectives are often in tension with each other, as measures to promote one objective can often have a negative impact on the others. For example, reducing inflation through restrictive monetary policies can sometimes increase unemployment, or stimulating growth may require measures (such as increased spending) that can lead to inflation (Mankiw, 2022).

Effective climate policies should be based on and seek to achieve a high level of public and institutional participation[11]; policies should be effective and efficient; for example, a carbon tax should achieve its purpose of providing a financial incentive for individuals, businesses and governments to reduce their carbon footprint and switch to cleaner energy sources; and they should ensure compliance with climate policy measures. Each of the three objectives of the magic triangle of climate policy is difficult to achieve on its own. At the global level, effective and efficient climate policies must address the fundamental injustice of climate change. The imposition of taxes may lack flexibility and increase incentives to circumvent such regulations. Participation may entail costs that are perceived as too high by the public. Compliance may be difficult to monitor and enforce. The design of compliance and cooperation mechanisms may be weak. Why are each of the goals of the magic triangle important?

Timothy Mitchell (2010: 4) emphasized in his monograph *Carbon Democracy* that there is limited, if any, reliable evidence that the

> presence of a civic culture – attitudes of trust, tolerance, mutual respect and other liberal virtues – facilitates the emergence of democracy. There is, in fact, no shortage of historical evidence to suggest the opposite. One can find repeated examples in the history of democratic struggles in the West of tolerant, educated, liberal political classes who were opponents of democratisation, fighting to prevent the extension of effective political rights to those who did not own property, to religious and racial minorities, to women, and to colonial subjects.

In light of these prospects, I maintain that participation in the political affairs of an existing democratic society engenders trust and a sense "of having an influence on political processes and, to a lesser degree, the confidence to participate in politics"; the combination of having an influence on politics and the feeling of having confidence to take part in political processes are defined as "political agency which is crucial in explaining variations in trust towards the national government" (OECD, 2024). Enhancing the sense and the active participation in political activities directed at climate issues is important, as the *OECD Trust Survey* shows. The transparency of politics and its effective compliance should enhance the democratic governance of climate issues. Trust in political institutions enhances support for environmental measures and regulations; the strength of the relationship may be weak, but nonetheless, Fairbrother, Sevä, and Kulin (2019) found, in statistical terms a strong interaction effect between political trust and beliefs about environmental problems.

Indeed, the 2024 OECD survey on public trust shows that

> trust in public institutions varies more depending on individuals' sense of political agency and partisanship than on their socio-economic and demographic characteristics. This underscores the well-established link between political trust and the feeling of having a say in policy decision making. [...] Moreover, partisanship, measured by whether an individual voted for the incumbent government in the last election, also plays a significant role. A comparison of the sizes of the trust gaps reveals that, on average, trust levels differ less among socio-economic groups and demographics, such as education, gender, and age, than variations based on feelings of political agency and partisanship. This trend, showing larger variations in trust levels by feelings of political agency and partisanship, holds true in all countries.

Similarly,

> the Trust Survey finds that a large share of those who have no trust in the national government and that feel they lack a political voice still feel confident in being able to participate in politics and indeed have engaged in various political activities.

The figure below shows the detailed results for each OECD country (Figure 7.1):

Broad-based participation is crucial for powerful climate policies (Willis, 2020: 81–99; Willis et al., 2022). Powerful climate policies require *policy mixes* that include dimensions other than environmental effectiveness. Annika Stechemesser and her colleagues (2024: 891), in an ex-post evaluation of climate policies in 41 countries for reducing emissions, noted that their

> findings highlight that successful policy mixes vary across sectors and that policy-makers should focus on sector-specific best practices when designing climate policy rather than following a one-size-fits-all approach. [...] we have identified pricing as a particularly effective policy in those sectors dominated by profit-maximizing firms—namely, industry—but also the electricity sector in developed economies.

Effective policies are in general strongly dependent on the conditions at hand. For example, "the lack of liberalized markets and existence of other price distortions can limit the effectiveness of price-based instruments" (Stechemesser, 2024: 891).

One of the framing conditions for successful public policy and political ascendancy is a *strong* socio-moral *political narrative*. A political narrative frames a policy in a story that resonates with people's values, experiences, and aspirations, transforming technical policy details into something relatable and meaningful. A well-crafted policy narrative can distinguish the essence of a policy into a clear and relatable story. Narratives tap into emotions, which, as noted, are often more powerful than facts alone in swaying public opinion. A political narrative can help people see where a policy wants to take society by connecting individual sacrifices or changes to a collective goal. Political narratives often draw on values or concerns that cut across political lines, creating a common story that unites diverse constituencies. A compelling narrative can energize grassroots movements, which are critical for sustained support (Lakoff, 2004; Tilly, 2006: Jones and McBeth, 2010).

At the global level, participation in climate policies signals the participation of all countries. Less developed countries have to be incentivized to participate out of a Win-Win perspective: Such countries should be able to reduce emissions simultaneously, adapt to climate change, and retain the possibility

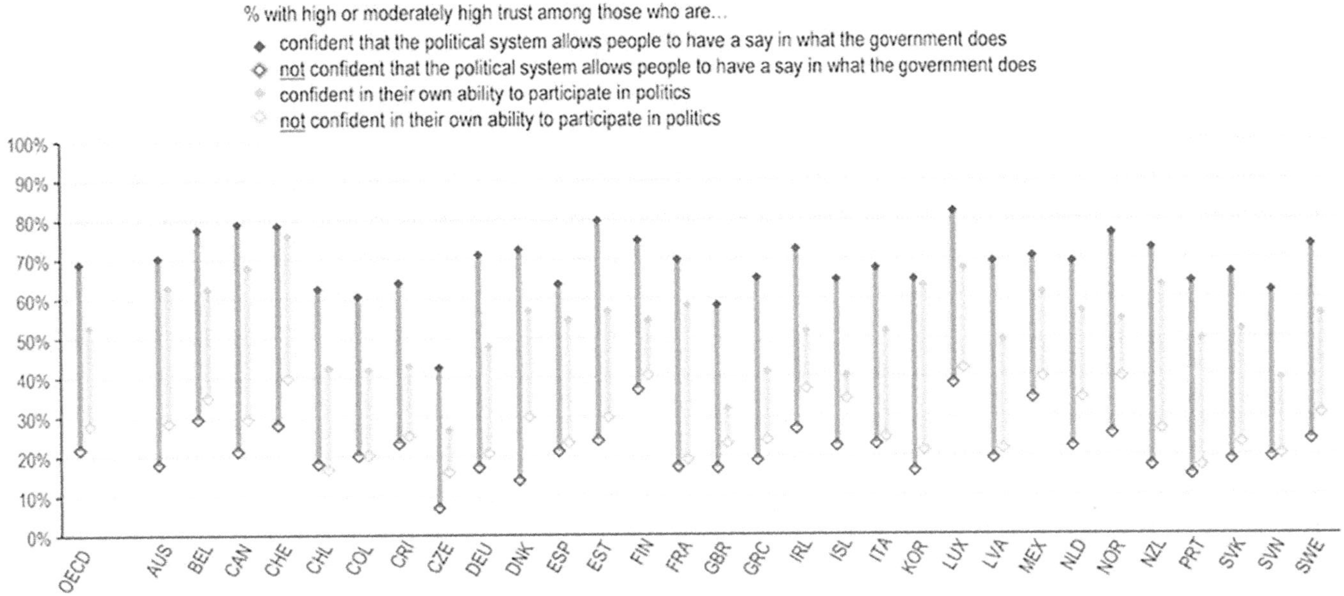

FIGURE 7.1 Share of population with high or moderately high levels of trust in the national government by feeling they have a say in what the government does (*grey*) and are confident in their ability to participate in politics (*yellow*), 2023.

Note: The figure presents the within-country distributions of responses to the question "On a scale of 0 to 10, where 0 is not at all and 10 is completely, how much do you trust the national government?" by respondents' feelings of having a say (blue) and confidence to participate in politics (yellow). Shown here is the proportion of respondents that have "high or moderately high trust" based on the aggregation of responses from 6–10 on the 0–10 response scale, grouped by whether people feel they have a say (blue): "How much would you say the political system in [COUNTRY] allows people like you to have a say in what the government does?" and feel confident to be able to participate in politics (yellow): "How confident are you in your own ability to participate in politics?". The OECD presents the unweighted average across countries.

Source: OECD Trust Survey 2023.

of enhancing the standard of living of their citizens. Whether countries with advanced economies will be prepared to generate the needed funds to realize these goals for developing nations remains to be determined. In a positive narrative, it is possible to advance the idea that "the relationship between democracy and climate change can be seen as a positive, mutually beneficial one, in which democracy finds solutions to climate change and our response to climate change strengthens democracy" (Brechin and Lee, 2024: 484).

Finally, in a speculative narrative, and as a novel condition for the possibility of broad political participation, the political scientist Mathias Risse (2024: 60–61) noted that

> it might well only be a question of time until some techno-optimists propose the large-scale involvement of AI in our collective-choice processes. Perhaps they will do so as a way of transferring Aristotle's [...] 'argument from the wisdom of the multitude' – according to which a group might display virtue that no individual features – into the context of twenty-first-century representative democracies [...] AI improves procedural legitimacy through greater participation, while the calibre of decision-making increases because voters are well informed [...] something like this might become more prominent in debates about AI and democracy.

The policy challenge of the concept of the Anthropocene and the urgent multiple demands of climate change risks of the future present in general and in relation to democratic governance in particular is one of significant tension in terms of the time scale. Ejsing, Tønder, Jensen and Hansen (2025) explored the temporal demands within the frame of the Anthropocene:

> Instead of engaging in the current debate about whether liberal democracies fare better than other regimes when it comes to climate policies, we argue that scepticism about democracy arises not only from concerns about political efficacy, but also because of a more radical shift in temporal orientation revealed by the arrival of the Anthropocene.

The authors make the case that liberal democracies, a reference to authoritarian regimes is missing, traditionally operate within linear and cyclical temporal frameworks, and are ill-equipped to address the rapid and complex ecological crises characterized by geological, biological, and human-induced changes. From a geological perspective, that is, from non-human timescales, the notion of the Anthropocene has compressed time, yet far less than the typical time frame of governance. From a human time perspective, anthropogenic climate change has also compressed the time scale for responses, challenging democratic institutions to react effectively.[12]

The following chapter explores the compatibility and necessity of democratic governance in confronting climate change within the Anthropocene. Unforgiving and merciless observers have a clear-cut answer to the compatibility of democratic climate policies: "Democracy affords a politics of procrastination that is quite deadly in the environmental arena" (Barber, 2012: 163). The following chapter continues to frame climate change as a "wicked problem" that requires persistent, flexible, and inclusive choices and responses. These choices and responses are ultimately linked to basic moral considerations rather than technocratic, top-down solutions. Despite criticisms, democracy—through participation, adaptability, and local knowledge—offers resilience and legitimacy that authoritarian models lack. Effective climate policy must mobilize civic engagement, prioritize justice, and reflect diverse societal voices. The chapter stresses that democracy fosters critical knowledge development, supports scientific innovation, and encourages collective responsibility. Legal and institutional challenges are addressed, including climate litigation and its growing significance. However, democracy's slow processes must be made more agile to meet urgent climate needs. Calls to "dare more democracy" highlight the need for enhanced civic agency and global cooperation. Ultimately, the chapter argues that sustainable, equitable responses to climate change require revitalized democratic governance rooted in transparency, trust, participation, and moral commitment. A *realistic* appraisal of calls for greater democracy must acknowledge the profound challenges posed by political polarization and entrenched economic and ideological interests, especially those resisting ecological transformation. These challenges also include motivated reasoning, misinformation, and resistance to scientific discourse within actual political systems. The literature examining multiple efforts to enhance democracy in the past present is filled with examples of the pitfalls of such heroic undertakings. However, the potential of the probable conditions for action in the future present that I have described could well determine the prerequisites for such a strengthening of democratic decision-making.

In the following chapter, I depend on the standard definition of the Anthropocene as a proposed geological epoch marking the period when human activity began significantly impacting Earth's ecosystems and geology, especially through climate change, pollution, and biodiversity loss, distinguishing it from previous natural epochs like the Holocene (cf. Hannah, 2021; Trenberth, 2022). From a societal frame, the Anthropocene era introduces a dynamic and unstable Earth system that is shaped by human activity. This challenges traditional institutions, such as states and markets, which rely on path dependency and resist transformation.

Notes

1 My criticism of the prevailing view of the climate problem is based on our *Hartwell Paper* (Prins et al., 2010).

2 Turnbull and Hoppe (2019) critically examine the concept of "wicked problems" in public policy, challenging the dichotomy between "wicked" and "tame" problems. The authors argue that the distinction between wicked and tame problems is unsustainable, as it oversimplifies the complexity of policy problems by failing to consider the relationships between policy workers and the stakeholders who deal with them. The counter-model proposed by Turnball and Hoppe reframes policy problems as existing on a continuum where their nature changes based on stakeholder perspectives and contextual factors. Ultimately, the distinction between a rigid analytical framework and more adaptive, reflective practices in policy analysis reiterates the notion that some problems are persistent problems, i.e. wicked problems.

3 Boykoff argues that effectively addressing climate change requires a multifaceted strategy rather than a single, universal solution. He emphasizes that different audiences and contexts require different communication approaches to foster meaningful and sustained engagement with climate issues. There is no "silver bullet" for communicating climate change; instead, a "silver buckshot" approach is needed, where strategies effectively reach different audiences in different contexts.

4 I draw on the longer discussion of "knowing and risks" in Stehr (2001) and Bechmann und Stehr (2000).

5 Friedrich Hayek (1944: 105) did acknowledge the dilemma of externalities for economic planning: he pointed out that in those instances where "certain harmful effects [...cannot...] be confined to the owner of the property in question [...] we have to resort to the substitution of direct regulation by authority where the conditions for proper functioning of competition cannot be created."

6 Most of the de-growth literature lacks basic evidence and is mostly opinionated (Savin and Jeroen, 2024).

7 See Dana Fisher's (2024; also Nyberg and Wright, 2025) observation that systemic societal changes toward effective policies are best served through the mass mobilization of the actors of civil society.

8 Max Weber uses the term "*elective affinity*" (in reference to Goethe's novel) metaphorically and sociologically to describe complex connections between ideas, values, social structures and economic developments. It is not about a biological or causal relationship, but rather a *structural affinity*, that is, a mutual fit or reciprocal attraction between different cultural or social phenomena.

9 I had already referred to Helmut Willke's (2014: 9) equally pessimistic observation about the political information of the average voter in democratic societies: "Democracy presupposes that citizens entitled to vote can assess and understand what they are voting on. Both constitutive conditions for democracy are increasingly proving to be fiction." Obviously, this is not a requirement in all practicing democracies.

10 An experimental study by Lukas Reinhardt and Harvey Whitehouse (Whitehouse, 2024: 206; also https://royalsocietypublishing.org/doi/epdf/10.1098/rsos.231632) inquiring into the question of what kind of speeches motivate climate action shows that "activists and leaders alike need to appreciate that reasoning based on scientific evidence and rational argument may not be enough to bring about the sweeping changes needed to combat climate change."

11 Among the institutions that are often overlooked is the insurance industry as an agent of change in climate policy issues. For example, insurance companies that refuse to underwrite environmental risks. The role of insurance companies also extends to other key policy areas, including risk assessment, lobbying, investment strategies and influencing climate resilience policies.

12 Critically, while Ejsing, Tønder, Brandt Jensen, and Hansen identify the problem, their proposed solutions remain somewhat abstract. The call for "multi-trajectory democratic politics" as a solution to divergent time scales is conceptually rich, but lacks concrete pathways for implementation; the critique of linear democratic time is valid, but the paper does not thoroughly explore alternative temporal frameworks that could better align democracy with climate action.

8

DEMOCRACY AND THE ANTHROPOCENE

> If we accept this distinction of the present future and future presents, we can define an open future as present future which has room for several mutually exclusive future presents.
>
> *Niklas Luhmann (1976: 140)*

Niklas Luhmann (1976: 141) added in the same volume, we "can think of degrees of openness and call *futurization* increasing and *defuturization* openness of a present future." In this chapter, I reduce the complexity of the future present and offer to address the following fundamental questions: What is good governance under exceptional circumstances? Is democratic governance effective governance? Why would a more democratic and egalitarian society be beneficial as a socio-political basis for dealing with extreme situations? Does a democratic society guarantee that the climate will be treated as a public good? As I noted at the beginning, democracy is not having its best time at the time of writing. It is therefore difficult, at times even uncomfortable, but important to argue from the opposite conviction, that democracy is crucial to maintaining a living balance with nature. Broad democratic conduct embedded in a democratic *state* will have to take on a significant role in a world of becoming. Democracies, with exceptions, have shown an impressive ability to overcome major tensions resulting from political, economic, and cultural change and to adapt to new circumstances. We should not underestimate the resilience of democratic systems in the face of challenges. Democracies can implement reforms in response to public demands, economic pressures, or global challenges. For example, during economic recessions, democracies can enact reforms to address systemic problems rather than ignore them. Decentralization often allows local governments to address regional issues

DOI: 10.4324/9781003682066-10

more effectively, increasing the adaptability of different levels of government. Similarly, Prantl et al. (2024) stress that effective Earth system governance must recognize multiple worldviews and knowledge systems, promoting epistemological parity in co-producing governance solutions. Overall, the resilience of democracies depends on their ability to strike a balance between continuity and proactive adaptability.

The virtues of democratic governance

The discourse of impatient scientists in their disenchantment with democracy privileges hegemonic actors such as world powers, states, transnational organizations, and multinational corporations. However, participatory strategies are rarely mentioned. The global damage limitation also takes precedence over local adaptation. "Global" knowledge triumphs over "local" knowledge. However, social trends seem to be working in the opposite direction. The ability of large social institutions to impose their will on citizens is declining (Stehr, 2001). Nevertheless, people are mobilizing around local concerns and efforts, including the consequences of climate change, to strengthen democratic governance. This is a very sensible approach. After all, the nature of climate change impacts depends on local conditions. As Mike Hulme (2009: 31) stresses, the "outcome of any climate shift is deeply dependent on the resilience and adaptability of local societies and institutions. There is no simple mapping between climate change and the fortune of civilizations." Given the world's emission records, "there is simply no way to avoid the fact that we will have to adapt to many of the consequences of warmer planet that has a changed climate" (Morgan and Dowlatabadi, 2023: 209). Similarly, there is no way but to reduce the emissions of greenhouse gases. What is good governance under these circumstances?

The discussion of options for a future present climate policy often supports the impression that the same failed climate policy measures must be retained and are the only correct approach; it is simply a matter of making these measures more effective and "reasonable." It follows that international negotiations must lead to an agreement on concrete but much broader emission reduction targets. Only a super-Kyoto can still help us, for example is attributed to Hans-Joachim Schellnhuber. Incremental policy measures are inadequate; a transformative, enforceable global framework is essential to mitigate the risks of climate change effectively. However, how the lofty goals of comprehensive emissions reduction can be practically and politically *enforced* remains in the midst of general declarations of intent and only sharpens the political skepticism of scientists.

The prevailing climate policy thrust has not been crowned with success at either the national or the global level. In contrast, everything that is set in motion worldwide is aimed at sustained economic growth, which prevents a reduction in emissions. An alternative model is needed: A model in which action can be enforced in ambivalent, uncertain, and unexpected circumstances. This type of model also recognizes that climate change is a wicked problem that can be approached only indirectly and requires persistence over time. This kind of model is found only through revitalized interaction, not less democratic interaction. The apprehension of climate as a collective good cannot be assured based on the pre-existence of internalized moral codes that function to deliver such a public good. In this sense, the free-rider problem[1] cannot be solved automatically.[2] As Barry Barnes (1988: 134) emphasized, "there is far too much free riding to vindicate a view of society as being based upon commitments to internalized norms, yet far too little to allow its reduction to the actions of so many independent, self-interested agents." Russell Hardin (1982) documents that the provision of collective goods is best assured through the intensive but exceptional public cooperation of people. This suggests that "interactions between people, and the effects that people have on each other in the course of such interactions, may be what largely account for the provision of collective goods — where they are indeed provided" (Barnes, 1988: 134).

My assumption is that climate policy must be compatible with democracy and rest on public support to be effective and durable (Lahsen, 2024); otherwise, the threat to civilization will go far beyond changes to our physical environment.[3] It is a form of

> action which does not have to run counter to self-interest, and which is perfectly compatible with a calculative, reflectively aware attitude. It seems plausible to suggest that this is why collective sanctioning is in practice the most important factor making for the actual production of collective goods even if it is not the only factor.
>
> *(Barnes, 1988: 139)*

Without such co-operation the free-rider problem will/cannot be overcome.

Climate change calls for complex solutions that require the global empowerment and knowledge of individuals, groups, and movements working on environmental issues. More democracy combined with political efforts to create a more just society could be the key to sustainable climate policy. More democracy is linked to greater political participation, especially among those who are typically on the margins of political participation today, e.g. the young and economically disadvantaged.[4] More democracy enables the mobilization of civil societies in the interest of encouraging and enacting climate policies for the future present. More democracy defends and enlarges places of internal refuge, for example, in the sense of compensating for loss from injustice and oppression (see Garston, 2024). Democracy demands that public officials publicly justify their decisions.[5]

A more egalitarian society "would not necessarily adopt rational ecological Democracy demands public officials to publicly justify their decisions policies, but it would be more likely to do so" (Best and Connolly, 1975: 59). If life chances are more evenly distributed and it is ensured that no one can escape the benefits and costs of solving a serious public problem,[6] one should expect that "the political system is very likely to produce collective responses to common hazards and burdens" (Best and Connolly, 1975: 59). The English political scientist David Runciman (2013a: 316) identifies two other clear, practical advantages of democracies over authoritarian governments faced with exceptional circumstances: "The first is their ability to pull together when the threat becomes too great to ignore [...]. The second is their ability to constantly experiment and adapt to the challenges they face."[7] Collective resilience and openness to experiment are virtues of democracies. However, openness to experiments and a changing course considering failure may result in complacency and overconfidence:

> The knowledge that democracies have of their long-term strengths does not tell them how to access those strengths at the right moment. If anything, it makes it harder. This is why climate change is so dangerous for democracies. It represents the potentially fatal version of the confidence trap.
>
> *(Runciman, 2013a: 318)*

Effective climate policies for the future present or a world of becoming rely on the development of knowledge and technologies for climate change policies. Democratic governance is the best insurance that such knowledge can develop unimpeded. Robert Merton's ([1942] 1957: 550/560) now more than eighty-year-old essay written during World War II on "Science and the democratic social structure" continues to apply: Effective scientists are aware of their dependence on types of social structure: "Ein Professor ist ein Mensch, der anderer Meinung ist" (a professor is a person who has a different opinion). Democratic countries lead in the development of novel technologies.

A war-like approach, on the other hand, has the opposite effect. A war-like approach reduces the complexity of social and political life insofar as war "nationalizes people's lives. Private activities [...] are largely shaped by collective constraints" (Rosanvallon, [2011] 2013: 183), as would be the case under authoritarian rule. Under modern conditions, the increased cognitive and social skills of citizens require their political participation for successful politics and good governance.

In addition, further denationalization of governance will help create new, diverse forms of social solidarity and obligations, strengthen local/regional responses to climate change, and improve the understanding of social interdependence. Furthermore, the autonomy of social institutions must be ensured and, if necessary, recreated to cross borders and combine supposedly different motives and practices of different social institutions, for example, by combining economic and moral incentives and increasing the complexity of needs.

The tendency to ascribe a crucial role to the uniqueness of knowledge (and information) in social behavior becomes clear when one asks *how much knowledge* one needs to accomplish a particular task, let alone how deeply and subtly one needs to know it. Curiosity about how much we need to know also extends to the question of what we do not need to know. First, this topic has rarely been studied systematically. Second, there is a tendency to assume that the resource of knowledge is somehow sufficient to carry out a particular transaction. A more adequate assumption would be that most decisions and actions are carried out with rather limited knowledge and information (cf. Akerlof, 1970; Smith, 2015) about future conditions for action and that actors are aware of how little knowledge they can typically mobilize in many situations. The pressure to act that characterizes everyday life ensures that decisions are made and that actions are taken despite the limited knowledge and information of most actors: the fact that we are often forced to act with limited knowledge is not a constitutive deficit of democracy. "Life cannot wait" (Durkheim, [1912] 1965: 479; see also Gehlen, [1940] 1988: 296–297). In most social contexts, the need to act takes precedence over the need to know.

The erosion of democracy may seem "convenient" to some, e.g. populists, but it is certainly an unnecessary suppression of social complexity. Friedrich Hayek ([1949] 1960: 25) pointed to a paradoxical development: the more science advances, the stronger the statement that we should "strive for a *more conscious and comprehensive control of all human activities.*" Hayek adds pessimistically: "This is why those who are intoxicated by the progress of knowledge so often become enemies of freedom."

There is no denying that democratic governance is *slow*, for example, compared with the speed at which decisions are made in the modern economy (see Stehr and Voss, 2019) and is less than inclusive. In the eyes of many citizens, including climate scientists, the slowness and deliberateness of decision-making create permanent dissatisfaction. Climate scientists, with their increasing warnings about the impending risks and dangers of the effects of climate change and their communication about the failure of politicians to heed these warnings, are not helping to reduce this unrest among citizens. It is therefore a major challenge for democracies to accelerate political decision-making and improve opportunities for participation in democratic decision-making, for example, in the workplace (see Herzog, 2019) and in the local political community. Citizens concerned about the impact of a warmer world have better opportunities and more efficient means to influence political decisions in democratic nations. Environmental interest groups have greater legitimacy in states with democratic governance and a better chance to compete with interests that favor efforts to dismantle environmental legislation. Democratic regimes are less preoccupied with external security in rivalry with internal environmental goals, and autocratic regimes are more active in concentrating on advancing military technology.

Anthony Giddens (2009: 74) subscribes generally to the idea that democracies not only permit "but positively encourage the open development of science, the very basis of our awareness of problems of global warming." Whether Giddens' claim continues to carry validity is doubtful since some autocratic regimes, for example, China, have quite an active scholarly community in environmental research, if only in their self-interest.

The empirical evidence on the comparative performance of democratic nations in terms of environmental issues is limited. However, the evidence is more encouraging than Jedediah Purdy's (2015: 267) bold assertion of the "awkwardness of calling for more democracy when democracy seems a formula for failure." Purdy (2015: 267) does not suggest that we should transcend democracy but demand "more democracy." A call for

> deepened democracy had better not amount to 'More of this, please!' [...] The Anthropocene question — what kind of world to make together — should be taken as a challenge to democracy. [...] A democracy that cannot do this will have marked itself as inadequate to its most basic problems.[8]

Selseng, Linnerud, and Holden (2022: 332) report covering 127 countries focusing on a wide subset of democratic qualities in addition to democracy itself that "the level of democracy and democratic qualities, both over time and across countries, has had no significant effect on CO_2 emissions from 1992–2014." The findings do not support the notion that democracy is inherently incompatible with climate change mitigation and thus do not substantiate the assertion that authoritarian climate governance is a necessary means of achieving sufficient mitigation outcomes. A more recent effort by Apra, Kumar Sedai, Kumar, and Nepal (2023: 73), using data from 150 countries over the period 1990–2010, estimates the marginal emission intensity (i.e. the change in per-capita carbon dioxide emissions for a unit change in per-capita income) in autocracies and democracies. Marginal emission intensity should depend on both per capita income and institutions. Their analysis suggests that

> autocracies have significantly higher turning points, and they emit more carbon dioxide per capita for a unit change in per-capita income than do democracies. In other words, autocracies emit more carbon dioxide to reach the same level of per-capita income vis-à-vis democracies.

Yacov Tsur's (2025: 2) study of the relationship between "key characteristics of democracy [that is, different democratic attributes such as popular voting, civil society participation, freedom of expression, or judicial constraint] and greenhouse gas emissions" for 150 countries over the last three decades, from 1990 to 2021, offers more detail. The findings of the study suggest that "Direct Popular Voting" is highly effective in reducing emissions of all types of greenhouse gases. Tsur explains that the correlation is likely due to the lack of political influence of interest groups in favor of heavy greenhouse gas emitters. The indicator "Civil Society Participation" is also highly effective, and "Political Civil Liberties" is the third most effective indicator. Since the indicator "Political Civil Liberties" is only one aspect of Liberal Democracy and entails fewer restrictions on regulation, its effectiveness score in reducing emissions is higher than that of Liberal Democracy. This suggests (Tsur, 2025: 2) that while individual and political liberties are important, "their influence on environmental outcomes is complex and multifaceted, often requiring a delicate balance between preserving freedoms and implementing effective environmental policies." The findings generally suggest

> that democracy increases emissions up to a certain level of GDP per capita, beyond which it reduces emissions. This critical level, referred to as the threshold GDP per capita, varies depending on the specific democracy indicators and types of emissions.
>
> *(Tsur, 2025: 10)*

All in all, these findings support the notion that democratic governance makes a difference, and that more democracy makes more difference.

The 2024 Environmental Performance Index (EPI) provides a data-driven summary of the state of sustainability around the world (Block et al., 2024). Using 58 performance indicators across 11 issue categories, the EPI ranks 180 countries in terms of climate change performance, environmental health, and ecosystem vitality. These indicators provide a gauge at the national scale of how close countries are to established environmental policy targets. The EPI offers a scorecard that highlights leaders' and laggards in environmental performance and provides practical guidance for countries that aspire to move toward a sustainable future.

In this context, the political ambitions of the European Green Deal (EGD) are noteworthy.[9] In principle, the EGD places significant emphasis on the roles of democracy and citizen participation, underscoring the idea that citizens can and should play a pivotal part in the transition to sustainability (Tosun et al., 2023).[10] However desirable the call for more citizen participation in climate action and more ways to promote democratic engagement in general, it meets limited social science knowledge in the sense of knowledge as capacity to act (Adolf and Stehr, 2017: 18–21). In the following section, I will explore the call for more democratic participation in climate change policy and its implementation, despite our limited knowledge.

Dare more democracy!

> Whatever is made by humans can be remade *by humans.*
>
> *Zygmunt Bauman (2007: 56)*

Willy Brandt's slogan "Dare more democracy" was formulated almost fifty years ago in 1969. Willy Brandt first offered the phrase on October 28, 1969, in his inaugural address as the newly elected Chancellor of Germany before the German Bundestag. This sentence became the guiding principle of his chancellorship and marked the beginning of the reform policy of the social-liberal coalition in the Federal Republic of Germany. The rule of thumb in the spirit of Willy Brandt should be, as expressed by Robert Merton ([1970] 1976: 82), that "the solution to the deficiencies of democracy is more democracy" (also Calhoun et al., 2020: 247). What is perhaps equally important to state following the call "dare more democracy" is the pragmatic observation by Ralf Dahrendorf (1997: 132) "that the permanent participation of all in everything is in fact a definition of total immobility." Innovation and participation are not natural co-conspirators.

More voices do not mean all voices all the time. Leadership that balances the legitimate interests of different stakeholders does not contradict democratic governance. More voices also do not mean the elimination of the state. More voices may well mean a continuation of government action with "mild" coercion, for example, "government spending [... that] leads to improved societal welfare and enhances [...] freedom" (Stiglitz, 2024: 71).

Brandt's slogan deserves to be evoked and repeated today, considering the massive problems posed by the future present of climate change and the growing calls for a strong state. The emergence of the global perspective in environmental politics, as relevant in this context, is sometimes accompanied by fears of a "democratic deficit" at the "local" political level and increasing numbers of decisions and regulations in the effort to mitigate and adapt to climate change have become burdens of increasing numbers of global actors and institutions (cf. Jasanoff and Martello, 2004: 346–347). Whether global warming strengthens democracy, as it should, or leads to its decline, it will certainly change democracy. Zygmunt Bauman's diagnosis of a transcending the falling apart of national power and politics in modern society, I have cited earlier, also implies efforts to relink power and politics but as I would argue not only of the global arena but also the local arena.

The authority wielded by scientific discourse within public debates over climate change and other contentious issues poses difficult questions about the scope and content of popular democracy, to the extent that such authority is typically beyond the Lay understanding of the general population, whose collective mobilization will be required to steer society in a more sustainable direction. In the absence of broad pressure for social transformation from below, the responses to global warming proposed by even the most courageous and free-thinking scientists will tend to be technical and managerial in nature, working within a constrained horizon of political possibility, and largely taking existing socio-economic arrangements as given. Such dynamics are overlooked by eco-socialists such as David Schwartzman (2009: 33), whose optimistic assessment of humanity's capacity to escape catastrophic climate change revolves around the civicminded efforts of a "committed scientific/technological intelligentsia," which can generate the innovations required to dematerialize production, tap into boundless renewable energy sources, and pave the way to a materially abundant form of "solar communism." Although Schwartzman's particular perspective is infected by a strong sense of utopianism, his underlying premise is correct: what is of social origin can be remade in society.

The call for more democracy is a sober reminder that increasing broad social participation is likely to be more than compensation for the sacrifices that humanity is likely to make in the future in the fight against the effects of climate change. More democracy, specifically democratic environmentalism, should ensure flexible approaches, rapid learning from mistakes (enhancing reflexivity, e.g. Pickering, 2019),[11] experimental designs, the application of new capacities for action at all levels of society, and greater legitimacy for policymaking (Blühdorn, 2013: 22).[12] However, we should not uphold illusions about mobilizing more permanently large segments of the public in favor of climate mitigation and adaptation. Ulrich Beck's (2010b: 254–255) sober, more than a decade old assertion continues to carry validity: "The discourse on climate politics so far is an expert and elitist discourse in which peoples, societies, citizens, workers, voters and their interests, views and voices are very much neglected." The asymmetrical discourse described by Beck encourages the public to take a passive role, reflecting a common approach to science (Aitken, 2012: 214).

The call to "dare more democracy" in the case of climate change policies is therefore not merely but also an appeal for an "environmental democracy" (cf. Fischer, 2017, 2018). The idea of environmental democracy as a normative concept is based on the idea that everyone affected by climate change should have a voice and an *equal right* in political decisions that arise as a result of the impact of climate change on their lives. The appeal for more democracy is related to the greater efficacy of the political *agency* of citizens. Daring more democracy linked to political agency presumably enhances social trust and social trust in turn enlarges the ability of citizens to "organize more quickly, initiate action, and sacrifice for the common good" (Brooks, 2020; Baber and Barlett, 2021b; Wolf, 2023).

Martin Wolf (2023: 23) observes that such trusting sentiments toward the socio-political system among the public can be quite fragile: the functioning

> of democratic capitalism also depends on the presence of certain virtues in the population at large and especially in elites. Neither politics nor the economy will function without a substantial degree of honesty, trustworthiness, self-restraint, truthfulness, or loyalty to shared political, legal, and other institutions. In the absence of these virtues, a cycle of mistrust will corrode social, political, and economic relations.

The societal conditions in recent decades in many developed countries that undermine trust and confidence in the socio-economic system are not attributed by many among the public to incompetence, the lack of foresight, or the absence of luck (Figure 8.1):

> these failings were the result not just of stupidity but of the intellectual and moral corruption of decision-makers and opinion formers at all levels—in the financial sector, regulatory bodies, academia, media, and politics. They also saw the resources of the state used to rescue both banks and bankers—the architects, as they saw it, of the disaster—while they (and those they loved) suffered large losses through foreclosure, unemployment, a prolonged period of stagnant or falling real wages, and fiscal austerity.
>
> *(Wolf, 2023: 115)*

Environmental policy in democracies benefits from political and knowledge transparency, such as the publication of policy mistakes, from a vibrant civil society, from free access to relevant informational sources, from ways limiting the commodification of knowledge, from free media, from the diffusion of power, from better informed citizens, from a larger number of actors who feel they have a real stake in policy making, from citizens being free to experimenting with their own solutions, from an effective, shared responsibility of citizens, from the legitimacy of power allowing more room for persuasion and appeals to adapt and mitigate, from the presence and voice of different generations with different stakes in the future (cf. Beckman, 2008), as well as members of different genders, from the fact that the priorities of governance can be the pursuit of multiple goals and active participation in international environmental agreements (cf. Burnell, 2012: 823). In fact, not only climate change policies and their effectiveness but also democratic governance itself could benefit from climate change (e.g. Fisher et al., 2015).

Ulrich Beck (2006, 2008) was one of the prominent observers of the global political stage and optimistically suggested that the politicization of climate change could act as a lever for the emergence of "cosmopolitan realism." Cosmopolitan realism successfully transcended the narrow nation-state

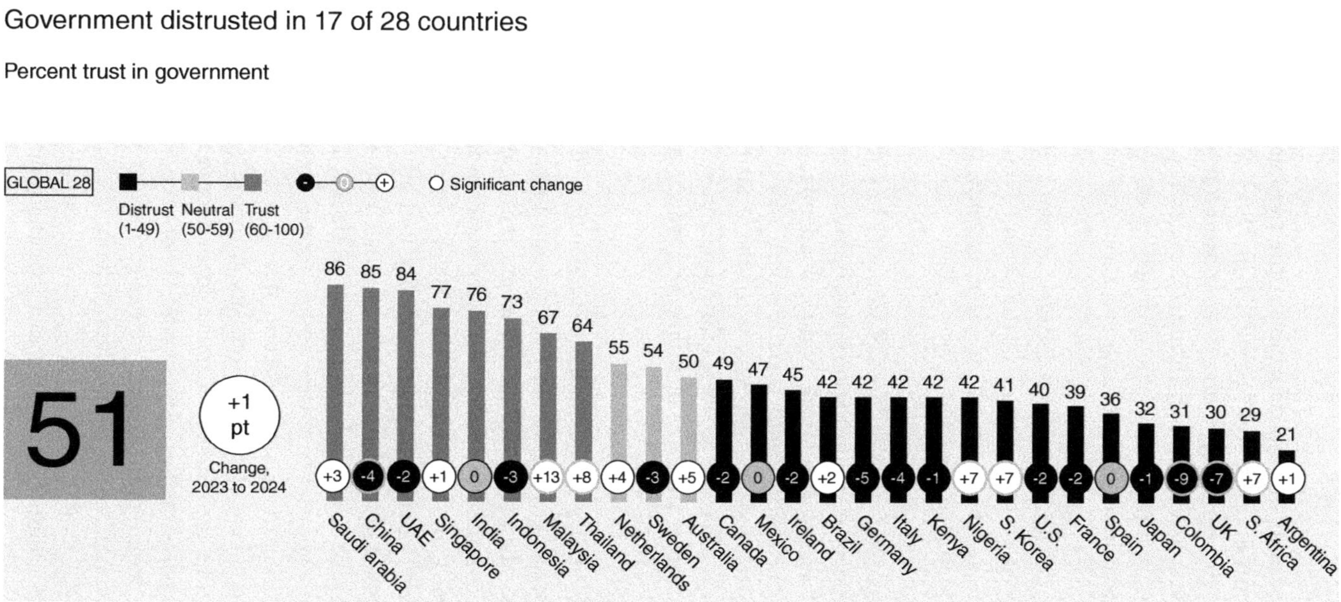

FIGURE 8.1 Government trust in 28 countries (in percent).

Source: Edelman Trust Barometer 2040; www.edelman.de/de/research/2024-edelman-trust-barometer, p. 42. Annual online survey in 28 countries, 32,000+ respondents, 1,150+ respondents per country. Field work was conducted from November 3rd to November 22nd, 2023.

frame of climate change politics. Ulrich Beck (2010a: 176) refers to novel transnational communities that could evolve in response to new global risks:

> At the beginning of the 21st century, in the face of the ongoing self-destruction of civilization, it might be possible to create imagined cosmopolitan communities that are constituted in the first place through the consent and co-determination of citizens, and thus to enable a democratic cosmopolitanism of climate change. These cosmopolitan communities of shared destiny should not negate national identities and states but rather re-establish them through cosmopolitan openness and cooperation in transformed and pooled sovereignty.[13]

The "politics of climate change is necessarily inclusive and global – it is cosmopolitical *realpolitik*" (Beck and van Loon, 2011: 124; also, Biermann, 2014; Abbott et al., 2016; Young, 2017).[14] As Ulrich Beck and his colleague recognize, the desirable scale for dealing with the ecological crisis is planetary.[15] The call for scaling up climate policies has become a normal demand and a strong voice on international political platforms. Scaling up climate policies means that an evolution of geopolitics has come into view.[16] However, despite the necessity, geopolitics remains wishful thinking—so far. And Geopolitics are changing independently of the evolution of anthropogenic climate change. The demand for and hope for geopolitics of the ecological crisis is countered by Niklas Luhmann's sobering (theoretical) observation (1980: 21) that "a system is complex when it can no longer link each of its elements to every other element." Furthermore, at least in theory, society's increased ability to influence itself should minimize the convergence of knowledge societies toward a common model (Stehr, 1994). However, the increased capacity for action brought about by scientific progress is filtered through social, political, and economic mechanisms that can reduce the potential diversity of societies.

Sovereignty: the dilemma of scales

It is widely recognized that climate change, mitigation, and adaptation occur at all scales: local, regional, national, and global. In some cases, the challenge is primarily at a single scale. While geoengineering can be carried out by a single country with planetary impact, effectively reducing greenhouse gas emissions requires the coordination of the efforts of many nations and less reliance on the sum of the economic interests of individual nation-states. Finally, climate adaptation takes place at the local level. The contemporary geopolitical field of climate policy is as *fragile* as regional and national climate policies. There is no global government. A global government that coerces and forces nations to follow agreed-upon global rules and regulations.

Climate policies are essentially interlinked with multiple political fields at all levels of action and often have unintended outcomes. For example,

> climate policy has come to serve many other political and social functions beyond its declared formal objective. Thus, undeclared political, religious, ethical, and wider life-styles and social purposes are being fulfilled which complicate the design and application of a formal policy process.
>
> *(cf. Hulme, 2009)*

Already, in the first decade of the century, the global constellation of institutions of over "200 international environmental agreements" suffered from a problem of what Held and Hervey (2009: 20) called " 'anarchic inefficiency,' featuring a diverse set of players whose roles are largely uncoordinated among each other." This has not changed in the intervening years but has most likely become more anarchic. Under the probable circumstances of the future present, it is difficult to be specifically confident that the transnational "solution" to the climate change dilemma will be "imagined cosmopolitan communities," as Ulrich Beck (2010a: 176) hoped to see evolve as an effective political response to climate change.

Ulrich Beck placed significant hope in the rise of a **global civil society**—NGOs, transnational movements, and informed citizens—to address climate change. However, the increasing influence of **state-centric power structures, corporate lobbying,** and **digital disinformation** has challenged the efficacy and reach of such non-state actors. His approach might now seem to understate the obstacles to genuine democratic global governance. Ulrich Beck imagined a world where national boundaries would be transcended by shared ecological risks. But in practice, **climate change negotiations remain mired in national interest,** sovereignty concerns, and unequal capacities. The **nation-state has proven more resilient** than Beck anticipated, and in the present, the cosmopolitan framework can seem outdated given the renewed **rise of nationalism** and the **fragmentation of global cooperation,** such as the U.S. withdrawal from the Paris Agreement under Trump. In the digital age, the idea of imagined communities has evolved. Beck's model didn't fully account for how **digital spaces fragment discourse** rather than unify it, leading to **polarization** rather than cosmopolitan consensus. Moreover, theoretically at least, the increased ability of society to act upon itself should minimize the convergence of knowledge societies toward some common model (Stehr, [1986] 1994). But the increased capacity of social action brought about by the advances in science is filtered through social, political, and economic mechanisms which may reduce the possible variety of societies.

In his 1984 lecture on "What is enlightenment?", Michel Foucault (1984: 46) strongly objects to any aspirations of a *global* political program:

> the historical ontology of ourselves must turn away from all projects that claim to be global or radical. In fact, we know from experience that the claim to escape from the system of contemporary reality to produce the overall programs of another society, of another way of thinking, another culture, another vision of the world, has led only to the return of the most dangerous traditions.

Past theoretical or practical grand political ambitions for global design have all turned into disasters. Foucault's position developed well before the global warming issue became a global affair, with an elective affinity with many strands of contemporary post-war theoretical thought that oppose a technocratic or rational design of modern society(s). Whether the extraordinary challenge of climate change is the reason for rethinking the position is itself a challenge (see Busk and Reynolds, 2023).

As Edmundson and Levy (2023: 36) summarized important and fragile elements of the political scaling problem of global warming,

> the international political community is comprised of structures and institutions that preserve the sovereign status of states. Throughout recent world history, efforts to achieve prosperity and international political order and stability have been ascribed priority over other political issues. Preserving states' rights to independent decision-making and interests has created an international political context that is poorly equipped for developing and implementing effective responses to global climate change.

The failure of the various COP conferences in recent decades has amply illustrated this dilemma.

Since the conclusion of the *Paris Climate Agreement* in 2015, major geopolitical shifts have occurred in recent years, particularly a significant increase in global tensions. Geopolitical discord has made it more difficult to scale up climate action. The incentives for cooperation have been massively reduced. China has emerged as the largest producer of renewable technology infrastructure; its control of resources such as batteries is needed to produce advanced technologies in the renewable energy transition, and the problem of adequate financial capital to underwrite climate policies around the world as costs continue to rise.

Finally, I turn to the often-underestimated role of social science and its various disciplines in the narrative of climate change and climate policy. The first premise is that social scientists (as often very visible users of language, cf. Schelsky, 1975; Tenbruck, 1984; Gergen, 1986: 153–155) have considerable influence on the cultural and symbolic repertoires of a society. Sociological language has influenced public discourse as well as numerous specialized intellectual communities and disciplines in many countries (cf. Bressler, 1967; Stehr, 1992). Social science knowledge is severely limited in its practical effectiveness only if one limits one's conception of what is effective to the restrictive notions propagated by the instrumentality model. Social science may well be most effective in practice in the form of knowledge that functions primarily as a means of practical *orientation and interpretation*, including, of course, for orientations directed to a future present. Niklas Luhmann ([1992] 1998: 63) is far gloomier when it comes to offering a picture of the future present: "But why should we describe the future? And how can we do so when what is to be described is not yet visible in the present?" The correct response to Luhmann's scepticism is to treat perceptions of the future present as "social facts": shared, constructed beliefs that influence social outcomes (cf. Boyd et al., 2015; Beckert and Suckert, 2021).

Social science and the fragility of the future

The persistent and disquieting question that science, especially the social sciences, confronts is how to respond, cope with, and perhaps assist in molding a future present that we know might well be quite dangerous, not least because of the consequences of climate change. However, regardless of which strategy might be recommended with the assistance of social science, the initial perplexing issue is of course whether we are able to anticipate the present of the future. In recent times, artificial intelligence has claimed predictive power for itself. However, data are available only for the present and the past.

Although much effort has been invested in reducing the contingencies of economic affairs and improving the possibilities of planning and forecasting, the economy of the knowledge society is, as much as the rest of the global society is, increasingly subjected to an increase in indeterminacy. While success may at times justify the high hopes of many that techniques and technologies will be developed to reduce, if not eliminate, much of the uncertainty from economic conduct, sudden and unexpected events almost invariably disconfirm, almost cruelly, such optimistic forecasts about the possibility of anticipating and therefore controlling future events. In fact, and paradoxically, one of the sources of the growing indeterminacy can be linked directly to the nature of the technological developments designed to achieve greater certainty. The new technology contributes to and accelerates the malleability of specific contexts because of its lower dedication (limitation) to functions. Technological developments add to the fragility of economic markets and the need for organizations operating in such a context to become more flexible to respond to greater mutability in demand and supply.

In the sphere of production, as a result, a new utopian vision arises, a vision that Charles Sabel (1991: 24) sketches in the following and deliberately enabling terms:

> Universal materializing machines replace product-specific capital goods; small and effortlessly re combinable units of production replace the hierarchies of the mass-production corporation; and the exercise of autonomy required by both the machines and the new organizations produces a new model producer which view of life confounds the distinction between the entrepreneurial manager and the socialist worker-owner.

Charles Sabel acknowledges that he has been seen as the major author and is therefore responsible for this utopian vision; however, he prefers to subscribe to a more "prudent version of these caricatures." This perspective accounts for the "diversity and similarity of efforts to adjust to the new competitive environment" (Sabel, 1991: 24–25).

Much of the standard discussion of these matters, at least until recently, has been animated by opposite expectations. Bell (1973: 26), for example, confidently asserts that the "development of new forecasting and 'mapping' techniques makes possible a novel phase in economic history—the conscious, planned advance of technological change, and therefore the *reduction of indeterminacy* about the economic future" (emphasis added). However, greater fragility, malleability, and volatility are not confined to the economy, the labor market, or the social organization of work and management, nor do they merely have "positive" effects on social relations and individual psyches. Greater vulnerability corresponds to greater fragility, and greater flexibility is linked to new regimes of exclusion.

A more recent iteration of efforts to anticipate the future present relies on the large data budget of artificial intelligence developments thereby raising the question of whether AI can be of any help in predicting the future (see Antón Barba-Kay, 2023: 95; Scheirer, 2024), on the other hand, is not persuaded that AI will be of much help: climate change is a "long-term" problem, and long-term problems have no place in the digital world: whereas

> our experience of being online is focused on the fragmented and overwhelming experience of now, the problem of climate change stems from our inability to act coherently in connection with the distant future. Long-term solutions are not within the scope of digital technology, the point of which is innovation and upheaval.

In Chapter 2, I mentioned the growing importance of the legal system in addressing climate change issues. The following section focuses more narrowly on this issue. The focus is on the legal liability of those seen as causing the climate crisis.

Climate change and climate liability

Climate change and extreme weather events, their consequences, and costs are increasingly becoming legal issues (cf. Kuh, 2021) on the national and international front.[17] How can we show that climate change must be taken seriously? It cannot be ruled out that the unpredictability of weather and seasons[18] will increase during global warming, and that the material power of climate and weather will increase. In view of the significance of the climate semantics described above as a basic attitude of modern society, however, the question arises as to what political and legal consequences should be drawn from this. Confidence in the normality of the climate is of existential importance. The perceived normality of the climate guarantees physical and psychological security.

The right to climate normality should be enshrined in the constitution. Such a constitutional amendment should make future climate policy much easier and defend it against possible doubts and skepticism. What better way to show that society is in harmony with nature? The climate is taken seriously by including it in the constitution. Moreover, and at least as importantly, security is a condition of freedom. That individual, political, and economic actors will address climate change should be strengthened by fundamental legal norms. A constitutional mandate is the basis for a warmer society.

More realistically, suing an emitter of greenhouse gases for damages or liability related to climate change is a growing area of legal interest. However, legal challenges are also complex and face several major legal hurdles. These depend on the country. Typically, lawsuits are filed in national courts and legal systems. Some of the main legal issues are: Most importantly, causation: Showing that a certain emitter's greenhouse gas (GHG) emissions cause specific problems with the climate, like floods, wildfires, and sea-level rise. It is hard to determine who is responsible because climate change is caused by the combined effect of global emissions over many years. This makes it difficult to identify a specific harm caused by a specific emitter. Standing is another hurdle. To stand, plaintiffs must show that they have been personally and concretely harmed by the defendant's actions. For example, in the United States, under Article III of the Constitution, plaintiffs must show that they have suffered an actual injury, that there is a cause-and-effect relationship between the injury and the defendant's actions, and that the injury can be redressed. Courts may decide that climate change is a problem that should be handled by politicians and regulators, not the courts. Finally, what legal framework may be used?

There are numerous ongoing lawsuits around the world against fossil fuel companies that allege that they are responsible for damage related to climate change. These lawsuits which set important precedents for climate accountability are typically brought by cities, states, environmental groups, and even individuals. The lawsuits usually focus on the argument that fossil fuel companies contributed to climate change by producing and promoting fossil fuels despite having been aware of the associated risks for decades. Key types of litigation include lawsuits alleging that fossil fuel companies have created a "public nuisance" by causing harm through climate change impacts such as rising sea levels, extreme weather, and other environmental impacts. Several U.S. cities and states, including New York City, San Francisco, and Rhode Island, have filed lawsuits seeking compensation for the costs of adapting to climate impacts. Similar cases have been filed outside the United States, including in the Netherlands, where Dutch courts ordered Shell to significantly reduce its carbon emissions. This ruling underscores corporate responsibility to address climate change. Moreover, in November of 2024, Shell successfully challenged the court ruling that it must reduce greenhouse gas emissions by 45% by 2030. The Court of Appeal ruled that such limits could not be set by the courts. Shell argued that corporate emission limits are a matter for politicians, not the judiciary. However, the court upheld the statement that companies such as Shell have an obligation to protect human rights. The ruling can still be appealed to the Dutch Supreme Court.

A fascinating legal case with potentially far-reaching implications was before a higher court in Germany. A Peruvian small farmer, Saúl Luciano Lliuya, who lives in Huarez, Peru, wants RWE (*Rheinisch-Westfälisches Elektrizitätswerk*), a large German power company and major emitter, to pay up. He lives under a melting glacier. He holds RWE partly responsible for his house being threatened by the melting glacier. The case concerns whether major greenhouse gas emitters are globally liable for the consequences of climate change. Lluya's case will be decided by the Hamm Higher Regional Court (OLG). The amount at stake is 17,000 euros. This amount corresponds to 0.47% of the costs for a dam to protect against the feared floods. RWE is liable for this share because the company is responsible for 0.47% of global man-made climate change. The case has been ongoing since 2015 and was initially dismissed by a lower court. The case before the Superior Court is based in part on a review of the evidence to determine whether Lluya's house is threatened. Specifically, the risk of a tidal wave or landslide as a result of the melting glacier. A battle of experts ensued. The costs of the litigation will be borne by a German environmental foundation (Germanwatch). According to Germanwatch, about 50,000 people live in the risk zone. The case could end up before the highest court in Germany. But on May 27, 2025, the lawsuit brought by a Peruvian farmer against the energy company RWE was dismissed by the Higher Regional Court (OLG) in Hamm. The OLG ruled that the appeal filed by small farmer and mountain guide Saúl Luciano Lliuya would be rejected. The court ruled that there was no possibility of appealing this decision. The environmental protection organization Germanwatch already considered the Higher Regional Court's decision to allow the evidence to be heard a success: "This is the only lawsuit in the world seeking corporate liability for climate risks that has made it to the discovery phase and has thus already made legal history."[19]

A major beneficiary of the fossil fuel industry is, of course, governments: In 2023, fossil fuel taxes and related revenues in the United States will generate approximately $138 billion annually for government entities at all levels.[20] These revenues are largely derived from excise taxes on petroleum products, which account for the largest share, followed by revenues from oil and gas extraction. Such "benefits" to the state go hand in hand with the climate-warming carbon emissions of the major fossil fuel companies. According to statistics from the *Carbon Majors* report, half of the world's climate-warming carbon emissions come from fossil fuels produced by 36 of the largest fossil fuel companies, led by Saudi Aramco, Coal India, ExxonMobil, Shell, and numerous Chinese firms. A study by the think tank that produced the report concludes:

> Despite global climate commitments, a small group of the world's largest fossil fuel producers are significantly increasing production and emissions. The research highlights the disproportionate impact these companies have on the climate crisis and supports efforts to enforce corporate responsibility.[21]

However, contributions from coal taxes are declining rapidly and are expected to continue to decline significantly, especially under policies that aim to reduce emissions by 2050. Germany's energy taxes, including fuel taxes, remain a core part of its environmental tax structure, which has historically generated more than €60 billion annually from taxes on energy, motor vehicles, and emissions. This strong revenue stream reflects both Germany's high tax rates on fossil fuels and its commitment to reducing dependence on carbon-intensive fuels. Another recent case involves the government of Hawaii, which, on June 20, 2024, settled a lawsuit with a group of young people who had sued the state's Department of Transportation over its use of fossil fuels. A *New York Times* article sums up the case:

> The Hawaii case alleged that, by depending on polluting fuels such as gasoline, the state's Department of Transportation was violating Hawaii's Constitution, which guarantees the right to a clean and healthful environment. The settlement, which was announced shortly before the case was to have gone to trial, requires the state to make a series of changes

to reduce planet-warming emissions. It also recognizes the constitutional rights of young people in the state to a clean and healthy environment.[22]

In 2023, a group in Montana, also organized by Our Children's Trust, won its lawsuit against the state of Montana using an argument like the one used in Hawaii. In Great Britain, in the summer of 2024, the highest court ruled that local councils and planning groups must consider the full environmental impact of new fossil fuel projects when deciding whether to approve them, a decision that could have far-reaching consequences and that climate activists hailed as a victory, perhaps even around the world.[23]

Christopher Callahan and Justin Mankin (2025) work on estimating climate liability, bridging climate science and law, offering flexible tools to hold corporations accountable for climate damages while underscoring the urgency of addressing historical emissions' inequitable impacts. Their work presents a novel framework for attributing climate damages directly to major fossil fuel companies ("carbon majors"), arguing that the scientific basis for climate liability claims is now robust. The authors claim (Callahan and Mankin, 2025: 897) their "results provide the first estimate of the global economic toll that individual fossil fuel companies have produced owing to the extreme heat caused by their emissions of carbon dioxide and methane." The new study's framework could make it easier to file claims related to climate change, similar to how other major damage and loss cases have been handled. For example, tobacco companies were held responsible for lung cancer, and pharmaceutical companies were held responsible for opioid addiction.

Using peer-reviewed methods, they quantify the economic losses caused by extreme heat linked to emissions from companies like Chevron, ExxonMobil, and Saudi Aramco over 1991–2020. Their "end-to-end" attribution approach connects corporate emissions to specific harms through three steps: (1) Using simple climate models to show how global temperatures are rising due to emissions, (2) converting global warming into local heat extremes using pattern scaling, and (3) estimating economic impacts using damage functions based on real data.

The study's key findings, based on an examination of emissions from 111 carbon majors, caused an estimated $28 trillion in global economic losses due to extreme heat, with a 90% confidence range of $12–$49 trillion. These losses have disproportionately impacted tropical, low-latitude regions. In April 2925, Pacific island states wrote to the governments of developed countries. They asked the governments to make big cuts in carbon. The statement says that rich countries are taking too long to make new plans to deal with the climate crisis. This is putting poor countries at greater risk. All governments are supposed to publish new plans this year on cutting greenhouse gas emissions, but so far, only a few have done so, and some of the plans submitted have not been enough to deal with the problem. Developed countries have not yet said how they will make sure that $1.3 trillion a year in climate money goes to poor countries by 2035.[24]

Individual companies' contributions to climate change are substantial; for example, Chevron's emissions alone likely resulted in between $0.79 trillion and $3.6 trillion in damages (from a single climate impact). The analysis goes further by linking emissions to specific historic extreme heat events, such as the 2003 European heatwave and associated losses, demonstrating causal ties between corporate emissions and localized economic harm.

Callahan and Mankin's study also addresses legal hurdles like causation by testing "but for" scenarios (e.g. "but for Chevron's emissions, would this damage have occurred?"). It also explores proportional liability, where companies' contributions to global warming align with their share of damages. The framework is modular, allowing adaptation to other hazards (e.g. floods) and scales (e.g. national or sectoral emissions). Callahan and Mankin (2025: 898) advocate for a scientific initiative to support climate litigation, akin to an extension of the *World Weather Attribution* project, and highlight inequities: emissions from Global North companies harm vulnerable, low-latitude populations. They note that legal success depends on courts reconciling fossil fuels' benefits with their externalities, but stress that science no longer impedes liability claims. While they acknowledge that legal systems must still grapple with balancing fossil fuel benefits against their costs, Callahan and Mankin conclude that scientific uncertainty is no longer a barrier to climate liability claims.

Despite the progress represented by Callahan and Mankin's approach to estimating climate liabilities, key questions remain unanswered, such as the time period over which emissions should be counted. Beyond the scope of attribution science, there are still controversial socio-legal and political challenges, such as the personal attribution of damages, the devaluation of the benefits of fossil fuel use, overcoming epistemic differences between legal scholars and attribution scientists, and the question of whether climate change is even a matter that should be dealt with by the courts.

However, it would be a mistake to assume that litigation about climate issues almost automatically favors individuals, groups or institutions that litigate on behalf of the environment.[25] A major case in point in which such litigation may favor the fossil fuel industry is judgments rendered by the investor-state dispute settlement system (ISDS). The ISDS mechanism emerged in the mid-20th century as part of efforts to promote and protect foreign investment. It is a mechanism included in many international investment agreements, such as bilateral investment treaties (BITs) and free trade agreements (FTAs). It allows foreign investors to bring claims against host states for alleged breaches of those agreements, typically through arbitration rather than domestic courts.

The primary purpose of ISDS is to protect foreign investors from discriminatory, unfair, or expropriatory actions by host governments.[26] However, as the recent *Guardian* investigation of 1400 ISDS cases revealed,

> there have been long-held concerns about ISDS creating 'regulatory chill' – where governments are scared off action on nature loss and the climate crisis by legal risks. Now, government ministers from a range of countries have confirmed to the Guardian that this 'chilling' is already in effect – and that fear of ISDS suits is actively shaping environmental laws and regulations.

Numerous rulings of the ISDS dispute settlement system point to rulings in favor of fossil fuel companies and "hostility" to environmental regulation. Whether such rulings against environmental concerns, as manifested in the demands of the Paris Agreement against the economic interests of investing corporations, are bound to favor mostly investment firms is a matter of debate. For the time being, however, the Guardian's investigation into ISDS cases revealed $84 billion in pay-outs from governments to fossil fuel companies. Since 1976, more than $120 billion in public money has been awarded to private investors across all industries. The average payout for a fossil fuel claim was $1.2 billion. Litigation can therefore also constitute a major hurdle in addressing climate change. Climate litigation in the future present is likely to increase significantly in volume, public visibility, and judgments obtained; it can, of course, help efforts to address climate change. In the longer term, however, it is the conjunction of legislation, litigation, economic imperatives, improved knowledge capacities, institutional reform, moral commitment, and more democracy that will be decisive. Fossil fuel interests may lobby and even succeed, in different jurisdictions, to obtain total immunity from litigation, for example, with Republican control of the White House and both houses of Congress.[27]

The government's rejection of climate science and its implications for climate policies, as evidenced during the second Trump administration in the United States, is primarily driven by economic considerations. Support for this position may be found in a report of the *U.S. Department of Energy* doubting the evidence amassed by climate science, contending that its consensual finding is too dire and overlooks the benefits of global warming (Climate Working Group, 2025). Specifically, the policy argument is that the costs of addressing climate change outweigh the benefits, effectively shifting the burden of risk to individuals. Ultimately, however, the law should protect people.

In the concluding remarks, I emphasize the need to reintegrate nature and climate change into democratic political discourse (see Grundmann and Stehr, 2000). Drawing on thinkers such as Hannah Arendt and Georg

Simmel, the text explores the human condition as being inherently tied to both nature and politics. According to Arendt's philosophy, while humans construct an artificial world, they remain deeply rooted in natural processes. Losing this connection threatens democratic stability. While climate change poses an unprecedented threat, as I have emphasized throughout, it puts pressure on governance systems and could tempt various societal agents toward authoritarian responses. However, technocratic or authoritarian models lack democratic legitimacy and long-term effectiveness. I therefore advocate deepening democracy by ensuring informed citizen participation and fair resource distribution as the most robust way to address climate challenges. Ultimately, the conclusion argues that a democratic, legally grounded approach is vital for equitably managing climate risks and ensuring the sustainability of society and the planet.

Notes

1 The idea of the free-rider problem was first systematically analyzed by the American economist Mancur Olson in the 1950s in his work *"The Logic of Collective Action"* (1965). Olson examined why individuals often do not participate in collective action even though they benefit from the results. However, the idea was not new. For example, David Hume described a similar phenomenon in his work *"A Treatise of Human Nature"* (1739), in which individuals rely on the efforts of others. In an essay on public goods, the economist Paul Samuelson (1954) discussed the problem of non-excludability, which is closely related to free-riding behavior. However, Mancur Olson is considered www.edelman.de/de/research/2024-edelman-trust-barometerthe person who made the problem prominent and formalized it in modern social and economic science.

2 The free-rider dilemma in the case of decarbonization is described in necessary detail by Dani Rodrik (2024: 402): "Whether greenhouse gases are produced in my country or yours makes no difference to global warming. If I impose carbon taxes in mine, you benefit as much as I do. Under these circumstances, individual nations are likely to under-invest substantially in providing for the collective global good and have strong incentive to free ride on other nations' contributions – sadly an all-too apparent reality in climate change. Transfer of financial resources and technology to developing countries so that they can decarbonize their economies and embark on a green transition is an especially important component of global public goods in climate, since these countries are not the primary culprits for historical emissions."

3 Public support for climate policies varies over time and from country to country, depending on the issues at the forefront of urgent public discussion, for example. The findings of an empirical survey by Bretter and Schulz (2025) emphasize the importance of worldviews and ideologies in influencing support for different policy instruments. They also show that support in the sampled Global South countries (Brazil, China, and South Africa) was much stronger than in the Global North countries (Germany, the UK, and the USA). Subsidies and information-based policies received the highest support in all countries. However, a single data

point is subject to great variation as political circumstances and/or meteorological conditions change.

4 Concrete advice on avoiding oligarchic tendencies in organizations can be found, for example, in Robert K. Merton's (1966) essay on "Dilemmas of democracies in the voluntary association."

5 Jocelyn Stacey (2019: i; emphasis added) elaborates in her *The Constitution of the Environmental Emergency* "how the requirement of public justification enables a deeper, more coherent understanding of how creative institutional design and environmental principles such as precaution and sustainable development enable the maintenance of the rule of law in light of the ongoing *environmental emergency*."

6 The systematic reduction of patterns of social inequality in modern societies improves democratic governance and political participation (Soci et al., 2014: 46).

7 Hans Jonas ([1979) 1984: 146) makes a similar observation about the systematic inability of authoritarian governments to overcome political mistakes.

8 What kind of democracy would be able to live up to such expectations? For Purdy (2015: 268), the answer can only be "a democracy capable of self-restraint. But, ironically, democratic self-restraint can come only from democratic self-assertion: a political community must be able to act effectively and decisively on hard questions in order to commit to accepting certain limitations."

9 The European Green Deal is a growth strategy launched by the European Union in 2019 to achieve climate neutrality by 2050. The package of policy initiatives aims to promote a green transition and support sustainable growth. A key target is to reduce greenhouse gas emissions by at least 55% by 2030 compared to 1990 levels. demonstrates stronger climate ambition.

10 Karin Bäckstrand (2025) explored how public and democratic legitimacy impact the implementation of the EU's climate and energy legislation—specifically the "Fit for 55" package—through a comparative study of Denmark and Sweden (cf. Rosamond and Dupon, 2025). Both countries are historically viewed as climate leaders with strong democratic institutions and public support for climate action. However, their recent implementation trajectories diverge significantly. Denmark demonstrates stronger climate ambition and more robust implementation of EU targets. In contrast, Sweden has seen a rollback in climate ambition under a Liberal-Conservative government supported by the populist Sweden Democrats. Bäckstrand emphasizes how institutional design, democratic innovation, and political leadership are crucial for effective climate governance (cf. also Olsson, 2022).

11 In general terms, reflexivity can be defined as the capacity of an agent, structure, process, or set of ideas to undergo change in response to reflection on their performance (Dryzek, 2016: 942; Pickering, 2019: 1145) importantly of course learning from and correcting mistakes.

12 The assertion that democracies **are uniquely capable of self-correction** by learning from and correcting policy mistakes resonates, of course, with Karl Popper's (1945: 124) strong commitment to democratic governance precisely for this reason: "*Democracy provides the institutional framework for the reform of political institutions. It makes possible the reform of institutions without using violence, and thereby the use of reason in the designing of new institutions and the adjusting of old ones. It cannot provide reason or intelligence where none exists, but it provides the only framework in which reason and intelligence can operate.*"

13 As Ulrich Beck (2008: 795) also describes it, "Cosmopolitanization does not mean uniformization and homogenization. People, their groups, communities, political organizations, cultures, civilizations will (and should) remain different, sometimes even unique. But to put it metaphorically, the walls between them must be replaced by bridges."

14 The large research undertaking of the "Earth System Governance Project" published in multiple volumes by Cambridge University Press (e.g. Biermann and Kim, 2020; Prantly et al., 2024) or its precursors (Biermann, 2014) do not contain any affirmative or critical reference to the extensive work of Ulrich Beck on cosmopolitical governance of the climate crisis. A case of the strong and persistent division of labor social science.

15 The demand (from the point of view of the social sciences) for a cosmopolitan policy approach is self-exemplifying. Are the social sciences cosmopolitan? "Does the sociological universerve presently function in a manner of a *unified* field on a world scale?" (Bourdieu, 1991: 374). The answer obviously is negative. Hence, Pierre Bourdieu (1991: 385) is only able to call for the creation of this "field of world sociology as an autonomous space of struggles in which truth may be produced on condition that only scientific weapons are used." Bourdieu's vision is a utopian plan; see my chapter on "global knowledge" in Stehr (2025b).

16 Ulrich Beck's expectations are less than two decades old. The promising developments he anticipated demonstrate how volatile geopolitical, political, and economic processes actually are. Beck had high hopes for the emergence of a global civil society comprising NGOs, transnational movements, and informed citizens to address climate change. However, the growing power of state-centric structures, corporate lobbying, and digital misinformation has undermined the effectiveness and influence of these non-state actors. In hindsight, his approach may appear to understate the obstacles to genuine democratic global governance. Beck imagined a world where national boundaries would be transcended by shared ecological risks. In practice, however, climate change negotiations remain mired in national interests, concerns about sovereignty, and unequal capacities. The nation-state has proven to be more resilient than Beck anticipated. In the present climate, the cosmopolitan framework may appear outdated, given the resurgence of nationalism and the fragmentation of global cooperation—for example, the US withdrawal from the Paris Agreement under Trump. In the digital age, the concept of imagined communities has evolved. However, Beck's model did not fully account for how digital spaces can fragment discourse rather than unify it, leading to polarization rather than cosmopolitan consensus.

17 A landmark, international court decision was issued by the *International Court of Justice*, the judicial branch of the United Nations, in the summer of 2025. The court issued "a strongly worded opinion on [...] saying that states must protect people from the 'urgent and existential threat' of climate change, a major moment for the global environmental movement and for the countries at greatest risk of harm. It was the first time that the court, which is the United Nations' highest judicial body, has weighed in on climate change. The unanimous opinion said that the failure of nations to take action to protect the climate system may constitute 'an internationally wrongful act.' It also found that protection of the environment is "a precondition" for ensuring human rights and cited government support of fossil-fuel production as a potential violation of these principles." However, the

U.N. court has no mechanism to enforce its findings. Nevertheless, its findings will likely resonate around the world in lower court decisions and spur more cases (see Karen Zraick and Marlise Simons, "Top U.U. court says countries must act on climate change," *New York Times*, July 23, 2025; www.nytimes.com/2025/07/23/climate/icj-hague-climate-change.html?searchResultPosition=1 .

18 Depending on the time of year and independent of global warming, the northern and southern hemispheres lean either toward or away from the sun. When the northern half of the globe leans closer toward the sun, the north experiences summer; and the southern half of the planet leans away from the sun, and is in winter. It is not known why Earth is angled this way. Some astronomers are convinced that the slope is responsible for the possibility of life on Earth.

19 In May 2025, another landmark case brought against the fossil fuel industry is the first wrongful death lawsuit: Leon v. ExxonMobil (https://climateintegrity.org/uploads/media/LeonvExxonComplaint.pdf). The argument "put forward in *Leon v. ExxonMobil* is relatively straightforward: Julie Leon's death was caused by unnaturally extreme heat; this heat was the result of climate change caused by Big Oil's fossil fuel products; these companies knew that their products would cause disasters like the Heat Dome and deaths like Julie's, but rather than warning the public about this lethal danger, Big Oil launched a campaign of deception to defraud the public about climate risks and block solutions that could have prevented Julie's death;" cited in https://newrepublic.com/article/195912/climate-leon-wrongful-death-lawsuit?utmsource=newsletter&utmmedium=email&utmcampaign=tnrdaily&vgoee=%2FUGxAk8%2Boq%2FztPXyRAwRUeVZNWpJ6CDcpQM%2F4haH1bwCLUk5yTk%3D%3A5KLW43kJlyG4QOTD80G9AMjz4mfLaizm.

20 On the other side of the ledger, the International Monetary Fund (IMF, 2023) reported that global fossil fuel subsidies reached **$7 trillion** in 2022, equivalent to **7.1% of global GDP**. This figure encompasses both **explicit subsidies**—such as direct government support—and **implicit subsidies**, which arise from undercharging for environmental costs like air pollution and climate change, as well as forgone consumption tax revenues. The subsidies are expected to continue into the future and perhaps even increase; see www.imf.org/-/media/Files/Topics/energy-subsidies/EXTERNALfuelsubsidiestemplate2023new.ashx.

21 Damian Carrington, "Half of world's CO_2 emissions come from 36 fossil fuel firms, study shows," *Guardian*, March 5, 2025; www.theguardian.com/environment/2025/mar/05/half-of-worlds-co2-emissions-come-from-36-fossil-fuel-firms-study-shows .

22 "Hawaii settles with young plaintiffs in climate case," *New York Times*, June 21, 2024; www.nytimes.com/2024/06/21/climate/hawaii-youth-climate-lawsuit.html.

23 "Oil projects must consider full climate impact, top U.K. court rules," *New York Times*, June 21, 2024; www.nytimes.com/2024/06/21/world/europe/uk-supreme-court-oil-ruling.html.

24 Cf. "Pacific island states urge rich countries to expedite plans to cut emissions," *Guardian*, April 26, 2025; www.theguardian.com/environment/2025/apr/26/pacific-island-states-urge-rich-countries-to-expedite-plans-to-cut-emissions.

25 Phoebe Weston and Patrick Greenfield, "Why fear of billion-dollar lawsuits stops countries phasing out fossil fuels," *Guardian*, March 6, 2025; www.theguardian.com/environment/2025/mar/06/isds-fear-of-billion-dollar-lawsuits-stops-countries-phasing-out-fossil-fuels-aoe.

26 Criticisms of ISDS includes lack of transparency: Arbitration is often conducted in private, raising concerns about accountability; Regulatory chill: Governments may be reluctant to enact regulations in the public interest (e.g. environmental or health measures) for fear of costly arbitration claims and imbalances: Critics argue that ISDS favors investors over states and undermines national sovereignty. The European Union has proposed a Multilateral Investment Court (MIC) to replace traditional ISDS with a more transparent and permanent system.

27 In the summer of 2025, Lee Zeldin, the administrator of the *Environmental Protection Agency* (E.P.A.), announced, for example that the Trump administration would revoke the scientific determination that underpins the government's legal authority to combat climate change. He announced that the E.P.A. intended to rescind the 2009 'endangerment finding', which concluded that greenhouse gases that warm the planet pose a threat to public health. The Obama and Biden administrations used this determination to impose strict limits on greenhouse gas emissions from cars, power stations, and other industrial sources of pollution; www.nytimes.com/2025/07/29/climate/epa-endangerment-finding-repeal-proposal.html?searchResultPosition=1. Representatives of the U.S. climate science community vigorously opposed the policy announcement: www.theguardian.com/us-news/2025/aug/01/trump-epa-climate-change-report.

CONCLUSIONS

Or bringing nature and climate change back into politics

> Although people must die, they are not born to die, but rather to begin something new.
>
> Hannah Arendt ([1958] 1972: 328)[1]

> But just as nature, with all its undeniable autonomy and harsh reality, is nevertheless included in the self, so on the other hand, this self, with all its freedom and self-centeredness, its opposition to mere nature, is nevertheless a part of it; that is precisely the overarching natural context, that it encompasses this being that is independent of itself, indeed often enough hostile, that what, according to its deepest sense of life, stands outside of itself must nevertheless be its element.
>
> Georg Simmel ([1908] 2009: 47)[2]

Hannah Arendt's statement underscores her conviction that human life is defined not by mortality, but by the ability and initiative to start anew. Georg Simmel was—and so was Hannah Arendt—a strong proponent of the idea that the essential human condition means that social beings are strongly incorporated into nature (see also Ghosch, 2021). In *The Human Condition*, Hannah Arendt ([1958] 1998: 126) saw humans as beings who exist both within and apart from nature. While humans are biological creatures subject to natural processes, they also transcend nature through their capacity to create a shared, enduring world and to engage in *political action*. For Arendt, the capacity to build a world apart from nature is a defining characteristic of humanity. This separation is necessary for the creation of politics, culture, and art—activities that define the human condition. While separation from nature is intrinsic to human life, complete alienation from it risks undermining

DOI: 10.4324/9781003682066-11

the balance between human biological existence and the artificial world we create:

> It is as though we had forced open the distinguishing boundaries which protected the world, the human artifice, from nature, the biological process which goes on in its very midst as well as the natural cyclical processes which surround it, delivering and abandoning to them the always threatened stability of human world.

Arendt's concern is not environmental or ecological in the modern sense but philosophical and political. She focuses on the existential and cultural consequences of losing the connection to the natural world, which she sees as essential to understanding and preserving the human condition. However, in the modern world, the severing of ties with nature may prove to be even more consequential than Hannah Arendt envisaged.

Societies have built a world of their own making. Are societies smart enough to keep the world they created from collapsing? Democracies live in perilous times. The resurgence of authoritarian movements in recent years, not least in response to policies to address the threat of anthropogenic climate change, is a stark reminder that constitutional democracy cannot be taken for granted.[3] However, it is not democracy as such that is failing. We are failing to be democratic enough:

> Our crises are rooted in a decades-long depoliticization, a collapse of political vision, energy, and faith, whose hinge was, most ironically, the democratic victories of 1988–1991, when the Soviet Union disowned one-party rule and dissolved its empire, and "velvet revolutions" swept away autocracies, tore down the Berlin Wall, and dissolved the Iron Curtain.
>
> *(Purdy, 2022: 71)*

In the future present, democracies should count on

> the most important sources of stability in a democracy [which] are the benefits that citizens experience from living in it: the freedom to pursue their own goals, the opportunity to participate in politics as equals, and access to a fair share of resources.
>
> *(Chandler, 2023: 326)*

The problem is not democracy but how democracy is run. The democracy we need is one that has more of the means to participate in the decisions of society, that is, knowledgeable citizens. Although democracies may be the worst process for deliberating and deciding on socio-moral issues, they are better suited for this task than all other forms of governance.

The magnitude of the expected changes due to climate change is not yet clear to us. *Nature* (December 4, 2014: 8) comments, "The extent of [...] climate change is worryingly uncertain. Even more uncertain are the physical, social and economic side effects of global warming. There is every reason to believe that they will be broadly harmful." The central question is no longer whether climate change has occurred. Rather, it is about what should be done about it. Climate change is the greatest threat humanity has faced in history. How much time is there? Presumably, there is very little time to act. This will put enormous pressure on governance. In addition, turning to the state appears to be, for many, the only viable option. Moreover, climate change will produce both winners and losers. The form of governance of the future present will determine not only who gains and who loses but also how large the collective between winners and losers will be.

If one were to suspend democratic debate and decision-making—as an expression of the citizen's freedom to create or change social relations—including full citizen participation and a government under the rule of law, in order to do what is necessary, one would either have to appoint experts as decision-makers or delegate power to policy-makers (who happen to believe a certain group of experts). Neither the first, the technocratic or socio-technical vision as a species, nor the idea of an authoritarian "climate state" (Christoff, 2022: 123) is attractive, let alone an effective problem solver. Given that authoritarian regimes promise to respond effectively to the dangers of climate change in a forward-looking manner, authoritarian governance is initially nothing more than a promise. Whether "virtuous leaders" actually prioritize environmental emergencies and use their powers to act without democratic procedures and legal constraints in the interest of exactly these rather than other priorities is uncertain (cf. Shahar, 2015). Equally unpredictable is the detailed path of the future course of the local, national, or global climate. As noted above, it cannot be ruled out that during global warming, weather extremes will become more intense, seasons will become unpredictable or even disappear, and climate and weather in general will increase their material power.

I have collected and put forward arguments that democracy needs to be strengthened, not abolished, and that constitutional democracy represents the best political basis for policies to tackle climate change. It is important to resist simplistic solutions to climate change (Willke, 2023). In discussing, researching, and understanding climate and climate change, we consider not only the complex interconnectedness of the climate system, but also the social processes, practices, and tensions through which science, society, nature, and climate interpenetrate, accompany, cover, and envelop each other (for such a theoretical perspective, see Stehr and Machin, 2019). The future present shaped by the political and economic order of society and nature requires the mobilization of all forms of knowledge, natural as much as social science knowledge, and the realization that democracy is the bulwark against climate change.

Climate change and extreme weather events are increasingly becoming legal issues. How can we show that climate change must be taken seriously? It cannot be ruled out that the unpredictability of the weather and the seasons will increase in the course of global warming, and that the material power of climate and weather will increase. In view of the significance of the climate semantics described above as a basic attitude of modern society, however, the question arises as to what political and legal consequences should be drawn from this. Confidence in the normality of the climate is of existential importance. The perceived normality of the climate guarantees physical and psychological security.

A final reference to the frequent and extensive reporting of extreme weather events as evidence for my argument, paradoxically, it has been argued, that normality manifests itself only in deviation, in contrast, and is thus simultaneously asserted: as the reporting on extreme weather events all too often proves, "I have lived here for 30 years and have never experienced weather like this," is the typical statement of a local resident in the face of an extreme event. The extreme is ordinary. Consistency is, as so often in everyday life, a virtue.

The right to climate normality should be enshrined in the constitution. Such a constitutional amendment should make future climate policy much easier and defend it against possible doubts and skepticism. It could also be expected that a constitutional amendment would encourage a higher level of public and institutional participation in climate-related societal activities, thereby enhancing the democratic foundation of democracies. Moreover, what better way is there to demonstrate that society is in harmony with nature? The climate will be taken seriously if it is included in the constitution. Climate becomes a public good. As a public good, climate makes everyone better off regardless of whether individuals or corporate actors contribute to the provision of the public good. A public good, because it is enshrined in the Constitution, is much more equitably distributed. The risks and costs of climate change are more equitably distributed. As such, the climate as a public good strengthens democracy. The individual, political, and economic will to address climate change should be strengthened by fundamental legal norms. A constitutional mandate as a basis for coping with a warmer society.

My appeal is not particularly unusual. There are hundreds of lawsuits related to global warming pending in many countries. The first climate action before the European Court of Human Rights (ECtHR) was successful. The Swiss senior citizens who brought the case argued, and the court ruled in favor of them on April 9, 2024, that their human rights were being violated by climate change at their age. The ruling has a signal effect beyond Switzerland. It underlines the fact that politicians must not only proclaim, but also act.

Similarly spectacular is a ruling by the Federal Constitutional Court (BVerfG) in March 2021, according to which fundamental rights convey a right to protection (a right to the future) against a one-sided shift of climate impact management into the future (see Janda, 2021).

In a ruling issued in May 2024, *the International Tribunal for the Law of the Sea* (ITLOS), the judicial organ of the *United Nations Convention on the Law of the Sea* (UNCLOS), an international treaty representing 169 countries, determined that greenhouse gases are pollutants that are damaging the marine environment and that states have a legal obligation to regulate them. The *Commission of Small Island States on Climate Change and International Law* (COSIC), a group of nine Caribbean and Pacific Island nations led by Antigua and Barbuda and Tuvalu, sought the opinion of the court. As is widely acknowledged, islands are particularly susceptible to the effects of climate change and have become disillusioned with the sluggish pace of international negotiations. The tribunal asserted that the *Paris Agreement* signed in 2016 lacked sufficient teeth, as it did not include penalties for non-compliance with the agreement's objectives. The Tribunal also emphasized that the world's wealthier nations must bear a greater financial burden in addressing climate change and implementing adaptation measures. After all, climate justice requires consideration of who will enjoy the enormous material wealth and freedoms of the fossil age and who will be the primary victims and pay the disproportionate price in the future present of the coal and oil age. It is likely that other international courts will issue rulings on climate change and climate justice in the future. Climate litigation can help. In the longer term, it is more democracy that empowers large segments of the public; it is self-organization, learning from feedback and dead ends, and transparency. (cf. Fiorini et al., 2023; Landemore, 2023).

The threat of catastrophic climate change risks and the increasingly evident dangers of disruption, one might suspect, persuade even the most resistant individual and collective actors around the world to take adaptive action to save humanity from the excessive consequences of a warming world and to ensure that the world of our making does not become a world of our unmaking, for the alternative is that nature will make itself heard even more loudly, forcing societies and their citizens to listen. Climate science cannot transcend politics, as evidenced by the persistent and vehement rejection of climate science in some political and economic fields. The foregrounding of the role of democracy in responding to the threats of climate change, as my discussion has shown, has also shown that the independent role of the political order and interests, and in the background of the economic order, will continue to be crucial in the world's response to climate change.

Notes

1 I have quoted Hannah Arendt from the German edition of *Human Condition:* [Dass] "Menschen zwar sterben müssen, aber deshalb noch nicht geboren werden, um zu sterben, sondern im Gegenteil, um etwas Neues anzufangen."
2 Translation amended.
3 The difference between authoritarians and dictators with respect to democracy is that authoritarians use democratic rules to get elected and, once elected, attempt to disassemble democracy and/or refuse to relinquish power, while dictators destroy democracy to gain power (de Sousa Santos, 2018).

REFERENCES

The most recent works by each author are listed last. If a work has been translated into another language, the original publication date is also given in square brackets. References to newspaper articles and their authors are usually not included here. The locations of these articles can be found in the relevant footnotes.

Abadi, Cameron (2024), *Climate Radicals. Why Our Environmental Politics Isn't Working*. Stanford, CA: Stanford University Press.

Abbott, Kenneth W., Jessica F. Green, and Robert O. Keohane (2016), "Organizational ecology and institutional change in global governance," *International Organization* 70: 247–277.Acemoglu, Daron and James A. Robinson (2006), "Economic backwardness in political perspective," *American Political Science Review* 100: 115–131.

Acemoglu, Daron, Simon Johnson and James A. Robinson (2001), "The colonial origins of comparative development: An empirical investigation," *American Economic Review* 91: 1369–1401.

Adler, Jonathan H. (2009), "Taking property rights seriously: The case of climate change," *Social Philosophy and Policy* 26: 296–316.

Adolf, Marian and Nico Stehr (2017), *Knowledge*. Second Edition. New York: Routledge.

Adorno, Theodor and Max Horkheimer ([1947] 1977), *The Dialectic of Enlightenment*. London: Verso.

Aitken, Mhairi (2012), "Changing climate, changing democracy: A cautionary tale," *Environmental Politics* 21: 211–229.

Ajzen, Icek (1985), "From intentions to actions: A theory of planned behavior," pp. 11–39 in Julius Kuhl and Jürgen Beckman (eds.), *Action-Control: From Cognition to Behavior*. Heidelberg: Springer.

Ajzen, Icek (2011), "The theory of planned behaviour: Reactions and reflections," *Psychology & Health* 26: 1113–1127.

Akerlof, George A. (1970), "The market for 'lemons': Quality, uncertainty, and the market mechanism," *The Quarterly Journal of Economics* 84: 488–500.

Akerlof, George and Robert Shiller (2010), *Animal Spirits: How Human Psychology Drives the Economy, and Why It Matters for Global Capitalism*. Princeton, NJ: Princeton University Press.

Akerlof, George, et al. (2019). Economists' statement on carbon dividends. *Wall Street Journal*, 17. www.wsj.com/articles/economists-statement-on-carbon-dividends-11547682910

Albert, Michael J. (2025), "Capitalism, complexity, and polycrises: Towards neo-Gramsian polycrises analysis," *Global Sustainability* 8: 1–11.

Albertone, Baptiste (2025), "Lost principles of a 'sustainable developmentalism," *Review of International Political Economy*, 32: 1–24.

Alonso, Sonia, John Keane, and Wolfgang Merkel (eds.) (2011), *The Future of Representative Democracy*. Cambridge: Cambridge University Press.

Alsberg, Paul (1922), *Das Menscheitsrätsel. Versuch einer prinzipiellen Lösung*. Second Edition. Dresden: Sybillen Verlag.

Anderson, Kevin and Alice Bows (2011), "Beyond 'dangerous' climate change: Emission scenarios for a new world," *Philosophical Transactions of the Royal Society A: Mathematical, Physical and Engineering Sciences* 369: 20–44.

Andre, Peter, Teodora Boneva, Felix Chopra, and Armin Falk (2025), "Globally representative evidence on the actual and perceived support for climate action," *Nature Climate Change* 14: 253–259.

Angrist, Noam, Kevin Winseck, Harry A. Patrinos, and Joshua S. Graff Zivin (2024), "Human capital and climate change," *NBER Working Paper* No. 31000.

Applebaum, Anne (2020), *Twilight of Democracy: The Seductive Lure of Authoritarianism*. New York: Doubleday.

Apra, Sinha, Ashish K. Sedai, Abhishek Kumar, and Rabindra Nepal (2023), "Are autocracies bad for the environment? Global evidence from two centuries of data," *The Energy Journal* 44: 47–78.

Arendt, Hannah ([1958] 1972), *Vita activa oder Vom tätigen Leben*. Erweiterte Neuausgabe. München: Piper.

Arendt, Hannah ([1958] 1998), *The Human Condition*. Second Edition. Chicago, IL: University of Chicago Press.

Arias, Sabrina B. and Christopher W. Blair (2024), "In the eye of the storm: Hurricanes, climate migration, and climate attitudes," *American Political Science Review* 118: 1593–1613.

Arnold, David (1996), *The Problem of Nature. Environment, Culture and European Expansion*. Oxford: Blackwell.

Aron, Raymond ([2013] 2023), *Liberty and Equality*. Princeton, NJ: Princeton University Press.

Athreye, Suma, Vinish Kathuria, Alessandro Martelli, and Lucia Piscitello (2023), "Intellectual property rights and the international transfer of climate change mitigating technologies," *Research Policy* 52: 104819.

Baber Walter F. and Robert V. Bartlett (2021a), "Democratic governance in the anthropocene: Equivocal, experimental, equitable, empowered, embedded," pp. 1–22 in Walter F. Baber and Robert V. Bartlett (eds.), *Democratic Norms of Earth System Governance*. Elements in Earth System Governance. Cambridge: Cambridge University Press.

Baber, Walter F. and Robert V. Bartlett (2021b), "Empowered democratic agency in the anthropocene: Reconciling people to nature and each other," pp. 47–76 in Walter F. Baber and Robert V. Bartlett (eds.), *Democratic Norms of Earth System Governance. Elements in Earth System Governance*. Cambridge: Cambridge University Press.

Bäckstrand, Karin (2025), "Democracy and the implementation of the European Green Deal: Comparing Denmark and Sweden," *Journal of European Integration* 47: 277–297.
Barba-Kay, Antón (2023), *A Web of Our Own Making. The Nature of Digital Formation*. Cambridge: Cambridge University Press.
Barber, Benjamin (2012), "Fixing climate change means fixing democracy," pp. 165–168-168 in Christopher Bieber, Benjamin Drechsel, and Ann-Katrin Lang (eds.), *Gefahr im Konflikt*. Claus Leggewie revisited. Bielefeld: transcript.
Barnes, Barry (1988), *The Nature of Power*. Urbana: University of Illinois Press.
Barnes, Elizabeth A., Noah S. Diffenbaugh, and Sonia I. Seneviratne (2024), "Combining climate models and observations to predict the time remaining until regional warming thresholds are reached," *Environmental Research Letters* 20. https://doi.org/10.1088/1748-9326/ad91ca
Barry, John (2012), "Climate change, 'the cancer stage of capitalism' and the return of Limits to Growth," pp. 129–152 in Mark Pelling, David Manuel-Navarrete, and Michael Redclift (eds.), *Climate Change and the Crisis of Capitalism*. London: Routledge.
Bauman, Zygmunt (2000), *Liquid Modernity*. Cambridge: Polity Press.
Bauman, Zygmunt (2007), *Liquid Times. Living in an Age of Uncertainty*. Cambridge: Polity Press.
Beck, R. A. (1993), "Climate, liberalism and intolerance," *Weather* 48: 63–64.
Beck, Thessa M., Dominik L. Schumacher, Hicham Achebak, Ana M. Vicedo-Cabrera, Sonia I. Seneviratne, and Joan Ballester (2024), "Mortality burden attributed to anthropogenic warming during Europe's 2022 record-breaking summer," *NPJ Climate and Atmospheric Science* 7: 245.
Beck, Ulrich ([1986] 1992), *Risk Society. Towards a New Modernity*. London: Sage.
Beck, Ulrich (2000), "Living your own life in a runaway world: Individualism, globalisation and politics," pp. 164–174 in Will Hutton and Anthon Giddens (eds.), *On the Edge. Living with Global Capitalism*. London: Jonathan Cape.
Beck, Ulrich (2006), *Cosmopolitan Vision*. Cambridge: Polity Press.
Beck, Ulrich (2008), "Reframing power in the globalized world," *Organization Studies* 29: 793–804.
Beck, Ulrich (2010a), "Entweder Kant oder Untergang: Wieviel Klimawandel verträgt die Demokratie," pp. 174–180 in Christopher Bieber, Benjamin Drechsel, and Ann-Katrin Lang (eds.), *Gefahr im Konflikt*. Claus Leggewie revisited. Bielefeld: transcript.
Beck, Ulrich (2010b), "Climate for change, or how to create a green modernity," *Theory, Culture and Society* 27: 254–266.
Beck, Ulrich and Joost van Loon (2011), "'Until the last ton of fossil fuel has burnt to ashes': Climate change, global inequalities and the dilemma of green politics," pp. 111–134 in David Held, Angus Fane-Hervey, and Marika Theros (Hrsg.), *The Governance of Climate Change. Science, Economics, Politics and Ethics*. Cambridge: Polity Press.
Becker, William S. (2023), 'Breaking policy gridlocks," pp. 133–165 in David W. Orr (ed.), *Democracy in a Hotter Time*. Cambridge: MIT Press.
Beckert, Jens, and Lisa Suckert (2021), "The future as a social fact. The analysis of perceptions of the future in sociology," *Poetics* 84: 101499.
Beckman, Ludvig (2008), "Do global climate change and the interest of future generations have implications for democracy?," *Environmental Politics* 17: 610–624.

Beer, Christopher T. (2022), "'Systems change not climate change': Support for a radical shift away from capitalism at mainstream U.S. climate change protest events," *The Sociological Quarterly* 63: 175–198.

Beeson, Mark (2010), "The coming of environmental authoritarianism," *Environmental Politics* 19: 276–294.

Greven, Michael Th. ([2009] 2020), "Zukunft oder Erosion der Demokratie? pp. 155–178 in Rüb, Friedbert W., Veith Selk und Rieke Trimçev (eds.), *Die Erosion der Demokratie*. Beiträge von Michael Th. Greven zur kritischen Demokratietheorie. Wiesbaden: Springer.

Bell, Daniel (1973), *The Coming of Post-Industrial Society: A Venture in Social Forecasting*. New York: Basic Books.

Bell, Daniel A. (2015), *The China Model. Political Meritocracy and the Limits of Democracy*. Princeton, NJ: Princeton University Press.

Bell, Daniel A. and Wang Pei (2020), *Just Hierarchy. Why Social Hierarchies Matter in China and the Rest of the World*. Princeton, NJ: Princeton University Press.

Belloc, Filippo und Ugo Pagano (2012), "Knowledge Enclosures, Forced Specializations and Investment Crisis," *European Journal of Comparative Economics* 9: 445–483.

Bellocca, Gian P., Pilar Poncelab, and Esther Ruiz (2024), "Extreme temperatures and the profitability of large European firms," https://e-archivo.uc3m.es/entities/publication/4b21bc5e-075a-46c0-903f-4ff0d1220760

Benninghaus, Hans (1976), *Ergebnisse und Perspektiven der Einstellungs--Verhaltens-Forschung*. Meisenheim am Glan: A. Hain.

Benson, Jonathan (2024), "Epistemic problems in Hayek's defence of free markets," *Economics & Philosophy* 41: 1–23.

Berger, Loïc and Massimo Marinacci (2020), "Model uncertainty in climate change economics: A review and proposed framework for future research," *Environmental and Resource Economics* 77: 475–501.

Berger, Peter und Thomas Luckmann (1967), *Die gesellschaftliche Konstruktion der Realität*. Frankfurt am Main: S. Fischer.

Bergquist, Parrish und Christopher Warshaw (2019), "Does global warming increase public concern about climate change?," *The Journal of Politics* 81: 686–691.

Bermeo, Nancy (2016), "On democratic backsliding," *Journal of Democracy* 27: 5–19.

Best, Michael H. and William E. Connolly (1975), "Market images and corporate power: Beyond the 'economics of environmental management'," pp. 41–74 in Kenneth M. Dolbeare (ed.), *Sage Yearbooks Public Policy Evaluation*. Beverly Hills, CA: Sage.

Biebricher, Thomas (2018), *The Political Theory of Neoliberalism*. Stanford: CA: Stanford University Press.

Biermann, Frank (2014), *Earth System Governance: World Politics in the Anthropocene*. Cambridge: MIT Press.

Block, S., J. W. Emerson, D. C. Esty, A. de Sherbinin, Z. A. Wendling (2024). *2024 Environmental Performance Index*. New Haven, CT: Yale Center for Environmental Law & Policy. epi.yale.edu.

Bloor, David ([1976] 1991), *Knowledge and Social Imagery*. Second Edition. Chicago, IL: University of Chicago Press.

Blühdorn, Ingolfur (2013), "The governance of unsustainability: Ecology and democracy after the post-democratic turn," *Environmental Politics* 22a: 16–36.

Boecher, Michael, Ulrike Zeigermann, Lars E. Berker, and Djaa Jabra (2022), "Climate policy expertise in times of populism–knowledge strategies of the AfD regarding Germany's climate package," *Environmental Politics* 31: 820–840.

Böhme, Gernot (1992), "Was ist Natur? Charaktere der Natur aus der Sicht der modernen Naturwissenschaft," pp. 56–76 in Gernot Böhme (ed.), *Natürlich Natur. Über Natur im Zeitalter ihrer technischen Reproduzierbarkeit*. Frankfurt am Main: Suhrkamp.

Bolin, Bert (2007), *A History of the Science and Politics of Climate Change. The Role of the Intergovernmental Panel on Climate Change*. Cambridge: Cambridge University Press.

Boltanski, Luc ([2009] 2011), *On Critique. A Sociology of Emancipation*. Cambridge: Polity.

Bonneuil, Christophe and Jean-Baptiste Fressoz (2017), *The Shock of the Anthropocene. The Earth, History and US*. London: Verso.

Bourdieu, Pierre ([1972] 1979), "Public opinion does not exist," pp. 124–130 in Seth Siegelaub (ed.), *Communication and Class Struggle* 1. New York: International General.

Bourdieu, Pierre ([1991] 1994), "Rethinking the state: Genesis and structure of the bureaucratic field." *Sociological Theory* 12: 1–18.

Boyd, David R., Nykquist, S., Borgström, S. und Stacewicz, P.A. (2015), "Anticipatory governance for social-ecological resilience," *Ambio* 44: 149-161.

Boykoff, Maxwell (2019), *Creative (Climate) Communications*. Cambridge: Cambridge University Press.

Braudel, Fernand ([1979] 1985), *Civilization and Capitalism 15th – 18th Century*. Volume 1: The Structure of Everyday Life. The Limits of the Possible. London: William Collins.

Braudel, Fernand ([1979] 1992), *Civilization and Capitalism, 15th-18th Century*. Berkeley: University of California Press.

Brechin, Steven R. and Seungyun Lee (2024), "Future governance under climate change," pp. 480–491 in Steven R. Brechin and Seungyun Lee (eds.), *Routledge Handbook of Climate Change and Society*. Second Edition. New York: Routledge.

Bremmer, Ian (2010), *The End of the Free Market. Who Wins the War between States and Corporations*. New York: Viking.

Bremmer, Ian (2022), *The Power of Crisis. How Three Threats – And Our Response – Will Change the World*. New York: Simon & Schuster.

Brent, Kerryn, Jeffrey McGee, Jan McDonald, and Eelco J. Rohling (2018), "International law poses problems for negative emissions research," *Nature Climate Change* 8: 444–553.

Bressler, Marvin (1967), "Sociology and collegiate education," pp. 45–77 in Paul F. Lazarsfeld, William H. Sewell and Harold L. Wilensky (eds.), *The Uses of Sociology*. New York: Basic Books.

Bretter, Christian and Felix Schulz (2025), "Public support for climate policies and its ideological predictors across countries of the Global North and Global South," *Ecological Economics* 233(2025): 108603.

Broecker, Wallace S. (1975), "Climatic change: Are we on the brink of a pronounced global warming?," *Science* 189(4201): 460–463.

Broecker, Wallace S. and Sidney Hemming (2001), "Climate swings come into focus." *Science* 294(5550): 2308–2309.

Brooks, David (2020), "America is having a moral revulsion," *The Atlantic*, October. www.theatlantic.com/ideas/archive/2020/10/collapsing-levels-trust-are-devastating-america/616581/

Brosch, Tobias (2021), "Affect and emotions as drivers of climate change perception and action: A review," *Current Opinion in Behavioural Sciences* 42: 15–21.

Brown, Mark B. (2014), "Climate science, populism, and the democracy of rejection," pp. 129–145 in Deserai A. Crow and Maxell T. Boykoff (eds.), *Culture, Politics and Climate Change. How Information shapes our Common Future*. New York: Routledge.

Brückner, Eduard (1890), *Klimaschwankungen seit 1700. Nebst Bemerkungen über die Klimaschwankungen der Diluvialzeit*. Wien and Olmütz: Hölzel.

Brückner, Eduard ([1889] 2000), "How constant is today's climate? [In wie weit ist das heutige Klima konstant?, Verhandlungen des VIII Deutschen Geographentages]," pp. 63–76 in Nico Stehr and Hans von Storch (eds.), *Eduard Bruckner – The Sources and Consequences of Climate Change and Climate Variability in Historical Times*. Dordrecht: Kluwer.

Burkhart, Ross E. and Michael S. Lewis-Beck I (1994), "Comparative democracy: The economic development thesis." *American Political Science Review* 88: 903–910.

Burnell, Peter (2012), "Democracy, democratization and climate change: Complex relationships," *Democratization* 19: 813–842.

Burnell, Peter (2014), "International support for action on climate change and democracy: Exploring complementarities," *Third World Quarterly* 35: 1216–1238.

Busk, Larry A. and Iaan Reynolds (2023), "Climate change and the irrational society," *Theory & Event* 26: 559–575.

Byrne, Brendan, Junjie Liu, Kevin W. Bowman, Madeleine Pascolini-Campbell, Abhishek Chatterjee, Sudhanshu Pandey, Kazuyuki Miyazaki, Guido R. van der Werf, Debra Wunch, Paul O. Wennberg, Coleen M. Roehl, und Saptarshi Sinha (2023), "Carbon emissions from the 2023 Canadian wildfires," *Nature* 633 (2024): 835–839. https://doi.org/10.1038/s41586-024-07878-z

Calhoun, Craig, Dilip P. Gaonkar, and Charles Taylor (2020), *Degenerations of Democracy*. Cambridge, MA: Harvard University Press.

Callahan, Christopher W. and Justin S. Mankin (2025), "Carbon majors and the scientific case for climate liability," *Nature* 640: 893–901.

Callon, Michel, Pierre Lascoumes, and Yannick Barthe (2009), "A quelle éducation citoyenne l'enseignement des «Thèmes de convergence» au collège en France, contribue-t-il?," *Studies in Science Education* 45.

Camus, Albert ([1953] 1957), *Heimkehr nach Tipasa*. Aus dem Französischen von Monique Lang. Zürich: Arche.

Caplan, Bryan (2007), *The Myth of the Rational Voter*. Princeton, NJ: Princeton University Press.

Caramani, Daniele (2017), "Will vs. reason: The populist and technocratic forms of political representation and their critique to party government," *American Political Science Review* 111: 54–67.

Carleton, Tamma, Esther Duflo, Kelsey Jack, and Guglielmo Zappalà (2024), "Adaptation to climate change," *NBER Working Paper* No. 33264, December 2024.

Cassady, Reed (2024), "The challenge of powering AI," *Wells Fargo Investments*, Clearbridge Investment. www.clearbridge.com/perspectives/institutional/2024/the-challenge-of-powering-ai

Chakrabarty, Dispeh (2009), "The climate of history: Four theses," *Critical Inquiry* 35: 197–222.

Chakrabarty, Dispeh (2015), "The anthropocene and the convergence of histories," pp. 45–56 in Clive Hamilton, Christophe Bonneuil, and François Gemenne (eds.), *The Anthropocene and the Global Environmental Crisis*. London: Routledge.

Chakrabarty, Dispeh (2017), "The politics of climate change is more than the politics of capitalism," *Theory, Culture & Society* 34: 25–37.

Chandler, Daniel (2023), *Free and Equal. What Would a Fair Society Look Like?* London: Penguin.

Chavez, Anthony E. (2015), "Exclusive rights to saving the planet: The patenting of geoengineering inventions," *Northwestern Journal of Technology and Intellectual Property* 13: 2–35.

Chesler, Angela, Debra Javeline, Kimberly Peh, and Shana Scogin (2023), "Is democracy the answer to intractable climate change?," *Global Environmental Politics* 23: 201–216.

Christoff, Peter (2022), "Crises, COVID, and the climate state," pp. 117–126 in Samuel Alexander, Sangeetha Chandrashekeran, and Brendan Gleeson (eds.), *Post-Capitalist Futures: Paradigms, Politics, and Prospects*. Singapore: Springer Singapore.

Christophers, Brett (2024), *The Price Is Wrong. Why Capitalism Won't Save the Planet*. London: Verso.

Churchman, C. West (1967), "Wicked problems," *Management Science* 14: B141–B142.

Clarke, Ben, Barnes Clair, Sparks Nathan, Toumi Ralf, Yang Wenchang, Giguere Joseph, Woods Placky Bernadette, Gilford Daniel, Pershing Andrew, Winkley Shel, Vecchi Gabriel A, Arrighi Julie, Roy Malini, Poole-Selters Leah, Van Sant Carolyn, Grieco Matthew, Singh Roop, Vahlberg Maja, Kew Sarah, Pinto Izidine, Otto Friederike, Hess Victoria, Gorham Ellen, Rodgers Sam, Philip Sjoukje, Kimutai Joyce (2024), "Climate change key driver of catastrophic impacts of Hurricane Helene that devastated both coastal and inland communities," www.worldweatherattribution.org/climate-change-key-driver-of-catastrophic-impacts-of-hurricane-helene-that-devastated-both-coastal-and-inland-communities/

Clarke, Ben, Friederike Otto, Rupert Stuart-Smith, and Luke Harrington (2022), "Extreme weather impacts of climate change: An attribution perspective," *Environmental Research: Climate* 1. https://doi.org/10.1088/2752-5295/ac6e7d

Clayton, Susan (2020), "Climate anxiety: Psychological responses to climate change," *Journal of Anxiety Disorders* 74: 102263.

Climate Working Group (2025), *A Critical Review of Impacts of Greenhouse Gas Emissions on the U.S. Climate*. Washington DC: Department of Energy.

Coeckelbergh, Mark (2021), "AI for climate: Freedom, justice, and other ethical and political challenges," *AI and Ethics* 1: 67–72.

Coeckelbergh, Mark, and Henrik S. Sætra (2023), "Climate change and the political pathways of AI: The technocracy-democracy dilemma in light of artificial intelligence and human agency," *Technology in Society* 75: 102406.

Coen, Deborah R. (2018), *Climate in Motion. Science, Empire and the Problem of Scale*. Chicago, IL: University of Chicago Press.

Coffé, Hilde, Sam Crawley, and Josh Givens (2025), "Growing polarisation: Ideology and attitudes towards climate change," *West European Politics* 48: 1–29. https://doi.org/10.1080/01402382.2024.2435727

Coleman, James (1984), "Introducing social structure into economic analysis," *The American Economic Review*, Papers and Proceedings of the Ninety-Sixth Annual Meeting of the American Economic Association 74: 84–88.
Collins, Harry M. and Robert Evans (2002), "The third wave of science studies: Studies of expertise and experience," *Social Studies of Science* 32: 235–296.
Connolly, William (1993), *TheTerms of Political Discourse.* New York: John Wiley & Sons.
Connolly, William E. (2019), *Climate Machines Fascist Drives and Truth*. Durham, NC: Duke University Press.
Connolly, William E. (2025), "The anthropocene as abstract machine." *Journal of Human Rights and the Environment* 16: 35–45.
Connolly, William E. and Bradley J. Macdonald (2015), "Confronting the anthropocene and contesting neoliberalism: An interview with William E. Connolly: Interviewer: Bradley J. Macdonald," *New Political Science* 37: 259–275.
Conti, Gregory (2014), "Liberalism and equality," *American Affairs* 8(3). https://americanaffairsjournal.org/2024/08/liberalism-and-equality/
Crippa, M., D. Guizzardi, F. Pagani, M. Banja, M. Muntean, E. Schaaf, W. Becker, F. Monforti-Ferrario, R. Quadrelli, A. Risquez Martin, P. Taghavi-Moharamli, J. Köykkä, G. Grassi, S. Rossi, J. Brandao De Melo, D. Oom, A. Branco, J. San-Miguel, and E. Vignati (2023), *GHG Emissions of All World Countries*. Luxembourg: Publications Office of the European Union. https://doi.org/10.2760/953332, JRC134504.
Crouch, Colin (2004), *Post-Democracy*. Cambridge: Cambridge University Press.
Crutzen, Paul J. and Eugene F. Stoermer (2000), "The anthropocene," *IGBP [International Geosphere-Biosphere Programme] Newsletter* 41: 17–18.
Dahl, Robert A. (1977), "On removing certain impediments to democracy in the United States," *Political Science Quarterly* 92: 1–20.
Dahrendorf, Ralf (1979), *Life Chances. Approaches to Social and Political Theory*. Chicago: University of Chicago Press.
Dahrendorf, Ralf (1980), "Effectiveness and legitimacy: On the 'governability' of democracies," *Political Quarterly* 51: 393–410.
Dahrendorf, Ralf (1997), *After 1989: Morals, Revolution and Civil Society*. Wiesbaden:Springer.
Darian-Smith, Eve (2022), *Global Burning. Rising Antidemocracy and the Climate Crisis*. Stanford, CA: Stanford University Press.
Darian-Smith, Eve (2024), "The challenge of political will, global democracy and environmentalism," *Environmental Policy and Law* 54: 117–126.
Davidson, Debra (2025), *Feeling Climate Change. How Emotions Govern Our Responses to the Climate Emergency*. New York: Routledge.
Davies, William (2024), "Antimarket," *London Review of Books* 46(7) (April 4).
De Bellaigue, Christopher (2023), "A world off the hinges," *New York Review of Books*, November 23.
de Sousa Santos, Boaventura (2018), *The End of the Cognitive Empire: The Coming of Age of Epistemologies of the South*. Durham, NC: Duke University Press.
Dean, Joshua F., Gemma Coxon, Yanchen Zheng, Jack Bishop, Mark H. Garnett, David Bastviken, Valier Galy (2025), "Old carbon routed from land to the atmosphere by global river systems," *Nature* 642: 105–111.
Delay, Tad (2024), *Future of Denial. The Ideologies of Climate Change*. London: Verso.

Department of Defense, Office of the Undersecretary for Policy (Strategy, Plans, and Capabilities) (2021), *Department of Defense Climate Risk Analysis*. Report Submitted to National Security Council.

Deutloff, Jakob, Hermann Held, and Timothy M. Lenton (2025), "High probability of triggering climate tipping points under current policies modestly amplified by Amazon dieback and permafrost thaw," *Earth System Dynamics* 16: 565–583.

Di Paola, Marcello and Dale Jamieson (2018), "Climate change and the challenges to democracy," *University of Miami Law Review* 72: 369–424.

Dorman, Peter (2022), *Alligators in the Arctic and How to Avoid Them. Science, Economics and the Challenge of Catastrophic Climate Change*. Cambridge: Cambridge University Press.

Douglas, Mary ([1970] 1975), "Environments at risk," pp. 230–248 in Mary Douglas (ed.), *Implicit Meanings. Essays in Anthropology*. London: Routledge and Kegan Paul.

Downs, Anthony (1972), "Up and down with ecology – The issue-attention cycle," *Public Interest* 28: 38–50.

Drahos, Peter (2020), "Responsive science," *Annual Review of Law and Science*16: 14.1–14.15.

Drahos, Peter (2021), *Survival Governance: Energy and Climate in the Chinese Century*. Oxford: Oxford University Press.

Drucker, Peter F. (1993), "The rise of the knowledge society," *The Wilson Quarterly* 17: 52–71.

Dryzek, John S. (1994), "Ecology and discursive democracy. Beyond liberal capitalism and the administrative state." pp. 176–197 in Martin O'Connor (ed.), *Is Capitalism Sustainable? Political Economy and Politics of Ecology*. New York: Guildford Press.

Dryzek, John S. (2016), "Institutions for the anthropocene: Governance in a changing Earth system," *British Journal of Political Science* 46: 937–956.

Dumaine, Carol (2022), "Redefining security. The recent national intelligence estimate on climate change illustrates how traditional national security approaches must be expanded to contend with global environmental change," *Issues in Science and Technology* 38: 81–83.

Dura, Tina, William Chilton, David Small, Andra J. Garner, Andrea Hawkes, Diego Melgar, Simon E. Engelhart (2025), "Increased flood exposure in the Pacific Northwest following earthquake-driven subsidence and sea-level rise," *PNAS Earth, Atmospheric, and Planetary Sciences* 122): e2424659122. https://doi.org/10.1073/pnas.2424659122

Durkheim, Èmile ([1893] 1933), *The Division of Labor in Society*. New York: The Free Press.

Durkheim, Èmile ([1912] 1965), *The Elementary Forms of Religious Life*. New York: Free Press.

Durkheim, Èmile ([1950] 1992), *Professional Ethics and Civic Morals*. London: Routledge and Kegan Paul.

Durkheim, Èmile (1960), *Montesquieu and Rousseau. Forerunners of Sociology*. Ann Arbor: University of Michigan Press.

Durkheim, Èmile ([1897] 1952), *Suicide: A Study in Sociology*. London: Routledge.

Edmundson, Beth and Stuart Levy (2023), *Climate Change and Order. The End of Prosperity and Democracy*. Houndmills: Palgrave Macmillan.

Ekiert, Grzegorz (2023), *Democracy and Authoritarianism in the 21st Century: A sketch*. Ash Institute for Democratic Governance and Innovation.

Elçi, Ezgi (2024), "Authoritarian populism," pp. 45–58 in Natasha Lindstaedt and Jeroen J. J. den Bosch (eds.), *Research Handbook on Authoritarianism*. Cheltenham: Edward Elgar Publishing.

Elias, Norbert (1977), "Towards a theory of social processes," *British Journal of Sociology* 48: 355–383.

Elias, Norbert ([1939] 1982), *The Civilizing Process*. Volume 2: Power and Civility. Oxford: Blackwell.

Elias, Norbert (1984), "Knowledge and power," pp. 251–292 in Nico Stehr and Volker Meja (eds.), *Society and Knowledge. Contemporary Perspectives on the Sociology of Knowledge*. New Brunswick, NJ: Transaction Books.

Elias, Norbert (1987), "The retreat of sociologists into the present," pp. 150–172 in Volker Meja, Dieter Misgeld, und Nico Stehr (eds.), *Modern German Sociology*. New York: Columbia University Press (now Routledge, 2023).

Englert, Alexander T. (2023), "We'll meet again," *aeon*. https://aeon.co/essays/kurt-godel-his-mother-and-the-argument-for-life-after-death

Eskander, Shaikh, Saam Fankhauser und Joana Setzer (2024), "Testing the ambition loop: do country-and company-level net-zero targets reinforce each other? A global comparison," *Journal of Comparative Policy Analysis: Research and Practice* 26: 1–17.

Esper, Jan M. T. and Ulf Büntge (2024), "2023 summer warmth unparalleled over the past 2,000 years," *Nature* 631: 94–97. https://doi.org/10.1038/s41586-024-07512-y

Fairbrother, Malcolm, Ingemar J. Sevä, and Joakim Kulin (2019), "Political trust and the relationship between climate change beliefs and support for fossil fuel taxes: Evidence from a survey of 23 European countries," *Global Environmental Change* 59: 102003.

Falchetta, Giacomo, Enrica De Cian, Ian S. Wing, and Deborah Carr (2024), "Global projections of heat exposure of older adults," *Nature Communications* 15(1): 3678. https://doi.org/10.1038/s41467-024-47197-5

Felli, Romain ([2016] 2021), *The Great Adaptation. Climate, Capitalism, Catastrophe*. London, Verso.

Fiorini, Ann, Gordon LaForge, and Anne-Marie Slaughter (2023), "Democratic governance for the long emergency," pp. 151–165 in David W. Orr (ed.), *Democracy in a Hotter Time. Climate Change and Democratic Transformation*. Cambridge: MIT Press.

Fiorino, Daniel J. (2022), "Climate change and right-wing populism in the United States," *Environmental Politics* 31(5): 801–819.

Fischer, Frank (2017), *Climate Crisis and the Democratic Prospect. Participatory Governance in Sustainable Communities*. Oxford: Oxford University Press.

Fischer, Frank (2018), "Environmental democracy: Participation, deliberation and citizenship," pp. 257–279 in Magnus Boström and Debra J. Davidson (eds.), *Environment and Society: Concepts and Challenges*. Heidelberg: Springer.

Fischer, Isabel, Claire Beswick, and Sue Newell (2021), "Rho AI – Leveraging artificial intelligence to address climate change: Financing, implementation and ethics." *Journal of Information Technology Teaching Cases* 11: 110–116.

Fisher, Dana R. (2024), *Saving Ourselves: From Climate Shocks to Climate Action*. New York: Columbia University Press.

Fisher, Dana R., Erika Svendsen, and James Connolly (2015), *Urban Environmental Stewardship and Civic Engagement: How Planting Trees Strengthens the Roots of Democracy*. New York: Routledge.

Forster, Piers M., Chris Smith, Tristram Walsh, William F. Lamb, Robin Lamboll, Christophe Cassou, Mathias Hauser, et al. (2025), "Indicators of Global Climate Change 2024: Annual update of key indicators of the state of the climate system and human influence," *Earth System Science Data* 17: 2641–2680.

Foucault, Michel ([1981–1985] 2005), "Die Maschen der Macht," pp. 224–244 in Michel Foucault (ed.), *Schriften*. Volume 4. Frankfurt am Main: Suhrkamp.

Foucault, Michel (1984), "What is enlightenment?" pp. 32–50 in Paul Rabinow (ed.), *Foucault Reader*. New York: Pantheon Books.

Frankenberg, Günter and Wilhelm Heitmeyer (2024), "Authoritarian developments as a threat to democracy and pluralism," pp. 2–59 in Günter Frankenberg and Wilhelm Heitmeyer (eds.), *Drivers of Authoritarianism. Paths and Developments at the Beginning of the 21st Century*. Cheltenham: Edward Elgar.

Frankopan, Peter (2023), *The Earth Transformed*. New York: Knopf.

Frederiksson, Per G. and Eric Neumayer (2013), "Democracy and climate change: Is history important?," *Ecological Economics* 95: 11–19.

Freier, Anna N. and Josua Schneider (2022), "'Die Arktis brennt' – Zur verlorenen Zukunft in Katastrophennarrativen von Klimaprotestbewegungen," pp. 95–122 in Julia Zilles, Emily Drewing, and Julia Janik (eds.), *Umkämpfte Zukunft. Zum Verhältnis von Nachhaltigkeit, Demokratie und Konflikt*. Bielefeld: Transkript Verlag.

Fremstad, Anders and Mark Paul (2022), "Neoliberalism and climate change: How the free-market myth has prevented climate action," *Ecological Economics* 197(2022): 107353.

Fritsch, Matthias (2023), "Climate change and democracy," pp. 1001–1026 in Gianfranco Pellegrino and Marcello DiPaola (eds.), *Handbook of the Philosophy of Climate Change*. Cham: Springer.

Fritz, Livia, Lucille Losi, Chad M. Baum, Sean Low, and Benjamin Sovacool (2015), "Between inflated expectations and inherent distrust: How publics see the role of experts in goiverning climate intervention technologies," *Environmental Science and Policy* 164: 104005. https://doi.org/10.1016/j.envsci.2025.104005

Froehlich, Halley E., Jamie C. Afflerbach, Melanie Frazier und Benjamin S. Halpern (2019), "Blue growth potential to mitigate climate change through seaweed offsetting," *Current Biology* 29: 3087–3093.

Fukuyama, Francis (1992), *The End of History and the last Man*. New York: The Free Press.

Fukuyama, Francis (2018), *Identity. Contemporary Identity Politics and the Struggle for Recognition*. London: Profile Books.

Fukuyama, Francis (2014), *Political Order and Political Decay: From the Industrial Revolution to the Globalization of Democracy*. New York: Farrar, Straus and Giroux.

Funtowicz, Silvio O. and Jerome R. Ravetz (1993), "Science for the post-normal age," *Futures* 25: 739–755.

Galbraith, John K. (1958), *The Affluent Society*. Boston, MA: Houghton Mifflin.

Gardiner, Stephen M. (2002), "The real tragedy of the commons," *Philosophy and Public Affairs* 30: 387–416.

Garston, Bryan (2024), "The liberalism of refuge," *Journal of Democracy* 35: 136–151.
Gaskill, Nicholas (2022), "Rorty against Rorty: Climate change, rug-pulling, and the rhetoric of philosophy," *Common Knowledge* 28: 380–401.
Gattuso, Jean-Pierre, Alexandre K. Magnan, Laurent Bopp, William W. L. Cheung, Carlos M. Duarte, Jochen Hinkel, Elizabeth Mcleod, et al. (2018), "Ocean solutions to address climate change and its effects on marine ecosystems," *Frontiers in Marine Science* 5: 337.
Gauchat, Gordon and Kenneth T. Andrews (2018), "The cultural-cognitive mapping of scientific professions," *American Sociological Review* 83: 567–595.
Gehlen, Arnold ([1940] 1988), *Man. His Nature and Place in the World*. New York: Columbia University Press.
Gergen, Kenneth J. (1986), "Correspondence versus autonomy in the language of understanding human action,"' pp. 136–62 in Donald W. Fiske and Richard A. Shweder (eds.), *Metatheory in Social Science. Pluralisms and Subjectivities*. Chicago, IL: University of Chicago Press.
Geroulanos, Stefanos (2024), *The Invention of Prehistory. Empire, Violence, and Our Obsession with Human Origins*. New York: Norton.
Gerrard, Michael B. and Tracy Hester (eds.) (2018), *Climate Engineering and the Law: Regulation and Liability for Solar Radiation Management and Carbon Dioxide Removal*. Cambridge: Cambridge University Press.
Ghosch, Amitav (2021), *The Nutmeg's Curse. Parables for a Planet in Crisis*. New York. Allen Lane.
Giddens, Anthony (1990), *The Consequences of Modernity*. Stanford, CA: Stanford University Press.
Giddens, Anthony (2009), *The Politics of Climate Change*. Cambridge: Polity Press.
Gigerenzer, Gerd and Rocio Garcia-Rettamero (2017), "Cassandra's regret: The psychology of not wanting to know," *Psychological Review* 124: 179–196.
Gilley, Bruce (2012), "Authoritarian environmentalism and China's response to climate change," *Environmental Politics* 21: 287–307.
Giuliano, Paola and Antonio Spilimbergo (2024), "Aggregate shocks and the formation of preferences and beliefs," *NBER Working Paper* No. 32669.
Gore, A. (2009). *Our Choice: A Plan to Solve the Climate Crisis*. New York: Melcher Media.
Gorin, Andrew L., Jeremy D. Shakun, Andrew G. Jones, Tori M. Kennedy, Shaun A. Marcott, Brent M. Goehring, Lucas K. Zoet, et al. (2024), "Recent tropical Andean glacier retreat is unprecedented in the Holocene," *Science* 385(6708): 517–521.
Gorz, André (2002), "Welches Wissen? Welche Gesellschaft?," pp. 14–35 in Heinrich-Böll-Stiftung (ed.), *Gut zu wissen. Links zur Wissensgesellschaft*. Münster Westphaelisches Dampfboot.
Gorz, André ([2008] 2009), *Auswege aus dem Kapitalismus*. Zürich: Rotpunktverlag.
Gouldner, Alvin W. (1980), "Is amnesia in sociology discontinuous, and the problem of permeable boundaries in culture," Paper presented at an International Conference on "The Political Realization of Social Science Knowledge, Institute for Advanced Studies, Vienna, June 18–20.
Granovetter, Mark (1985), "Economic action and social structure: The problem of embeddedness," *American Journal of Sociology* 91: 481–510.
Greene, Chad A., Alex S. Gardner, Michael Wood, and Joshua K. Cuzzone (2024), "Ubiquitous acceleration in Greenland Ice Sheet calving from 1985 to 2022," *Nature* 625: 523–528.

Greider, Thomas and Lorraine Garkovich (1994), "Landscapes: The social construction of nature and the environment," *Rural Sociology* 59: 1–24.

Greven, Michael T. ([2009] 2020), "Zukunft oder Erosion der Demokratie?," S. 155–178 in Friedbert W. Rüb, Veith Selk, and Rieke Trimçev (eds.), *Die Erosion der Demokratie.* Wiesbaden: Springer.

Grundmann, Reiner (1991), *Marxism and Ecology*. Oxford: Oxford University Press.

Grundmann, Reiner and Nico Stehr (1997), "Klima und Gesellschaft, soziologische Klassiker und Außenseiter. Über Weber, Durkheim, Simmel und Sombart," *Soziale Welt* 48: 85–100.

Grundmann, Reiner and Nico Stehr (2000), "Social science and the absence of nature: Uncertainty and the reality of extremes," *Social Science Information* 39: 155–179.

Grundmann, Reiner und Nico Stehr (2010), "Climate Change: What Role for Sociology? A Response to Constance Lever-Tracy," *Current Sociology* 58: 897–910.

Grundmann, Reiner and Nico Stehr (2012), *The Power of Scientific Knowledge. From Research to Public Policy*. Cambridge: Cambridge University Press.

Grzegorz, Ekiert (2023), *Democracy and Authoritarianism in the 21st Century: A Sketch*. Ash Institute for Democratic Governance and Innovation, 2023. https://ash.harvard.edu/wp-content/uploads/2023/12/democracy_and_authoritarianism_in_the_21st_century-_a_sketch.pdf

Haas, Peter (2004), "When does power listen to truth? A constructivist approach to the public policy process," European Public Policy 11: 569–592.

Habermas, Jürgen ([1968] 1971), *Knowledge and Human Interests*. Boston, MA: Beacon Press.

Habermas, Jürgen ([1969] 1971) *Toward a Rational Society*. London: Heinemann.

Hacker, Jacob S., Alexander Hertel-Fernandez, Paul Pierson, and Kathleen Thelen (2022), "The American political economy: A framework and agenda for research," pp. 1–48 in Jacob S. Hacker, Alexander Hertel-Fernandez, Paul Pierson, and Kathleen Thelen (eds.), *American Political Economy*. Oxford: Oxford University Press.

Haerpfer, Christian W., Patrick Bernhagen, Christian Welzel, and Ronald F. Inglehart (eds.) (2019), *Democratization*. Second Edition. Oxford: Oxford University Press.

Hajek, André und Hans-Helmut König (2020), "Climate anxiety in Germany," *Public Health* 212: 89–94.

Hajek, André und Hans-Helmut König (2023), "Climate anxiety and mental health in Germany," *Climate* 11(8): 158.

Haldane, Andrew ([2009] 2013), "Rethinking the financial network," pp. 243–278 in Stephan Jansen, Eckhard Schröter, and Nico Stehr (eds.), *Fragile Stabilität – stabile Fragilität*. Wiesbaden: Springer VS.

Hannah, Michael (2021), "The anthropocene and the earth system," pp. 11–28 in Michael Hannah (ed.), *Extinctions: Living and Dying in the Margin of Error*. Cambridge: Cambridge University Press.

Hannigan, John (2025), *Rethinking Environmental Sociology Moral Claims and Moral Meanings*. Cheltenham: Edward Elgar Publishing.

Harari, Yuval N. (2024), *Nexus: A Brief History of Information Networks from the Stone Age to AI*. London: Signal.

Hardin, Garrett (1968), "The tragedy of the commons. The population problem has no technical solution; it requires a fundamental extension in morality," *Science* 162: 1243–1248.

Hardin, Russell (1982), *Collective Action.* Baltimore, MD: John Hopkins University Press.
Hayek, Friedrich A. (1945), "The use of knowledge in society," *The American Economic Review* 35: 519–530.
Hayek, Friedrich A. v. (1944), *The Road to Serfdom.* London: George Routledge & Sons.
Hayek, Friedrich A. v. ([1949] 1960), *The Constitution of Liberty.* London: Routledge.
Heikonen, Sara, Matias Heino, Mika Jalava, Stefan Siebert, Daniel Viviroli und Matti Kummu (2025), "Climate change threatens crop diversity at low latitudes," *Nature Food* 6: 331–342 https://doi.org/10.1038/s43016-025-01135-w
Heilbroner, Robert L. (1974), *An Inquiry into the Human Prospect.* New York: Norton.
Heilbroner, Robert L. ([1974] 2013), *An Inquiry into the Human Prospect.* New York: Norton.
Heinberg, Richard (2024), "From climate crisis to polycrisis," *World Literature Today* 98: 32–36.
Heitmeyer, Wilhelm (2018), *Autoritäre Versuchungen. Signaturen der Bedrohung I.* Berlin: Suhrkamp.
Held, David (2022), "The overlapping crises of democracy, globalization, and global governance," pp. 307–309 in J. Tully, K. Cherry, F. Forman, et al. (eds.), *Democratic Multiplicity: Perceiving, Enacting, and Integrating Democratic Diversity.* Cambridge: Cambridge University Press.
Held, David and Angus F. Hervey (2009), "Democracy, climate change and global governance," Policy Network Paper. London: Policy Network.
Held, David and Angus F. Hervey (2011), "Democracy, climate change and global governance. Democratic agency and the policy menu ahead," pp. 89–110 in David Held, Angus Fane-Hervey, and Marika Theros (eds.), *The Governance of Climate Change. Science, Economics, Politics and Ethics.* Cambridge: Polity Press.
Hennis, Wilhelm, Peter G. Kielmannsegg, and Ulrich Matz (eds.) (1977), *Regierbarkeit. Studien zu ihrer Problematisierung.* Stuttgart: Klett-Cotta.
Herder, Johann G. (1794), *Ideen zur Geschichte der Menschheit.* Carlsruhe: Christian Gottlieb Schmieder.
Heritage Foundation (2023), *Mandate for Leadership. The Conservative Promise.* Washington, DC: The Heritage Foundation.
Herzog, Lisa (2019), *Die Rettung der Arbeit.* München: Hanser Berlin.
Herzog, Lisa (2024), *Citizen Knowledge. Markets, Experts, and the Infrastructure of Democracy.* Oxford: Oxford University Press.
Hitchcock, Dian R. and James E. Lovelock (1967), "Life detection by atmospheric analysis," *Icarus* 7: 149–150.
Hobsbawn, Eric (2007), *Globalisation, Democracy and Terrorism.* London: Abacus.
Hobson, Christopher (2012), "Addressing climate change and promoting democracy abroad: Compatible agendas? *Democratization* 19: 974–992.
Homer-Dixon, Thomas (2002), *The Ingenuity Gap.* Toronto: Vintage.
Horn, Eva (2024), *Klima. Eine Wahrnehmungsgeschichte.* Frankfurt am Main: S. Fischer.
Horowitz, Irving L. (2010), "Introduction to the transaction edition. Man and society in calamity: Demythologizing Pitirim Sorokin," pp. ix–xviii in Pitirim Sorokin (ed.), *Man and Society in Calamity.* New Brunswick, NJ: Transaction Publishers.

Huber, Bettina, Menno Boldt, Wendell Bell, and James A. Mau (1971), *Sociology of the Future: Theory, Cases and Annotated Bibliography*. New York: Russel Sage Foundation.

Hulme, Mike ([2009] 2014), "Why we disagree about climate change (2009)," pp. 230–243 in Steve Rayner and Mark Caine (eds.), *The Hartwell Approach to Climate Policy*. London: Routledge.

Hulme, Mike (2009), *Why we Disagree about Climate Change*. Understanding Controversy, Inaction and Opportunity. Cambridge: Cambridge University Press.

Hulme, Mike (2011), "Reducing the future to climate: A story of climate determinism and reductionism," *Osiris* 26: 245–266.

Hultgren, John (2025), *The Smoke and the Spoils. Anti-Environmentalism* and Class Struggle in the United States. Cambridge, Massachusetts: MIT Press.

Huntington, Ellsworth (1907), *The Pulse of Asia*. Boston, MA: Houghton & Mifflin.

Huntington, Ellsworth ([1915] 1924), *Civilization and Climate*. Third Edition, Revised and Rewritten with Many New Chapters. New Haven, CT: Yale University Press.

Huntington, Ellsworth (1945), *Mainsprings of Civilization*. New York: John Wiley and Sons.

Innerarity, Daniel (2021), "No democracy without comprehension: Political unintelligibility as a democratic problem," *Polity* 53: 264–287.

Innis, Harold A. (1951), *The Bias of Communication*. Toronto: University of Toronto Press.

International Monetary Fund (2023), *IMF Fossil Fuel Subsidies Data: 2023 Update*. Washington, DC: International Monetary Fund.

Iverson, Torben and David Soskice (2019), *Democracy and Prosperity: Reinventing Capitalism through a Turbulent Century*. Princeton, New Jersey: Princeton University Press.

Jamieson, Dale (2014), *Reason in a Dark Time. Why the Struggle against Climate Change Failed – And What It Means for Our Future*. New York: Oxford University Press.

Janda, Constanze (2021), "Sozialstaat for future," *Zeitschrift für Rechtspolitik* 5: 149–153.

Jasanoff, Sheila (1990). *The Fifth Branch: Science Advisors as Policymakers*. Cambridge, MA: Harvard University Press.

Jasanoff, Sheila (2012), *Science and Public Reason*. London and New York: Routledge.

Jasanoff, Sheila and Marybeth L. Martello (2004), "Knowledge and governance," pp. 335–350 in Sheila Jasanoff and Marybeth L. Martello (eds.), *Earthly Politics. Local and Global in Environmental Governance*. Cambridge: MIT Press.

Jonas, Hans (Ed.) ([1979] 1984), "The imperative of responsibility," in *Search of an Ethics for the Technological Age*. Chicago, IL and London: University of Chicago Press.

Jones, Michael D. and Mark K. McBeth (2010), "A narrative policy framework: Clear enough to be wrong?," *Policy Studies Journal* 38: 329–353.

Kaack, Lynn H., Priya L. Donti, Emma Strubell, and David Rolnick (2020), *Artificial Intelligence and Climate Change*. Heinrich-Böll-Stiftung Ecology. Berlin: Heinrich-Böll-Stiftung.

Kaase, Max (2007), "Perspectives on political participation," pp. 783–796 in Russell Dalton and Hans D. Klingemann (eds.), *Oxford Handbook of Political Behaviour*. Oxford: Oxford University Press.

Kahan, Dan M., Hank Jenkins-Smith, and Donald Braman (2011), "Cultural cognition of scientific consensus," *Journal of Risk Research* 14: 147–174.
Kahn, Matthew E. (2016), "The climate change adaptation literature," *Review of Environmental Economics and Policy* 10: 166–178.
Kahneman, Daniel (2011), *Thinking, Fast and Slow*. New York: Farrar, Straus, Giroux.
Kames, Henry ([1788] 2007), *Sketches of the History of Man*. Indianapolis, IN: Liberty Fund.
Kapczynski, Amy, and Joel Michaels (2024), "Administering a Democratic Industrial Policy," *Harvard Law & Policy Review* 18: 279–344.
Keane, John (2020), *The New Despotism*. Cambridge, MA: Harvard University Press.
Keane, John (2022), "Thoughts on uncertainty," *Journal of Social and Political Philosophy* 1(1): 1–13.
Keith, David W. and Hadi Dowlatabadi (1992), "A serious look at geoengineering." *Eos, Transactions American Geophysical Union* 73(27): 289–293.
Keller, Evelyn F. (2017), "Climate science, truth and democracy," *Studies in History and Philosophy of Biological and Biomedical Sciences* 64: 106–122.
Kellogg, Ryan (2024), "The end of oil," *NBER Working Paper* No. 33207, November 2024.
Kemp, Luke (2025), *Goliath's Curse*. The History and the Future of Societal Collapse. London: Penguin.
Keuschnigg, Marc and Fabian Kratz (2018), "Thou shalt recycle: How social norms of environmental protection narrow the scope of the low-cost hypothesis," *Environment and Behavior* 50: 1059–1091.
Keynes, John M. ([1919] 1920), *The Economic Consequences of the Peace*. New York: Harcourt, Brace, and Howe.
Keynes, John M. ([1942] 2013), *The Collected Writings of John Maynard Keynes*. Volume 27. Activities 1940–1946. Edited by Donald Moggridge. Cambridge: Cambridge University Press.
Kheir-Mataria, Wafa A. E. and Sungsoo Chun (2025), "Climate change and women's cancer in the MENA region: Assessing temperature-related health impacts," *Frontiers in Public Health* 13: 2025. https://doi.org/10.3389/fpubh.2025.1529706
Kitcher, Philip (2011), *Science in a Democratic Society*. Amherst, MA: Prometheus Books.
Klein, Naomi (2014), *This Changes Everything: Capitalism vs. the Climate*. New York: Simon & Schuster.
Klinenberg, Eric (2002), *Heat Wave. A Social Autopsy of Disaster in Chicago*. Chicago: The University of Chicago Press.
Krämer-Hoppe, Rike (2021), "The climate protection order of the federal constitutional court of Germany and the north-south divide," *German Law* 22: 1393–1408.
Kreuder-Sonnen, Christian (2019), *Emergency Powers of International Organizations: Between Normalization and Containment*. Oxford: Oxford University Press.
Kuh, Katrina F. (2021), "Litigating government (in)action on climate change," pp. 109–134 in Karl S. Coplan Shelby D. Green, Katrina Fischer Kuh, Smita Narula, Karl R. Rábago,and Radina Valova (eds.), *Climate Change Law. An Introduction*. Cheltenham: Edward Elgar.
Kuklinski, James D. (1990), "Information and the study of politics," pp. 391–395 in John A. Forejohn and James H. Kuklinski (eds.), *Information and the Democratic Processes*. Urbana and Chicago, IL: University of Illinois Press.

Kurzweil, Ray (2024), *The Singularity Is Nearer. When We Merge with AI.* New York: Viking.

Kuttner, Robert (2018), *Can Democracy Survive Global Capitalism?* New York: WW Norton & Company.

Lahson, Myanna (2024), "Steering signification for sustainability," *Global Sustainability* 7: 1–10 Lahsen.

Lahsen, Myanna and Jesse Ribot (2022), "Politics of attributing extreme events and disasters to climate change," *WIREs Clim Change* 13: e750.

Lakoff, George (2004), *Don't Think of an Elephant! Know Your Values and Frame the Debate*, p. 148. Green Publishing: Chelsea.

Lally, Roisin, Cristina P. Bonfiglioli, Jan K. B. Friis, Don Ihde, Babettet Babich, Patricia Glazebrook, Dana S. Belu, Brenden Mahoney, Galit Wellner, Danielk Bradley, Thomas M. Jeannot, and Lars Botin (2019), *Sustainability in the Anthropocene: Philosophical Essays on Renewable Technologies.* New York: Rowman & Littlefield.

Lamb, Hubert H. (1959), "Our changing climate, past and present," *Weather* 14: 299–318.

Landemore, Héléne (2023), "No decarbonization without democratization: To save the climate, open democracy," pp. 37–50 in David W. Orr (ed.), *Democracy in a Hotter Time.* Cambridge: MIT Press.

Landes, David (1998), *The Wealth and Poverty of Nations. Why Some Are So Rich and Some Are So Poor.* New York: W.W. Norton.

Latour, Bruno ([1999] 2004), *Politics of Nature: How to Bring the Sciences into Democracy.* Cambridge, MA: Harvard University Press.

Latour, Bruno (2015), "Telling friends from foes in the time of the anthropocene," pp. 145–155 in Hamilton, Clive, Christophe Bonneuil, and François Gemenne (eds.), *The Anthropocene and the Global Environmental Crisis.* New York: Routledge.

Latour, Bruno ([2015] 2017), *Facing Gaia. Eight Lectures on the New Climatic Regime.* Cambridge: Polity Press.

Latour, Bruno ([2017] 2018), *Down to Earth. Politics in the New Climate Regime.* Cambridge: Polity Press.

Lazar, Nomi C. (2009), *States of Emergency in Liberal Democracies.* Cambridge: Cambridge University Press.

Le Roy Ladurie, Emmanuel ([1967] 1971), *Times of Feast, Times of Famine. A History of Climate since the Year 1000.* New York: Doubleday & Company.

Leggewie, Claus (2010), "Gibt es ein Leben nach der Demokratie? Herausforderungen des Westens durch die Naturgefahr," pp. 149–158 in Christopher Bieber, Benjamin Drechsel, and Ann-Katrin Lang (eds.), *Gefahr im Konflikt.* Claus Leggewie revisited. Bielefeld: transcript.

Lenton, Timothy M. (2011), "Early warning of climate tipping points," *Nature Climate Change* 1: 201–209.

Lessig, Lawrence (2006), *Code 2.0.* New York: Basis Books.

Levitsky, Steven and Daniel Ziblatt (2018), *How Democracies Die. What History Reveals about the Future.* London: Viking.

Levitsky, Steven and Lucan Way (2015), "The myth of democratic recession," *Journal of Democracy* 26: 45–58.

Lewandowsky, Stephan, Ullrich K. H. Ecker, John Cook, Sander van der Linden, Jon Roozenbeek, and Naomi Oreske (2023), "Misinformation and the epistemic integrity of democracy," *Current Opinion in Psychology* 54: 101711. www.cogsciwa.com

Li, Pengfei, Jianyi Yang, Mohammad A. Islam, and Shaolei Ren (2023), "Making AI less "thirsty": Uncovering and addressing the secret water footprint of AI models," Cornell University arXiv:2304.03271v3.

Lindblom, Charles (1995), "Markets and democracy – Obliquely," *PS: Political Science & Politics* 28: 684–688.

Lippmann, Walter ([1922] 1997), *Public Opinion*. New York: Routledge.

Lipset, Martin M. ([1960] 1980), *Political Man. The Social Basis of Politics*. New York: Doubleday.

Lipset, Martin M., Martin A. Trow, and James S. Coleman ([1956] 1962), *Union Democracy: The Internal Politics of the International Typographical Union*. New York: Doubleday & Company.

Lipset, Seymour M. (1959). "Some social requisites of democracy: Economic development and political legitimacy," *American Political Science Review* 53: 69–105.

Liu, Huan and Otwin Renn (2025), "Polycrisis and systemic risk: Assessment, governance, and communication," *Internationl Journal Disaster Risk Science* 28: 1–24. https://doi.org/10.1007/s13753-025-00636-3

Livingstone, David N. (2020), "Climate and civilization," pp. 58–74 in Michael Boyden (ed.), *Climate in American Literature and Culture*. Cambridge: Cambridge University Press.

Livingstone, David N. (2024), *The Empire of Climate. A History of an Idea*. Princeton, NJ: Princeton University Press.

Lockwood, Matthew (2018), "Right-wing populism and the climate change agenda: Exploring the linkages." *Environmental politics* 27: 712–732.

Lorenz-Spreen, Philipp, Lisa Oswald, Stephan Lewandowsky, and Ralph Hertwig (2023), "A systematic review of worldwide causal and correlational evidence on digital media and democracy," *Nature Human Behaviour* 7: 74–101.

Lovelock, James (2009), *The Vanishing Face of Gaia*. New York: Basic Books.

Lowe, Adolph (1971), "Is present-day higher education learning relevant?," *Social Research* 38: 563–580.

Luce, Edward (2017), *The Retreat of Western Liberalism*. London: Abacus.

Luckmann, Thomas (1970), "On the boundaries of the social world," pp. 73–100 in Maurice Natanson (ed.), *Phenomenology and Social Reality*. The Hague: Martinus Nijhoff.

Luckmann, Thomas (1974), "Natur – eine geschichtliche Gegebenheit?" pp. 193–213 in Landeshauptstadt Stuttgart (ed.), *Zum Naturbegriff der Gegenwart*. Kongressdokumentation zum Projekt "Kultur im Kopf". Stuttgart, 21–26. Juni 1993. Stuttgart: problemata frommann-holzboog.

Luhmann, Niklas (1976), "The future cannot begin: Temporal structure in modern society," *Social Research* 43: 130–152.

Luhmann, Niklas ([1973] 1979), *Trust and Power. With Introduction by Gianfranco Poggi*. New York: Wiley.

Luhmann, Niklas ([1986] 1989), *Ecological Communication*. Chicago, IL: University of Chicago Press.

Luhmann, Niklas (1980), "Gesellschaftliche Struktur und semantische Tradition," pp. 9–70 in Nijklas Luhmann (ed.), *Gesellschaftsstruktur und Semantik. Studien zur Wissenssoziologie der modernen Gesellschaft*. Volume 1. Frankfurt am Main: Suhrkamp.

Luhmann, Niklas (1988), *Die Wirtschaft der Gesellschaft*. Frankfurt am Main: Suhrkamp.
Luhmann, Niklas (1991), *Soziologie des* Risikos. Opladen: Westdeutscher Verlag.
Luhmann, Niklas ([1992] 1998), *Observations on Modernity*. Stanford, CA: Stanford University Press.
Luhmann, Niklas ([1997] 2012), *Theory of Society*. Volume 1. Stanford, CA: Stanford University Press.
Lukács, Georg ([1923] 1970), *Geschichte und Klassenbewußtsein. Studien über marxistische Dialektik*. Neuwied: Luchterhand.
Luo, Ming, Sijia Wu, Tao Pwei, ZhenLiu, Xiaoyyu Wang, Guicai Ning, Ting On Chan, Yuannjian Yang, and Wei Zhang (2024), "Anthropogenic forcing has increased the risk of longer-traveling and slower-moving large contiguous heatwaves," *Science Advances* 10(13). https://doi.org/10.1126/sciadv.adl1598
MacDuffie, Allen (2024), *Climate of Denial. Darwin, Climate Change, and the Literature of the Long Nineteenth Century*. Stanford, CA: Stanford University Press.
Machen, Ruth and Eric Nost (2021), "Thinking algorithmically: The making of hegemonic knowledge in climate governance," *Transactions of the Institute of British Geographers* 46: 555–569.
Maier, Charles S. (2023), *The Project-State and Its Rivals. A New History of the Twentieth and Twenty-First Centuries*. Cambridge, MA: Harvard University Press.
Malm, Andreas (2020), *Corona, Climate, Chronic Emergency: War Communism in the Twenty-First Century*. London: Verso.
Mankiw, N. Gregory (2022), *Macroeconomics*. Eleventh Edition. London: Macmillan.
Mann, Michael (2013), *The Sources of Social Power*. Volume 4: Globalizations, 1945–2011. Cambridge: Cambridge University Press.
Marcuse, Herbert ([1989/1990] 2009), "Ökologie und Gesellschaftskritik," pp. 165–174 in Peter-Erwin Jansen (ed.), *Herbert Marcuse Nachgelassene Schriften*, Band 6: Ökologie und Gesellschaftskritik. Springe: zu Klampen Verlag.
Marquand, David (2004), *The Decline of the Public. The Hollowing Out of Citizenship*. Cambridge: Polity Press.
Marquart-Pyatt, Sandra T., Aaron M. McCright, Thomas Dietz, and Riley E. Dunlap 2014), "Politics eclipses climate extremes for climate change perceptions," *Global Environmental Change* 29: 246–257.
Marsh, George P. ([1864] 1965), *Man and Nature*. Cambridge, MA: Harvard University Press.
Marshall, George (2014), *Don't Even Think about It. Why Our Brains Are Wired to Ignore Climate Change*. New York: Bloomsbury.
Martin, Geoffrey J. (1973), *Ellsworth Huntington. His Life and Thought*. Hamden, CT: The Shoestring Press.
Maschewski, Felix and Anna-Verena Nosthoff (2022), "Pandemic solutionism: The power of Big Tech during the COVID-19 crisis," *Digital Culture & Society* 8: 43–66.
McCright, Aaron M. and Riley Dunlap (2011), "The politicization of climate change and polarization in the American public's views of global warming, 2001–2010," *The Sociological Quarterly* 52: 155–194.
McDonald, Susan (2009), "Changing climate, changing minds," *International Journal of Sustainable Communities* 4: 45–63.

McKibben, Bill (2016), "We're under attack by a powerful enemy – And our only hope it to mobilize like we did in WWII," *New Republic* (September): 22–32.

Merton, Robert K. ([1942] 1957), "Science and the democratic social structure," pp. 550–561 in Robert K. Merton (ed.), *Social Theory and Social Structure*. Revised and Enlarged Edition. New York: Free Press.

Merton, Robert K. (1966), "Dilemmas of democracy in the voluntary associations," *American Journal of Nursing* 66: 1055–1061.

Merton, Robert K. ([1970] 1976), "The ambivalence of organization leaders," pp. 73–89 in Robert K. Merton (ed.), *Sociological Ambivalence and Other Essays*. New York: Free Press.

Michels, Robert ([1911] 1949), *Political Parties: A Sociological Study of the Oligarchical Tendencies of Modern Democracy*. New York: Free Press.

Miller, Nolan H., David Molitor, and Eric Zou (2024), "The nonlinear effects of air pollution on health: Evidence from wildfire smoke," *NBER Working Paper* No. 32924.

Mills, M. Anthony and Price S. Claire (2025), "The strange new politics of science," *Issues in Science and Technology* (Spring): 40–48.

Mishra, Manoranjan, Tamoghna Acharyya, Rajkumar Guria, Nihar R. Rout, Celso A. G. Santos, Richarde M. da Silva, Sanjay Srivastava, Ranjana Kumari, and Alok K. Pradhan, (2024) "Lightning-related fatalities in India (1967–2020): (2024), a detailed overview of patterns and trends," *Environment, Development and Sustainability*: 1–30. https://doi.org/10.1007/s10668-024-05216-x

Mitchell, Timothy (2010), *Carbon Democracy. Political Power in the Age of Oil*. London: Verso.

Mittiga, Ross (2022), "Political legitimacy, authoritarianism, and climate change," *American Political Science Review* 116: 998–1011.

Monnet, Éric ([2021] 2024), *Balance of Power. Central Banks and the Fate of Democracies*. Chicago, IL: University of Chicago Press.

Moore, Frances C. (2024), "Learning, catastrophic risk and ambiguity in the climate change era," *NBER Working Paper* No. 32684.

Morgan, Granger and Hadi Dowlatabadi (2023), "We have no choice but to adapt," pp. 208–218 in Granger Morgan and Hadi Dowlatabadi (eds.), *Interdisciplinary Research on Climate and Energy Decision Making*. New York: Routledge.

Morin, Edgar and Kern A. Brigitte (1993), *Terre-patrie*. Paris: Editions du Seuil.

Morozov, Evgeny (2022), "Critique of techno-feudal reason," *New Left Review* 133/134: 89–126.

Morozov, Evgeny (2024), "AI and the techno-utopian path not taken," *LeMonde Diplomatique*, August 2024. https://mondediplo.com/2024/08/07ai-cold-war

Mounk, Yascha (2018), *The People vs. Democracy. Why Our Freedom Is in Danger and How to Save It*. Cambridge, MA: Harvard University Press.

Musil, Robert ([1930] 1995), *The Man without Qualities*. Volume 1. New York: Alfred A. Knopf.

Natanson, Maurice (1962), "Introduction," in Alfred Schutz (ed.), *Collected Papers I: The Problem of Social Reality* xi-xlvii. The Hague: Martinus Nijhoff.

Nath, Ishan (2025), "Climate change, the food problem, and the challenge of adaptation through sectoral reallocation," *Journal of Political Economy* 133: 1705–1756.

Neal, Timothy N., Ben R. Newell, and Andy Pitman (2025), "Reconsidering the macroeconomic damage of severe warming," *Environmental Research Letters* 20: 044029.
Nelkin, Dorothy and Michael Pollack (1977), "The politics of participation and the nuclear debate in Sweden, the Netherlands, and Austria," *Public Policy* 25: 333–357.
Neumann, Franz (1957), *The Democratic and the Authoritarian State.* Edited and with a Preface by Herbert Marcuse. New York: Free Press.
Newell, Peter and Mathew Paterson (2010), *Climate Capitalism. Global Warming and the Transformation of the Global Economy.* Cambridge: Cambridge University Press.
Nordhaus, William (2015), *The Climate Casino. Risk, Uncertainty, and Economics for a Warming World.* New Haven, CT: Yale University Press.
Nordhaus, William (2020), 'The climate club. How to fix a failing global effort," *Foreign Affairs* 99: 10–17.
Norgaard, Kari (2011), "The psychology and sociology of denial," pp. 399–413 in John Dryzek, Richard Norgaard, and David Schlosberg (eds.), *Oxford Handbook of Climate Change and Society.* New York: Oxford University Press.
Norris, Pippa (2011), *Democratic Deficit. Critical Citizens Revisited.* Cambridge: Cambridge University Press.
Norris, Pippa (2019), "Varieties of populist parties," *Philosophy and Social Criticism* 45: 981–1012.
Norris, Pippa and Ronald Inglehart (2019), *Cultural Backlash. Trump, Brexit, and Authoritarian Populism.* Cambridge: Cambridge University Press.
Nyber, Daniel and Christopher Wright (2025), "Confronting the climate crisis: Fossil fuel hegemony and the need for decarbonization, degrowth, and democracy," *Journal of Management Studies.* 10.1111/joms.13198
Oatley, Thomas (2024), "The dual economy, climate change, and the polarization od American politics," *Socio-Economic Review* 22: 1355–1380.
OECD (2024), *OECD Survey on Drivers of Trust in Public Institutions – 2024.* Paris: OECD. www.oecd.org/en/publications/oecd-survey-on-drivers-of-trust-in-public-institutions-2024-results9a20554b-en/full-report.html
Offe, Claus (1987), "The utopia of the zero-option. Modernity and modernization as normative political criteria," *Praxis International* 7: 1–24.
Ogburn, William F. (1943), "Review of Clarence A. Mills, *Climate Makes Man.*" *American Journal of Sociology* 48: 784–787.
Olmstead, Albert T. (1912), "Climate and history." *Journal of Geography* 10: 163–168.
Olsson, David (2022), "From technocracy to democracy: Ways to promote democratic engagement for just climate change adaptation and resilience building," *Sustainability* 14: 1433.
Ophuls, William (1977), *Ecology and the Politics of Scarcity.* San Francisco, CA: W H Freeman.
Oreskes Naomi and Erik M. Conway (2014), *The Collapse of Western Civilization. A View from the Future.* New York: Columbia University Press.
Oschlies, Andreas, Markus Pahlow, Andrew Yool, and Richard J. Matear (2010), "Climate engineering by artificial ocean upwelling: Channeling the sorcerer's apprentice." *Geophysical Research Letters* 37. https://doi.org/10.1029/2009GL041961

Ostrom, Elinor (2009), "A polycentric approach for coping with climate change." *Policy Research Working Paper 5095*. Washington, DC: World Bank.
Ostrom, Elinor (2010), "The challenge of self-governance in complex contemporary environments," *Journal of Speculative Philosophy* 24: 316–332.
Otto, Friederike (2017), "Attribution of weather and climate events," *Annual Review of Environment and Resources* 42: 627–646.
Otto, Friederike (2023a), "Attribution of extreme events to climate change," *Annual Review of Environment and Resources* 48: 813–828.
Otto, Friederike (2023b), "Extreme weather events as teachable moments: Catalyzing climate change learning and action through conversation," *Environmental Communication* 17: 828–843.
Paehlke, Robert (2005), "Democracy and Environmentalism: Opening a Door to the Paehlke Administrative State? pp. 25-43 in Robert Paehlke and Douglas Torgerson (eds.), *Managing Leviathan*. Environmental Politics and the Administrative State. Second Edition. Peterborough, Ontario: Broadview Press.
Page, Benjamin I. and Martin Gilens (2017), *Democracy in America? What Has Gone Wrong and What We Can Do about It*. Chicago, IL: University of Chicago Press.
Park, R. Jisung, A. Patrick Behrer, and Joshua Goodman (2021), "Learning is inhibited by heat exposure, both internationally and within the United States," *Nature Human Behaviour* 5: 19–27.
Parsons, Talcott (1963), "On the concept of political power," *Proceedings of the American Philosophical Society* 107: 232–262.
Partzsch, Lena (2025), "Not one inconvenient truth, but many: Emergency narratives and climate change," *Journal of Political Power* 18: 77–92.
Passmore, John (1974), *Man's Responsibility for Nature*. London: Duckworth.
Paxton. Robert O. (2004), *The Anatomy of Fascism*. New York: Alfread A. Knopf.
Perrow, Charles (2010), "Organizations and global warming," pp.59–77 in Constance Lever-Tracey (ed.), *Routledge Handbook of Climate Change and Society*. New York: Routledge.
Peters, B. Guy (2017), "What is so wicked about wicked problems? A conceptual analysis and a research program," *Policy and Society* 36: 385–396.
Peters, Glen P., Robbie M. Andrew, Tom Boden, Josep G. Canadell, Philippe Ciais, Corinne Le Quéré, Gregg Marland, Michael R. Raupach, and Charlie Wilson (2013). "The challenge to keep global warming below 2°C," *Nature Climate Change* 3: 4–6.
Petersen, Michael B. and Lene Aarøe (2013), "Politics in the mind's eye: Imagination as a link between social and political cognition," *American Political Science Review* 107: 275–293.
Peterson, Lauri (2021), "Silver lining to extreme weather events? Democracy and climate change mitigation," *Global Environmental Politics* 21: 23–53.
Pickering, Jonathan (2019), "Ecological reflexivity: Characterising an elusive virtue for governance in the anthropocene," *Environmental Politics* 28: 1145–1166.
Pielke, Roger Jr. (2010), *The Climate Fix. What Scientists and Politicians Won't Tell You about Global Warming*. New York: Basic Books.
Piketty, Thomas (2020), *Capital and Ideology*. Cambridge, MA: Harvard University Press.
Pohjolainen, Pasi, Iida Kukkonen, Pekka Jokinen, Wouter Poortinga, Charles Adedayo Ogunbode, Gisela Böhm, Stephen Fisher, and Resul Umit (2021), "The

role of national affluence, carbon emissions, and democracy in Europeans' climate perceptions." *Innovation: The European Journal of Social Science Research* 37: 515–533. https://doi.org/10.1080/13511610.2021.1909465

Popper, Karl (1945), *The Open Society and Its Enemies*. Volume 1. London: Routledge & Kegan Paul.

Povitkina, Marina (2018), "The limits of democracy in tackling climate change," pp. 145–155 in Clive Hamilton, Christophe Bonneuil, and François Gemenne (eds.), *The Anthropocene and the Global Environmental Crisis*. London: Routledge.

Prantl, Jochen, Ana F. Barros-Platiau, Cristina Y. A. Inoue, Joana C. P. T. L. Ribeiro, Eduardo Viola, and Getulio Vargas (2024), Building Capabilities for Earth System Governance. Cambridge: Cambridge University Press.

Prewitt, Kenneth (2010, "Introduction: Limits to knowledge? No easy answer," *Social Research: An International Quarterly* 77: 901–904.

Prins, Gwyn, Isabel Galiana, Christopher Green, Reiner Grundmann, Mike Hulme, Atte Korhola, Frank Laird, Ted Nordhaus, Roger Pielke Jr., Steve Rayner, Daniel Sarewitz, Michael Shellenberger, Nico Stehr, und Hiroyuki Tezuka (2010), *Hartwell Paper*. London: London School of Economics.

Przeworski, Adam (2019), *Crisis of Democracy*. Cambridge: Cambridge University Press.

Przeworski, Adam and Fernando Limongi (1997), "Modernization: Theories and facts," *World Politics* 49: 155–183.

Purdy, Jedediah (2015), *After Nature. A Politics for the Anthropocene*. Cambridge, MA: Harvard University Press.

Purdy, Jedediah (2022), *Two Cheers for Democracy. Why Democracy Is Flawed, Frightening – And Our Best Hope*. New York; Basic Books.

Qiu, Minghao, Christopher W. Callahan, Iván Higuera-Mendieta, Lisa Rennels, Bryan Parthum, Noah S. Diffenbaugh, and Marshall Burke (2025), "Valuing wildfire smoke related mortality benefits from climate mitigation," *NBER Working Paper* No. 33829.

Radkau, Joachim (2017), *Geschichte der Zukunft. Prognosen, Visionen, Irrungen in Deutschland von 1945 bis heute*. München: Hanser.

Rayner, Steve (2006), "Wicked problems: Clumsy solutions – Diagnoses and prescriptions for environmental ills," ack Beale Memorial Lecture, UNSW.

Riemens, Rianne (2024), "Fixing the earth: Whole-systems thinking in Silicon Valley's environmental ideology," *Internet Histories* 8: 294–311.

Riris, Philip, Fabio Silva, Enrico Crema, Alessio Palmisano, Erick Robinson, Peter E. Siegel, Jennifer C. French et al. (2024), "Frequent disturbances enhanced the resilience of past human populations," *Nature 629*: 1-6.https://doi.org/10.1038/s41586-024-07354-8

Risse, Mathias (2024), *Political Theory of the Digital Age. Where Artificial Intelligence Might Take Us*. Cambridge: Cambridge University Press.

Rittel, Horst and Melvin M. Webber (1973), "Dilemmas in the general theory of planning," *Policy Sciences* 4: 154–59.

Robert, Reich (2022), "America's Billionaire Class Is Funding Anti-Democratic Forces." The Guardian.

Rodrik, Dani (2024), "Reimagining the global economic order," *Review of Keynesian Economics* 12: 396–407.

Romps, David M., Jacob T. Seeley, David Vollaro, and John Molinari (2014), "Projected increase in lightning strikes in the United States due to global warming," *Science* 346(6211) (2014): 851–854.

Rorty, Richard (2004), "Post-democracy," *London Review of Books* 26: 10–11.

Rosa, Hartmut (2017), *Beschleunigung: Die Veränderung der Zeitstrukturen in der Moderne*. Berlin: Suhrkamp.

Rosamond, Jeffrey and Claire Dupont (2025), "Investigating the politics of democratic legitimacy in EU climate policymaking: The special case of fit for 55," *Journal of European Integration* 47: 237–255.

Rosanvallon, Pierre (2006), *Democracy Past and Future*. New York: Columbia University Press.

Rosanvallon, Pierre ([2011] 2013), *The Society of Equals*. Cambridge, MA: Harvard University Press.

Roszak, Theodore (1969), *The Making of the Counterculture. Reflections on the Technocratic Society and Its Youthful Opposition*. New York: Doubleday.

Runciman, David (2011), "Can democracy cope?," *The Political Quarterly* 82: 536–545.

Runciman, David (2013), *The Confidence Trap. A History of Democracy in Crisis from World War I to the Present*. Princeton, NJ: Princeton University Press.

Runciman, David (2018), *How Democracy Ends*. London: Profile Books.

Russell, Jacob H. and Dennis Patterson (2025), *The Weaponization of Expertise. How Elites Fuel Populism*. Cambridge: MIT Press.

Rüttenauer, Tobias (2023), "More talk, no action? The link between exposure to extreme weather events, climate change belief and pro-environmental behaviour," *European Societies* 26: 1246–1270. https://doi.org/10.1080/14616696.2023.2277281

Sabel, Charles (1991), "Moebius-strip organizations and open labor markets: Some consequences of the reintegration of conception and execution in a volatile economy," pp. 23–62 in Pierre Bourdieu and James Coleman (eds.), *Social Theory for a Changing Society*. Boulder, CO: Westview Press.

Sachs, Ignacy (1977), "Civilization project and ecological prudence," *Alternatives* 3: 1–18.

Sachs, Jeffrey, John W. McArthur, Guido Schmidt-Traub, Margaret Kruk, Chandrika Bahadur, Michael Faye, and Gordon McCord (2004), "Ending Africa's poverty trap," *Brookings Papers on Economic Activity* 1: 117–240.

Sætra, Henrik S. (2020), "A shallow defence of a technocracy of artificial intelligence: Examining the political harms of algorithmic governance in the domain of government," *Technology on Society* 62: 101283. https://doi.org/10.1016/j.techsoc.2020.101283

Saito, Kohei (2023), *Marx in the Anthropocene: Towards the Idea of Degrowth Communism*. Cambridge: Cambridge University Press.

Samuelson, Psul A. (1954), "The pure theory of public expenditure," *Review of Economics and Statistics* 36(4): 387–89.

Sandel, Michael J. (2020), *The Tyranny of Merit. What's Become of the Common Good*. London: Penguin.

Sarewitz, Daniel (2010), "Normal science and the limits on knowledge: What we seek to know, what we choose not to know, what we don't bother knowing," *Social Research* 77: 997–1010.

Sartori, Giovanni (1987), *The Theory of Democracy Revisited*. London: Chatham House.
Sato, Misato, Glen Gostlow, Catherine Higham, Joana Setzer, and Frank Venmans (2024), "Impacts of climate litigation on firm value," *Nature Sustainability*. https://doi.org/10.1038/s41893-024-01455-y
Scheffran, Jürgen (2025), "Planetary boundaries, polycrisis and politics in the anthropocene: Climate pathways, tipping cascades and transition to sustainable peace in integrative geography," pp. 339–444 in Hans G. Brauch (ed.), *Towards Rethinking Politics, Policy and Polity in the Anthropocene: Multidisciplinary Perspectives*. Cham: Springer Nature Switzerland.
Scheirer, Walter J. (2024), *A History of Fake Things on the Internet*. Stanford, CA: Stanford University Press.
Schelsky, Helmut (1961), *Der Mensch in der wissenschaftlichen Zivilisation*. Wiebaden: VS Verlag für Sozialwissenschaften.
Schelsky, Helmut (1975), *Die Arbeit tun die anderen*. Klassenkampf und Priesterherrschaft der Intellektuellen. 2. Auflage. Opladen: Westdeutscher Verlag.
Schmitter, Philippe C. (1985), 'Neo-corporatism and the state," pp. 32–62 In Wyn Grant (ed.), *The Political Economy of Corporatism. Sociology, Politics and Cities*. London: Palgrave.
Schnaiberg, Allan (1980), *The Environment: From Surplus to Scarcity*. New York: Oxford University Press.
Schneider, Etienne, Alina Brad, Ulrich Brand, Mathias Krams, and Valerie Lenikus (2023), "Historical-materialist policy analysis of climate change policies," pp. 110–126 in Christoph Scherrer, Ana Garcia, and Joscha Wullweber (eds.), *Handbook on Critical Political Economy and Public Policy*. Cheltenham: Edward Elgar.
Schon, Donald A. ([1963] 1967), *Invention and the Evolution of ideas*. London: Tavistock.
Schöngart, Sarah, Zebedee Nicholls, Roman Hoffmann, Setu Pelz, and Carl-Friedrich Schleussner (2025), "High-income groups disproportionately contribute to climate extremes worldwide," *Nature Climate Change* 15: 627–633. https://doi.org/10.1038/s41558-025-02325-x
Schumpeter, Joseph A. (1942), *Capitalism, Socialism and Democracy*. New York: Harper-Collins.
Schwartz, Stephan A. (2024), "Humanity's precognition: Climate change and the decline of democracy," *Explore* 21(2025): 103104. www.sciencedirect.com/journal/explore
Schwartzman, David (2009), "Ecosocialism or ecocatastrophe," *Capitalism, Nature, Socialism* 20: 6–33.
Scott, James C. (1998), *Seeing Like a State. How Certain Schemes to Improve the Human Condition Have Failed*. New Haven, CT: Yale University Press.
Seefried, Elke (2015), *Zukünfte. Aufstieg und Krise der Zukunftsforschung*. Berlin: de Gruyter.
Selseng, Torbjørn, Kristin Linnerud, and Erling Holden (2022), "Unpacking democracy – The effects of different democratic qualities on climate change performance over time," *Environmental Science and Policy* 128: 326–335.
Semple, Ellen C. ([1911] 2023), *Influences of Geographic Environment. On the Basis of Ratzel's System of Anthropo-Geography*. London: Greenbie Press.
Semple, Ellen C. (1931), *The Geography of the Mediterranean Region. Its Relation to Ancient History*. New York: Henry Holt and Company.

Serres, Michel ([1992] 1995), *The Natural Contract*. Ann Arbor: The University of Michigan Press.

Shahar, Dan C. (2015), "Rejecting eco-authoritarianism, again," *Environmental Values* 24: 345–366.

Shahvandia, Mostafa K., Surendra Adhikarib, Mathieu Dumberryc, Siddhartha Mishrad, and Benedikt Sojaa (2024), "The increasingly dominant role of climate change on length of day variations," *PNAS Earth, Atmospheric, and Planetary Sciences* 121(30): 1–7.

Shakespeare, Tom (2015), *Disability Research Today*. London: Routledge.

Shearman, David and Joseph Wayne-Smith (2007), *The Climate Change Challenge and the Failure of Democracy*. London: Praeger.

Simmel, Georg ([1907] 1989), *Philosophie des Geldes*. Gesamtausgabe Band 6. Frankfurt am Main: Suhrkamp.

Simmel. Georg ([1911] 1981), „On the concept and the tragedy of culture," pp. 27-46 in Peter Etzkorn (ed.), *Georg Simmel*. The Conflict in Modern Culture and Other Essays. New York: Teachers College Press.

Simmel, Georg (1906), "The sociology of secrecy and of secret societies," *American Journal of Sociology* 11: 441–498.

Simmel, Georg ([1908] 1992), *Soziologie. Untersuchungen über die Formen der Vergesellschaftung*. Gesamtausgabe: Volume 11. Frankfurt am Main: Suhrkamp.

Simmel, Georg ([1908] 2009), *Sociology. Inquiries into the Construction of Social Forms*. Leiden: Brill.

Skidelsky, Robert and Edward Skidelsky (2012), *How Much Is Enough? Money and the Good Life*. New York: Other Press.

Skocpol, Theda (2013), "Naming the problem: What it will take to counter extremism and engage Americans in the fight against global warming," Prepared for the Symposium on The Politics of America's Fight against Global, Co-sponsored by the Columbia School of Journalism and the Scholars Strategy Network February 14, 2013, Harvard University.

Smil, Vaclav (2019), *Growth: From Microorganisms to Megacities*. Cambridge: MIT Press.

Smith, Charles (2015), *What the Market Teaches Us. Limitations of Knowing and Tactics for Doing*. Oxford: Oxford University Press.

Soci, Anna, Anna Maccagnan, and Daniela Mantovani (2014), "Does inequality harm democracy? An empirical investigation on the UK," 6th International Scientific Conference on Economic and Social Development and 3rd Eastern European ESD Conference: Business Continuity, Vienna 24–25 April 2014.

Soener, Matthew (2021), "Between green growth and de-growth: Locating the roots of climate change in capitalist accumulation," pp. 47–60 In Marlène Benquet and Théo Bourgeron(eds.), *Accumulating Capital Today*. London: Routledge.

Sombart, Werner (1938), *Vom Menschen. Versuch einer geisteswissenschaftlichen Anthropologie*. Berlin: Buchholz & Weisswange.

Sorg, Christoph and Jan Groos (2025), "Rethinking economic planning," *Competition & Change* 29: 3–16.

Sorokin, Pitirim (1928), *Contemporary Sociological Theories*. New York: Harper & Brothers.

Sorokin, Pitirim ([1942] 2010), *Man and Society in Calamity*. New Brunswick, NJ: Transaction Publishers.

Stacey, Jocelyn (2019), *The Constitution of Environmental Emergency*. Oxford: Hart Publishing.

Stechemesser, Annika, Nicolas Koch, Ebba Mark, Elina Dilger, Patrick Klösel, Laura Menicacci, Daniel Nachtigall, Felix Pretis, Nolan Ritter, Moritz Schwarz, Helena Vossen, and Anna Wenzel (2024), "Climate policies that achieved major emission reductions: Global evidence from two decades," *Science* 385: 884–893.

Steffen, Will, Jacques Grinevald, Paul Crutzen, and John McNeill (2011), "The anthropocene: Conceptual and historical perspectives," *Philosophical Transactions: Mathematical, Physical and Engineering Sciences* 369: 842–867.

Stehr, Nico (1978), "Man and the environment: A general perspective," *Archives for Philosophy of Law and Social Philosophy* 74:1–17.

Stehr, Nico (1991), "The power of scientific knowledge — And its limits," *Canadian Review of Sociology and Anthropology* 29: 460–482.

Stehr, Nico (1992), *Practical Knowledge. Applying the Social Sciences*. London: Sage.

Stehr, Nico ([1986] 1994), *Knowledge Societies*. London: Sage.

Stehr, Nico (1997), "Trust and climate," *Climate Research* 8: 163–169.

Stehr, Nico (2000), *Die Zerbrechlichkeit moderner Gesellschaften. Die Stagnation der Macht und die Chancen Individuums*. Weilerswist: Velbrück Wissenschaft.

Stehr, Nico (2001), *The Fragility of Modern Societies. Knowledge and Risk in the Information Age*. London: Sage.

Stehr, Nico (2005), *Knowledge Politics. Governing the Consequences of Science and Technology*. Boulder, CO: Paradigm Publishers.

Stehr, Nico (2008), *Moral Markets*. Boulder, CO: Paradigm Publishers.

Stehr, Nico (2015a), "Democracy is not an inconvenience," *Nature* 525: 449–450.

Stehr, Nico (2015b), "Enabling knowledge," pp. 75–88 in Peter Meusburger, Derek Gregory, and Laura Suarsana (eds.), *Geographies of Knowledge and Power*. Wiesbaden: Springer.

Stehr, Nico (2016), *Information, Power, and Democracy: Liberty Is a Daughter of Knowledge*. Cambridge: Cambridge University Press.

Stehr, Nico (2020), "The atmosphere of democracy: Knowledge and political action," pp. 69–91 in J. Glückler, G. Herrigel, and M. Handke (eds.), *Knowledge for Governance. Knowledge and Space*. Voloume 15. Cham: Springer, 2020.

Stehr, Nico (2022a), "In-between: The Simultaneity of the Non-simultaneous," *Social Epistemology* 36: 40–424.

Stehr, Nico (2022b), "The simultaneity of the non-simultaneous," *Social Epistemology* 36: 407–424.

Stehr, Nico (2023), *Knowledge Capitalism*. New York: Routledge.

Stehr, Nico (2025), *Society and Knowledge in a Turbulent Era*. Cheltenham: Edward Elgar.

Stehr, Nico and Amanda Machin (2016a), "Inequality in modern society: Causes, consequences," pp. 3–36 in Nico Stehr and Amanda Machin (eds.), *Understanding Inequality: Social Costs and Benefits*. Wiesbaden: Springer VS.

Stehr, Nico and Amanda Machin (2016b), "Trusting the climate: Catastrophe vs. stability," *Society* 53: 573–580.

Stehr, Nico und Amanda Machin (2019b), *Gesellschaft und Klima. Entwicklungen, Umbrüche, Herausforderungen*. Weilerswist: Velbrück Wissenschaft.

Stehr, Nico and Dustin Voss (2019), *Money. A Social Theory of Modernity*. New York: Routledge.

Stehr, Nico and Hans von Storch (1999), "Climate works: An anatomy of a disbanded line of research," pp. 137–185 in Heidrun Kaupen-Haas and Christian Saller (eds.), *Wissenschaftlicher Rassismus. Analysen einer Kontinuität in den Human- und Naturwissenschaften*. Frankfurt am Main: Campus.

Stehr, Nico and Hans von Storch (2000), "Global warming in perspective. Contemporary concern about climate change has an age-old resonance," *Nature* 405: 615.

Stehr, Nico and Hans von Storch (2023), *Science in Society: Climate Change and Climate Policies*. Singapore: World Scientific Publishers.

Stehr, Nico and Marion Adolf (2018), *Ist Wissen Macht? Wissen als gesellschaftliche Tatsache*. Weilerswist: Velbrück Wissenschaft.

Steiner, Philippe ([2005] 2011), *Durkheim and the Birth of Economic Sociology*. Princeton, NJ: Princeton University Press.

Stern, Nicholas (2015), *Why Are We Waiting? The Logic, Urgency, and Promise of Tackling Climate Change*. Cambridge: MIT Press.

Stevenson, Hayley (2021), "Reforming global climate governance in an age of bullshit," *Globalizations* 18: 86–102.

Steves, Franklin and Alexander Teytelboym (2023), "Political economy of climate change policy," Department of Economics, University of Oxford and Smith School for Enterprise and the Environment.

Stiglitz, Joseph E. (2024), *The Road to Freedom. Economics and the Good Society*. New York: W.W. Norton.

Stokes, Chris R., Jonathan L. Bamber, Andrea Dutton, and Robert M. DeConto (2025), "Warming of +1.5 °C is too high for polar ice sheets," *Communicatios Earth and Environment* 6 351. https://doi.org/10.1038/s43247-025-02299-w

Stone Jr., Brian, Carina J. Gronlund, Evan Mallen, David Hondula, Marie S. O'Neill, Mayuri Rajput, Santiago Grijalva, et al. (2023), "How blackouts during heat waves amplify mortality and morbidity risk," *Environmental Science & Technology* 57: 8245–8255.

Strange, Susan (1996), *The Retreat of the State: The Diffusion of Power in the World Economy*. Cambridge: Cambridge University Press.

Sundquist, Göran and Linda Soneryd (2024), "Science and democracy introduction: Environmental problems, science and democracy," pp. 504–510 in Christine Overdevest (ed.), *Elgar Encyclopedia of Environmental Sociology*. Northampton, MA: Edward Elgar.

Sunstein, Cass R. (2017), *#repubic. Divided Republic in the Age of Social Media*. Princeton, NJ: Princeton University Press.

Sunstein, Cass R. (2021), *This Is Not Normal. The Politics of Everyday Expectations*. New Haven, CT: Yale University Press.

Surdea-Hernea, Vlad (2025), "Attack on climate: The persistent effects of authoritarianism on climate policy equilibria." *Economics of Transition and Institutional Change* 33: 719–739. https://doi.org/10.1111/ecot.12446.

Susskind, Daniel (2024), *Growth. A History and a Reckoning*. Cambridge, MA: Harvard University Press.

Swyngedouw, Erik. (2009), "The political economy and political ecology of the hydro-social cycle." *Journal of Contemporary Water Research & Education* 142: 56–60.

Swyngedouw, Erik (2010), "Apocalypse forever? Post-political populism and the spectre of climate change." *Theory, Culture and Society* 27: 213–232.

Swyngedouw, Erik (2011), "Interrogating post-democracy: Reclaiming egalitarian political spaces," *Political Geography* 30: 370–380.

Swyngedouw, Erik (2022), "The unbearable lightness of climate populism," *Environmental Politics* 31(5): 904–925.

Tam, Kim-Pong, Hoi-Wing Chan, und Susan Clayton (2023), "Climate change anxiety in China, India, Japan, and the United States," *Journal of Environmental Psychology* 87: 101991.

Tarde, Gabriel ([1890] 2003), *The Laws of Imitation*. Second Edition. New York: Henry Holt & Company.

Tenbruck, Friedrich H. (1977), "Grenzen der staatlichen Planung," pp. 134–149 in Wilhelm Hennis, Peter G. Kielmansegg, and Ulrich Matz (eds.), *Regierbarkeit. Studien zu ihrer Problematisierung*. Band 1. Stuttgart: Klett-Cotta.

Tenbruck, Friedrich H. (1984), *Die unbewältigten Sozialwissenschaften*. Graz: Styria.

Tilly, Charles (2006), *Why? What Happens When People Give Reasons...and Why*. Princeton, NJ: Princeton University Press.

Tocqueville de, Alexis (2010), *Democracy in America*. Historical-Critical Edition of *De la democratie en Amerique*. Edited by Eduardo Nolla. Translated from the French by James T. Schleifer. Volume 1. Indianapolis, IN: Liberty Press.

Tol, Richard S. J. (2024), "A meta-analysis of the total economic impact of climate change," *Energy Policy* 185: 113922.

Tollefson, Jeff (2016), "2015 breaks heat record: Pacific Ocean warming helped to make last year the hottest in history," *Nature* 529: 450–451.

Tooze, Adam (2021), "Ecological Leninism," *London Review of Books* 43 (November 18).

Töpfer, Klaus and Günther Bachmann (2010), "One man – One vote – One carbon footprint: Knowledge for sustainable development," pp. 49–61 in Roeland J. in't Veld (ed.), *Knowledge Democracy. Consequences for Science, Politics, and Media*. Heidelberg: Springer.

Tormey, Simon (2015), *The End of Representative Politics*. Cambridge: Polity Press.

Tosun, Jale, Jan Pollex, and Laurence Crumbie (2023), "European climate pact citizen volunteers: Strategies for deepening engagement and impact," *Policy Design and Practice* 6: 344–356.

Toynbee, Arnold J. (1949), *Studie zur Weltgeschichte. Wachstum und Zerfall der Zivilisationen*. Hamburg: Claassen & Goverts.

Toynbee, Arnold J. (1973), "Foreword," pp. ix–x in Geoffrey J. Martin (ed.), *Ellsworth Huntington. His Life and Thought*. Hamden, CT: The Shoestring Press.

Treen, Kathie M. d. I., Hywel T. P. Williams, and Saffron J. O'Neill (2020), "Online misinformation about climate change," *Wiley Interdisciplinary Reviews: Climate Change* 11: e665.

Trenberth, Kevin E. (2022). "Earth and climate system," pp. 1–13 in *The Changing Flow of Energy through the Climate System*. Cambridge: Cambridge University Press.

Tsur, Yacov (2025), "The diverse impacts of democracy on greenhouse gas emissions," *Ecological Economics* 227: 108411. https://doi.org/10.1016/j.ecolecon.2024.108411

Turnbull, Nick and Robert Hoppe (2019), "Problematizing 'wickedness': A critique of the wicked problems concept, from philosophy to practice," *Policy and Society* 38: 315–337.

van der Sluijs, Jeroen P., Rinie van Est, and Monique Riphagen (2010). "Beyond consensus: Reflections from a democratic perspective on the interaction between

climate politics and science," *Current Opinion in Environmental Sustainability* 2: 409–415.
Van Dijk, Teun A. (2014), *Discourse and Knowledge. A Socio-cognitive Approach.* Cambridge: Cambridge University Press.
Van Westen, René, Michael Kliphuis, and Henk A. Dijkstra (2024), "Physics-based early warning signal shows that AMOC is on tipping course," *Science.Advances* 10(6). https://doi.org/10.1126/sciadv.adk1189
Verschraegen, Gert (2023), "Epistemic sociology: Luhmann's theory of science and knowledge," pp. 73–92 in Ralf Rogowski (ed.), *The Anthem Companion to Niklas Luhmann.* London: Anthem Press.
Victor, David G. (2009), "On the regulation of geoengineering," pp. 325–339 in Dieter Helm and Cameron Hepburn (eds.), *The Economics and Politics of Climate Change.* Oxford: Oxford University Press.
Victor, David G., M. G. Morgan, Jay Apt, John Steinbruner, and Katharine Ricke (2009), "The geoengineering option: A last resort against global warming?," *Foreign Affairs* 88: 64.
Wallerstein, Immanuel, et al. (1996), *Open the Social Sciences.* Report of the Gulbenkian Commission on the Restructuring of the Social Sciences. Stanford, CA: Stanford University Press.
Wallerstein, Immanuel M. (2004), *The Uncertainties of Knowledge.* Philadelphia, PA: Temple University Press.
Wan, Jessica S., Chih-Chieh Jack Chen, Simone Tilmes, Matthew T. Luongo, Jadwiga H. Richter, and Katharine Ricke (2024), "Diminished efficacy of regional marine cloud brightening in a warmer world." *Nature Climate Change* 14: 1–7.
Wang, Yan (2024), "Could working less promote environmental concern? A cross-national intercohort analysis," *Social Science Research* 121: 103038.
Weber, Elke U. (2006), "Experience-based and description-based perceptions of long-term risk: Why global warming does not scare us (yet)," *Climatic Change* 77: 103–120.
Weber, Max ([1918] 1994), "Parliament and government in Germany under a new political order," pp. 130–271 in Peter Lassman and Ronald Spiers (eds.), *Weber. Political Writings.* Cambridge: Cambridge University Press.
Weinstein, Jay and Nico Stehr (1999), "The power of knowledge: Race science, race policy, and the Holocaust," *Social Epistemology* 13: 3–36.
Weyland, Kurt (2024), "The populist threat to democracy," pp. 1–29 in Kurt Weyland (ed.), *Democracy's Resilience to Populism's Threat Countering Global Alarmism.* Cambridge: Cambridge University Press.
WGBU (2012), "World in transition – A social contract for sustainability." The German Advisory Council on Global Change. www.wbgu.de/en/publications/publication/world-in-transition-a-social-contract-for-sustainability
Whitehouse, Harvey (2024), *Inheritance. The Evolutionary Origins of the Modern World.* Cambridge, MA: Harvard University Press.
Williamson, Oliver E. (1975), *Markets and Hierarchies. Analysis and Antitrust Implications.* New York: Free Press.
Williamson, Oliver E. (1985), *The Economic Institutions of Capitalism. Firms, Markets, Relational Contracting.* New York: Free Press.
Williamson, Oliver E. (1995), "Transaction cost economics and organization theory," *Industrial and Corporate Change* 2: 107–156.

Willis, Rebecca (2020), *The Democratic Challenge of Climate Change*. Bristol: Bristol University Press.

Willis, Rebecca, Nicole Curato, and Graham Smith (2022), "Deliberative democracy and the climate crisis," *Wiley Interdisciplinary Reviews: Climate Change* 13: e759.

Willke, Helmut (2014), *Demokratie in Zeiten der Konfusion*. Berlin: Suhrkamp.

Willke, Helmut (2016), *Dezentrierte Demokratie*. Berlin: Suhrkamp.

Willke, Helmut (2023), *Klimakrise und Gesellschaftstheorie. Zu den Herausforderungen und Chancen globaler Umweltpolitik*. Frankfurt am Main: Campus.

Witt, Michael A. and Gordon Redding (2014), "China: Authoritarian Capitalism," pp. in Michaela A. Witt and Gordon Redding (Eds.), *The Oxford Handbook of Asian Business Systems*. Oxford: Oxford University

Wolf, Martin (2023), *The Crisis of Democratic Capitalism*. New York: Penguin Press.

Wolf, Martin, (2004), *Globalization Works*. New Haven. Connecticut: Yale University Press.

Wong, Joel, Matthias Münnich, and Nicolas Gruber (2024), "Column-compound extremes in the global ocean," *AGU Advances* 5(3) (2024): e2023AV001059.

Yearley, Steven (1992), *The Green Case: A Sociology of Environmental Issues, Arguments and Politics*. London: Routledge.

Young, Oran R. (2017), *Governing Complex Systems: Social Capital for the Anthropocene*. Cambridge: MIT Press.

Young, Rachel and Solomon Hsiang (2024), "Mortality caused by tropical cyclones in the United States," *Nature* 635: 121–128. https://doi.org/10.1038/s41586-024-07945-5

Zakaria, Fareed ([2004] 2007), *The Future of Freedom. Illiberal Democracy at Home and Abroad*. New York: W.W. Norton.

Zanocco, Chad, Hilarly Boudet, Roberta Nilson, und June Flora (2019), "Personal harm and support for climate change mitigation policies: Evidence from 10 U.S. communities impacted by extreme weather," *Global Environmental Change* 59: 101984. https://doi.org/10.1016/j.gloenvcha.2019.101984

Zhang, Sharon (2023), "Report on ChatGPT model's emissions offers rare glimpse of AI's climate impacts," *Truthout*, April 3, 2023.

Žižek, Slavoj (2010), *Living in the End Times*. London: Verso.

Zuleeg, Fabian (2024), "European cognitive dissonance," *Intereconomics* 59: 122–123.

Zürn, Michael (2024), "Authoritarian forms of rule in the twenty-first century," pp. 60–83 in Günter Frankenberg and Wilhelm Heitmeyer (eds.), *Drivers of Authoritarianism. Paths and Developments at the Beginning of the 21st Century*. Cheltenham: Edward Elgar.

NAME INDEX

SUBJECT INDEX

www.ingramcontent.com/pod-product-compliance
Ingram Content Group UK Ltd.
Pitfield, Milton Keynes, MK11 3LW, UK
UKHW020153200926
475702UK00016B/2680
* 9 7 8 1 0 4 1 1 5 9 5 8 2 *